P9-DTS-201

THE MODERN POETIC SEQUENCE: THE GENIUS OF MODERN POETRY

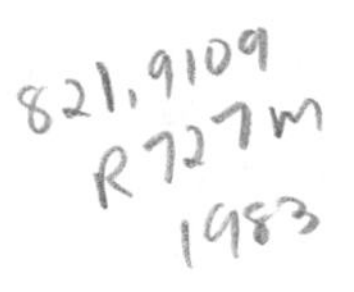

The Modern Poetic Sequence: *The Genius of Modern Poetry*

M. L. Rosenthal
Sally M. Gall

New York Oxford
OXFORD UNIVERSITY PRESS
1983

Library of Congress Cataloging in Publication Data

Rosenthal, M. L., 1917–
The modern poetic sequence

Includes index.
1. English poetry—20th century—History and criticism. 2. American poetry—History and criticism.
I. Gall, Sally M. II. Title.
PR601.R58 1983 821'.91'09 82-14529
ISBN 0-19-503170-9

Portions of this book, in their original form, first appeared in *Agenda* (London), *American Book Review, American Poetry Review, Basil Bunting: Man and Poet* (ed. Carroll F. Terrell), *Contemporary Literature, Massachusetts Review, Modern Poetry Studies, New American Review, New York Times Book Review, Robert Lowell: A Tribute* (Pisa—ed. Rolando Anzilotti), and *Sagetrieb*, and were presented in papers for the Modern Language Association (1977, 1979) and for the Ezra Pound International Conference (1979, 1981).

Printing (last digit): 9 8 7 6 5 4 3 2 1

Printed in the United States of America

for

Victoria Rosenthal

and

W. Einar Gall

workings *in their practice* many times over. And still today, like the sequence, it remains news—the open secret that poems keep telling but that remains a surprise.

Among sequence-makers two names spring up at once—Yeats for the British Isles, Pound for the United States—as rivals in inventive prolificacy. They have received space here accordingly; but their colleagues, major and minor, have also been treated as fully as possible. We have not been exhaustive. Not every sequence of quality could be discussed in these pages. But once the form has been isolated, we trust our critical colleagues will find further examples aplenty; and no doubt, too, they will re-examine the works we have fixed on in their own fashions. Moreover, we are speaking of a vital, thriving species, not an extinct one. Poets give every indication they will continue to write sequences and experiment with them; the form will surely develop further, and unpredictably, as time passes. Meanwhile, we have tried to identify the more centrally indicative sequences in our language, and also to characterize them in their own right, while noting significant perspectives they suggest: their implicit poetics.

Since the earliest sequences are American, we devote a chapter each to Whitman and Dickinson after our first chapter, which sets out some historical and theoretical considerations. The sequence-formation in Dickinson's fascicles is a perfect instance of the emergence of a form because "its time has come"—this despite the absence of a supporting body of theory or a goodly company of fellow artists, all working in related ways. *Ecce haec miracula!* Chapter Four then shifts to Britain, primarily to Hardy's elegiac sequence *Poems of 1912–13*. His work and that of other English poets lead naturally to Yeats's sequences, discussed in Chapters Five and Six. There are overlappings of time thereafter; we need to bear in mind, for example, that Yeats's first fully achieved sequences, "Meditations in Time of Civil War" and "Nineteen Hundred and Nineteen," are contemporary with Pound's early cantos and Eliot's *The Waste Land;* and much later, while Williams's *Paterson* was coming out in installments, Book by Book, Pound was still writing cantos—however different in character from his earliest ones.

Eliot, Pound, and Williams occupy the central third of the book (Chapters Seven through Ten), with Williams's *Paterson* providing a useful transition to our discussion, in Chapters Eleven and Twelve, of British and American "neo-regionalist" sequences. We have coined this term because these sequences are deeply anchored in specific locales and their history and cultural and familial memories; they range in time from Crane's and MacDiarmid's work of the 1920s to recent writing by Kinsella, Montague, and Hill. The final three chapters, moving from Stevens's and Auden's meditative sequences to the Confessional ones of

Lowell, Plath, and others and then to a wide stream of poetry by Guthrie, Hughes, Kinnell, Riding, Rich, and various additional figures, are grouped under the rubric "Poetry of Psychological Pressure." This description is applicable to all the sequences we take up in the book, but is especially stressed in these closing chapters. One reason is that the impact of the Confessional poets has opened more room for the psychological generally. Another is that our study as a whole has led more and more toward a sense of the importance of this factor. For instance, the power of the psychological dimension in politically engaged poetry of the past few decades reveals itself surprisingly in the course of our analyses.

The abundance of sequences, large and small, and their formal variety are matters of the most striking interest to criticism and theory. This is because of the challenge to greatness intrinsic in the genre: its demand for improvisation but for rigor as well, and for both emotional intensity and distancing, personal immediacy and wider engagement—in short, for all that yoking of opposites, associative brilliance and subtlety, and starkly unsentimental but profoundly humane interest we expect in great art. To deal with them at all adequately, we have had to be highly selective in the representative works discussed. We have also been led to mingle poetic history with critical description and evaluation and with essential theoretical orientation where it has seemed most needed—not only in the first chapter but wherever, along the way, the character of a sequence or of a related group of sequences has seemed to call for it.

Relevant theory grows out of direct engagement and sufficient empathy with literary works. They are great teachers if allowed to be—that is, if we open ourselves to the way a work enters upon its journey and moves from beginning to end: quite simply, from its initial self-presentation through its unfoldings and shifts of direction or subjective emphasis to whatever volatile equilibrium it seems to reach. Such attention has led us to some basic positions of evaluative, or dynamic, criticism and far from any approach depending on notions of "voice" or "persona." In a large sense, it has led to poetic criticism "in a new key," to borrow Susanne K. Langer's phrase for her valuable thinking about the arts. In the specific context of the present study, it can, we believe, enable poetic criticism to see the whole, complex mass of modern poetry in a unified purview. That purview is of the struggle of our poetry, almost independently of the poets' will, to realize itself through the emergence of the lyrical sequence and thereby triumph over all tendencies to render it obsolete, precious, and palatable only in tiny doses.

Suffern, New York M.L.R.
November 1982 S.M.G.

Contents

Part One
The Rise of the Modern Sequence

Chapter One

The Modern Sequence and Its Precursors

1. The "New" Genre and Organic Form

A "new" genre, the modern poetic sequence, has evolved over the past century and a half or longer. It has emerged so naturally, so without fanfare, as hardly to have been noticed. Yet, once its existence has been pointed out, and a name proposed for it, 'tis very like a camel or a mountain or a whale or—to return to things literary—an epic poem.

Its presence becomes abundantly obvious, for the modern sequence is the decisive form toward which all the developments of modern poetry have tended. It is the genre which best encompasses the shift in sensibility exemplified by starting a long poetic work "I celebrate myself, and sing myself," rather than "Sing, Goddess, the wrath of Achilles." The modern sequence goes many-sidedly into who and where we are *subjectively;* it springs from the same pressures on sensibility that have caused our poets' experiments with shorter forms. It, too, is a response to the lyrical possibilities of language opened up by those pressures in times of cultural and psychological crisis, when all past certainties have many times been thrown chaotically into question. More successfully than individual short lyrics, however, it fulfills the need for encompassment of disparate and often powerfully opposed tonalities and energies.

It is striking that the presence of this genre, the outgrowth of poets' recognition and pursuit of "new thresholds, new anatomies" (Hart Crane's visionary exclamation in a somewhat different context), has gone unappreciated for the most part. It is especially striking that experts in the very works that represent it so overwhelmingly—such works as Walt

Whitman's *Song of Myself,* the first great unmistakable exemplar of the form, and Ezra Pound's *Cantos*—should have missed the fact that they were confronting something new *artistically,* a creation of the genius of modern poetry to be closed with only through the dynamics of individual works.

How could this have happened? One explanation, we believe, is the character of poetic evolution itself. It is easy to detect superficial signs of newness: departures from traditional rhyme and meter, the absence of explanatory or narrative links between images or other evocative centers, the use of a vocabulary and subject matter hitherto taboo or considered unsuitable for poetry. But it is another matter to see the bearing of such signs. If not simply disregarded, they are usually thought odd or peripheral, although they may mark vast shifts of psychic direction and of the axes of aesthetic sensibility. Consider that Whitman is still a problematic figure to many people, and that Romantic poetry is still disdained by certain "classicists" who, as William Butler Yeats noted, would hardly know where to turn should "their Catullus" suddenly appear in the flesh. But for all of us it is, by definition, hard to recognize other people's originality and even harder to give it our complete empathy.

There is a further turn on the problem: critical recalcitrance. One needs to recall how seldom a literary work is actually treated without preconceptions, as a work of art that will tell us (if we allow it to do so) what its living character is. Now a work of art is made by a human being and reflects his or her empirical life and psychological set and times and general stock of ideas—these are points hardly in question. But as a work of *art* it is a construct, improvised to hold—however precariously—some sort of equilibrium or balance against disorder: especially, against its own internal disorder. The tendency to annihilate the thing that's made to a mere thought, however complex, or to a biographical shade, however humanly interesting, is the great and dominant retrogression of our criticism. We stress this position not out of any desire to renew familiar polemics, but for the good practical reason that it must be constantly sustained and reconceived if we are to see what is going on aesthetically in a literary work.

Critical recalcitrance results from the fact that "rational" scholarship and criticism too easily ignore the improvisatory, tentative nature of artistic experience, which is initimately related to the volatile richness of ordinary human awareness. But poets cannot ignore it; it is the primary condition of their art, and the key to the need for a formal resolution that a poem presses to fulfill. As William Carlos Williams tells us at the start of his sequence *Paterson,* they must work "with the bare hands." Their only recourse is

To make a start,
out of particulars
and make them general, rolling
up the sum, by defective means—
Sniffing the trees,
just another dog
among a lot of dogs. . . .

Thus the poet's task. Not every poet would put it in so deceptively humble a way. Yet even Yeats, with his aristocratic yearnings, wrote that the dreams and images of his poetry had their source in "the foul rag-and-bone shop of the heart." Commonplaces of poetic consciousness, these perspectives are hardly reflected anywhere in our poetic criticism; yet one might expect critics to use all their antennae to detect a poem's essential effort and idiom and then—by whatever "defective means"—to follow through as best they could. If, for example, one finds apparent fragmentation in a poem, one should want to get in touch with the fragments, get the tonal character of each, and open oneself to their internal relationships. Or if one finds apparently "unpoetic" elements, one must be ready for a possible widening of poetic range and the discovery of unexpectedly valuable dimensions. In short, what is called for is a critical method and theory congruent with and reflexive from poetic practice—a poet's poetics.

Criticism, however, has been drifting into various autotelic and self-adoring postures, rarely allowing itself to be instructed by the art of writers. Hence the strange blindness to what counts: here, to the towering presence of a genre that has dominated the ambition of every great modern poet. ("Blindness" may be too gentle a word in an era in which a number of critics begin to discount the value of artistic achievement altogether.) A cultivated empathy with the dynamic life of works can encourage the ability to discern new organic forms and aesthetic tendencies. Artistic change, in response to cultural change, goes on all the time independently of a critic's ability to recognize it. But the critic can at least try to be open to the most telling art, to its idiosyncratic energy and character, to the austere, driven intensity that Rebecca West named the "beautiful stark bride of Blake." The alerted state that devotion to such a modern muse entails draws language into discovery. Obviously the process precedes conscious formulation. By the same token, it leads to the only kind of "experimentation" that matters in art.

The whole development of the modern sequence, inseparable from that of modern poetry, has involved this sort of inevitable yet unformulated experimentation. These sequences have been written in the spirit of our changed world, meeting the needs of sensibilities impatient with posturing and keyed to rigorous emotional accuracy. Of course it is sheer dogmatism to rule out any form or genre as obsolete ("the novel is dead"), and it is always possible that some genius will find a way of writing a long, continuous narrative poem or logically or thematically developed one that will satisfy the most knowledgeable and sensitized of contemporary readers. But a fatal ennui with such efforts does seem to have set in, and not even the gifts of a poet as truly remarkable as Wallace Stevens have been able to achieve more than a momentary, suspect remission.

Like the first modulations toward the writing of sequences, critical inklings of the major direction of modern practice came early and piecemeal and—naturally—from poets. The most notable is Poe's remarkable insight in his 1846 essay "The Philosophy of Composition" (developed more fully in "The Poetic Principle" shortly afterwards):

> What we term a long poem is, in fact, merely a succession of brief ones—that is to say, of brief poetical effects. It is needless to demonstrate that a poem is such, only inasmuch as it intensely excites, by elevating, the soul; and all intense excitements are, through a psychal necessity, brief. For this reason, at least one half of the "Paradise Lost" is essentially prose—a succession of poetical excitements interspersed, *inevitably,* with corresponding depressions—the whole being deprived, through the extremeness of its length, of the vastly important artistic element, totality, or unity, of effect.[1]

Scholars in their serried ranks (Miltonists especially) have wittily or somberly demolished Poe's logic here, to their own satisfaction. What they have not dealt with is his unerring pointing of the new direction of sensibility. Poe may have been mistaken in any given assertion, but he was dead right in what is after all his essential position: that the traditional ways to structure the long poem no longer satisfy the modern poet.

A poem depends for its life neither on continuous narration nor on developed argument but on a progression of specific qualities and intensities of emotionally and sensuously charged awareness. A successful long poem, and the modern sequence pre-eminently, is made up of such centers of intensity. Its structure resides in the felt relationships among

1. James A. Harrison, ed., *The Complete Works of Edgar Allan Poe* (New York: 1902), XIV, 195–96. An interesting precedent for these ideas may be found in John Stuart Mill's essay "Alfred de Vigny" (1838), in his *Dissertations and Discussions* (London: 1867—second edition), I, 326–27. Mill holds it "impossible that a feeling so intense as to require a more rhythmical cadence than that of eloquent prose, should sustain itself at highest elevation for long together."

them. Narration and argument are useful poetically only as they provide certain kinds of dynamic structuring of the centers of intensity and tones—of suspense, expectation, thoughtfulness, or whatever—to go with them (Poe's "succession" of "brief poetical effects"). Chronological and rational ordering are but two among many possible structural devices subsumed in a work's lyrical structure. The problem cannot be solved by a debate over *Paradise Lost.* It is a matter of the fundamental character of lyrical structure: a way of making and viewing poetic constructs that concentrates on something other than logical and narrative and thematic links.

We shall return to the basic considerations of lyrical structure shortly. We should point out here that it is intimately related to the general conception of an organic rather than a mechanical approach to structure that has been with us for a long time, particularly in reference to the less subtle formal characteristics of a work—its rhyme scheme or stanza form or adherence to certain dramatic "laws." The sources of ideas about organic form lie in antiquity; in modern times they come to us by way of German and English Romantic theory, epitomized in Coleridge's defense of Shakespeare against the notion that he was a wild, undisciplined genius, "a sort of African nature, fertile in beautiful monsters." Borrowing his language directly from one of A. W. Schlegel's *Dramatic Lectures,* Coleridge said that such a mistake arises when we confound

> mechanic regularity with organic form. The form is mechanic when on any given material we impress a predetermined form not necessarily arising out of the properties of the material . . . The organic form, on the other hand, is innate; it shapes as it develops itself from within, and the fullness of its development is one and the same with the perfection of its outward form. Such is the life, such the form.[2]

In the United States the first transmitter of the idea of organic form was Emerson. His essay "The Poet," published two years before Poe's "The Philosophy of Composition," opened directly on the great poetry to come. Emerson's most striking formulation was derived from European ideas but absolutely original in its call for artistic exploration: our

2. Terence Hawkes, ed., *Coleridge's Writings on Shakespeare* (New York: G. P. Putnam's Sons, 1959), p. 68. Coleridge's *Shakespeare Lectures,* published in 1818, were delivered between 1808 and 1814. The passage in question (from Coleridge's 1808 lecture notes) uses A. W. Schlegel's distinction between "*mechanische*" and "*organische*" form and comes close to exact translation. See Schlegel's *Über dramatische Kunst und Litteratur* (Heidelberg: 1817—second edition), III, p. 8. For bibliographical guidance to the Schlegel text and to the Mill reference in the previous note we are indebted to René Wellek's *A History of Modern Criticism 1750–1950* (New Haven: Yale University Press, 1955), II, pp. 48, 354, and 359, and III (1965), pp. 135 and 315–16.

poets must find the form in which to "chaunt our own times and social circumstances." He understood the necessity for loosening the grip of tight, conventional formal structures. Although he—as a poet—lacked the nervy genius and the powerful pressures to make a reckless break himself, he helped create a climate less hostile to experiment than before. Charles Olson's 1959 pronouncement, echoing Robert Creeley, that "form is never more than an extension of content" was preceded by 115 years in Emerson's vivid and elegant declaration:

> For it is not metres, but a metre-making argument that makes a poem,—a thought so passionate and alive that like the spirit of a plant or an animal it has an architecture of its own, and adorns nature with a new thing. . . . The poet has a new thought; he has a whole new experience to unfold; he will tell us how it was with him, and all men will be the richer in his fortune. For the experience of each new age requires a new confession, and the world seems always waiting for its poet.

It was—of course, it *would* be—an embarrassing and perhaps disreputable poet like Whitman who would respond to this thrilling challenge. It couldn't be the personally and poetically inhibited Emerson, who entertained the most daring ideas but whose verbal behavior was modest and respectable. Whitman's *Song of Myself* is the first realized modern poetic sequence, to be followed by *Calamus* and *Drum-Taps* as his other prime contributions to the genre. And not far from this poetic epicenter was the virtually secret work of the great American "underground" poet, Emily Dickinson, mistress of a realm of confessional lyricism whose volcanic beauty did not erupt into general view until well after her death. Her fascicles—the hand-threaded packets (or booklets) in which many of her most powerful poems are arranged—belong in any consideration of the modern sequence, despite their erratic success as integrated works of art.

We shall return to Whitman and Dickinson, as pioneers in the art of the sequence, in our next chapters. The genre's full flourishing, however, came only in the 1920s, with T. S. Eliot's *The Waste Land,* Pound's early cantos and *Hugh Selwyn Mauberley,* Yeats's Irish Civil War sequences, Crane's "Voyages," and Hugh MacDiarmid's *A Drunk Man Looks at the Thistle.* Since that time, the outstanding poets writing in English have used the sequence to accommodate the complexities and passions of contemporary experience. When we speak of this dominating form we do not forget that any individual sequence has, in Emerson's words, its own "architecture." There are, for instance, obvious differences among the works just named and among the others we shall adduce, such as Crane's *The Bridge,* Williams's *Paterson,* Basil Bunting's *Briggflatts,* Robert Lowell's *Life Studies,* Olson's *The Maximus Poems,*

Ramon Guthrie's *Maximum Security Ward,* Austin Clarke's *Mnemosyne Lay in Dust,* Ted Hughes's *Crow,* and Galway Kinnell's *The Book of Nightmares.* These are works of varied quality, but all are high points in their authors' development and serious efforts within the genre: neither "long poem" nor "linked series" but "sequence." They exemplify a compelling process, the result of sheer, psychically powerful need on each poet's part to mobilize and give direction to otherwise scattered energies.

For reasons we have earlier suggested, we cannot assume general knowledge of the nature and achievement of this new genre. Nor can we assume general knowledge of the critical approach, emphasizing lyrical structure (but not confining that term to the short lyric poem), necessary for insight into the character of the sequence—its balancings of stress and interplay among its centers of passionate preoccupation. Indeed, this vital structuring principle in poetry has been almost as neglected as its major modern genre.

The modern sequence, then, is a grouping of mainly lyric poems and passages, rarely uniform in pattern, which tend to interact as an organic whole. It usually includes narrative and dramatic elements, and ratiocinative ones as well, but its structure is finally lyrical. Intimate, fragmented, self-analytical, open, emotionally volatile, the sequence meets the needs of modern sensibility even when the poet aspires to tragic or epic scope. (The intimate character, the strong sense we have of a highly subjective impulse of lyrical energy at work, makes it helpful to refer to the poem's "speaker" or "protagonist." Since this usage is more a convenience of critical discourse, very easily misleading, than a key to aesthetic structure, we must remember that we are not dealing with a literal speech or monologue or poeticized discourse but with something like a piece of music.) The intimate, self-conscious tonality is often present from the very start: "I celebrate myself, and sing myself," "Bring me to the blasted oak," "April is the cruellest month," "So name her Vivian. I, scarecrow Merlin." It establishes an initial poetic pitch more than a theme or character. The pitch may change; so may the level of diction, the points of reference, and the contexts of evocation. The poem creates its own dynamics, in fact, whether or not it has a specified dramatic speaker.

Yet a dramatic speaker is often its *vehicle,* not controlling its movement but frequently presented as under a pressure felt as tragic. We may be confronted by a speaker *in extremis.* A piercing example is the mortal agony of Guthrie's "Today Is Friday" in *Maximum Security Ward:*

> You could taste it being fed intravenously through a
> skein of tubes into your most plausible dreams
> It was happening   It was going on as suavely

as if it were a rank of drop-forges
smashing diamonds to dust as fast as
they could be fed to them.

Or, in a less acutely personal key of suffering but still in the realm of tragedy, the poem is oppressed by what Delmore Schwartz called "the burden of consciousness," the alienated music of "civilization and its discontents." Yeats gives us many instances, as in "The Stare's Nest by My Window"—part of "Meditations in Time of Civil War" and, in its way, not really less desperate than Guthrie's lines:

We are closed in, and the key is turned
On our uncertainty; somewhere
A man is killed, or a house burned,
Yet no clear fact can be discerned:
Come build in the empty house of the stare.

A barricade of stone or of wood;
Some fourteen days of civil war;
Last night they trundled down the road
That dead young soldier in his blood:
Come build in the empty house of the stare.

Seeking to locate the elements of its oppressed yet volatile state, the associative pressure in such works stirs up sunken dimensions of consciousness and memory from the depths, moving through confusions and ambiguities towards a precarious balance. The process of association and of modulation among shifting intensities is both psychological and cultural in its contexts of reference. It involves the feeling of obsolescence, the need to recover an identifying past. Here enters the heroic or epic aspect of the sequence: its effort to pit primal values and personal, historical, and artistic memory or vision against anomie and desolation. Thus, Guthrie's title ("Today Is Friday") refers distantly to the Crucifixion, and Yeats powerfully suggests natural values to be clung to in the face of war's perversity. Similarly, Bunting and Clarke bring moments of secure childhood bliss into their sequences to help counteract psychological or moral distress; Hughes invents a drastic and cynical Creation "myth" that mocks and parodies the Bible but also suggests a source of strength in the mere will to survive (epitomized negatively in the appalled poem "That Moment" and in minimally positive terms in "Littleblood"); and Bunting, Williams, David Jones, and others use regional history and dialect to make a music of affirmation of identity.

In short, the modern sequence has evolved out of a serious need for an encompassing poetry, one completely involved with what our lives really mean subjectively. That need reflects the ultimate pressure on modern sensibility to understand itself and to regain what Olson called the "human universe." The pressure, right or wrong, is to reconceive reality in humanly reassuring ways rather than in chillingly impersonal ones. It is felt inescapably by poets, even when they hardly realize why they write as they do, as a pressure to remember and deepen their sense of our human reality while rejecting any hubristic, anthropomorphic self-deception. (See Hart Crane's poem "To Brooklyn Bridge" for a very pure projection of this pressure.) The struggle against depression and loss of morale in this context was waged powerfully by great poets in the past in work like Donne's "A Nocturnall upon St. Lucie's Day," Wordsworth's "Tintern Abbey," and Tennyson's *Maud: A Monodrama.* These titles, in their chronological order, will suggest that certain pressures are a constant, cutting across differences of time and poetic genre. At the same time, the poems themselves may suggest an increasing emphasis, as poetry approaches its modern phase, on a complex music of feeling involving a number of radiant centers, progressively liberated from a narrative or thematic framework.

2. Lyrical Structure and the Genius of Modern Poetry

Here we must attempt a few basic observations on this liberated lyrical structure. Let's say, first, that *its object is neither to resolve a problem nor to conclude an action but to achieve the keenest, most open realization possible.* This realization is, naturally, rooted in a work's initial pressures but goes beyond them in scope. By initial pressures we mean the human occasion for the poem, its set of awareness, its situation (the felt reality within the poem), its condition of sensuous or emotional apprehension—whatever constitutes an emotional center energizing the poem, which moves towards a state of equilibrium that balances, resolves, or encompasses these pressures.

This state of equilibrium is generally a momentary one, among varied and volatile states of feeling, awareness, and active emotional or intellectual discovery. We may call it an equilibrium among affects in process. It provides a sense of encompassment or transcendence, because the poem has, as it were, reached a height of responsiveness to all pressures acting on it. The ability to hold in balance conflicting and logically irreconcilable energies, and to identify their presence and intensity, is felt as mastery over contradiction, mastery by poetic conversion into a pattern of unruly but mobilized affects.

It follows that the higher the work's level of responsiveness, and the more manifold the pressures acting on it yet held in balance, the more satisfying the lyrical structure will be and the more powerful the illusion of transcendence. A short, concentrated poem, it is true, may also synthesize complexly or subtly related tonalities despite its surface simplicity:

> Westron winde, when will thou blow,
> The smalle raine downe can raine?
> Crist, if my love wer in my armis,
> And I in my bed againe.

The whole of this little song is charged with longing, yet each line resonates on its own. The first line—an impatient wish, whatever the reason, for a change in the weather—is followed by the ambiguously related vision of the "smalle raine," with a certain gentle ardor accompanying both lines. Then comes the expletive, and the song swells into a lover's frustrated outburst.

But we are here primarily concerned with poems on a larger scale, in which such a piece as "Westron Winde," though lovely and self-contained, might form a single affect. It is instructive to consider a long poem like Browning's magnificent "The Englishman in Italy"—292 lines in five unnumbered sections—as a model of lyrical structure. The brief opening and close provide a dramatic and intellectual frame. The middle 273 lines are so richly loaded with intertwined sensuous effects, gathering and multiplying from moment to moment, that the poem seems an epic of riotous openness. For the cultivated "Englishman" who is the vehicle of the poem's dynamics, the daily world of Italian life in the country region outside Sorrento is all splendor and surprise. In the twelve-line opening section we see him filled with solicitude for little Fortù, who is frightened by the sirocco:

> Fortù, Fortù, my beloved one,
>   Sit here by my side,
> On my knees put up both little feet!
>   I was sure, if I tried,
> I could make you laugh spite of Scirocco.
>   Now, open your eyes,
> Let me keep you amused till he vanish
>   In black from the skies,
> With telling my memories over
>   As you tell your beads;
> All the Plain saw me gather, I garland
>   —The flowers or the weeds.

To amuse the child, the speaker spins out a vivid account of his adventures in observation. The long second section (116 lines) is a continuous explosion of the senses that begins with an exclamation and continues in high excitement and with some unsqueamish whimsy:

> Time for rain! for your long hot dry Autumn
>   Had net-worked with brown
> The white skin of each grape on the bunches,
>   Marked like a quail's crown,
> Those creatures you make such account of,
>   Whose heads,—speckled white
> Over brown like a great spider's back,
>   As I told you last night,—
> Your mother bites off for her supper.
>   Red-ripe as could be,
> Pomegranates were chapping and splitting
>   In halves on the tree . . .

From these passages (the first strophe and the next twelve lines of the poem) we can see much of the whole poem's quality. The exuberance of all that is to come is foreshadowed here in the tones of robustly active kindness and in the eagerness to pour out fresh "memories" that are at once compared to a rosary and a garland. This comparison is an active one, full of word-play—with *telling* beads and with *making* a garland. (The word "garland" is a verb here.) The rhythmic movement is contained within the simple rhyme scheme *xaxaxbxb* . . ., and the meter sustains the entranced tone and overflowing spirits by its flexible energy—alternating trimeter and dimeter lines in a mixture of iambic and anapestic, with feminine endings for most of the longer lines. The form allows for luxurious expansiveness while militating against mere prolixity.

"The Englishman in Italy" is such a delight one could spend pages over it. But our point here is primarily to suggest the kind of dynamics, and considerations of formal analysis, involved in lyrical structure: the shifting of affects and the means of modulation as the poem changes key. The 100-line third section, as glorious though not as wild in its detail as the second, takes us on a muleback journey to the top of a neighboring hill. The arrival may be compared to moments of buoyant exultation, tempered by surprise and awe, in Dante's *Divine Comedy*. From all the swarmingly physical, teeming, appetite-filled life on the plain below we rise, as on the higher reaches of the Purgatorial Mount—but not self-importantly, for the description of the mule provides a humorously delicate touch:

> Over all trod my mule with the caution
>   Of gleaners o'er sheaves,

Still, foot after foot like a lady,
  Till, round after round,
He climbed to the top of Calvano,
  And God's own profound
Was above me, and round me the mountains,
  And under, the sea,
And within me my heart to bear witness
  What was and shall be.
Oh, heaven and the terrible crystal!
  No rampart excludes
Your eye from the life to be lived
  In the blue solitudes.

The sudden leap into the language of revelatory terror is a triumph of virtuosity, one that gave every sort of clue to later poets. It is comparable to the passages in Browning's *Sordello* (Book One, lines 374–84 and 392–439) that Pound reprints in his *ABC of Reading* and that anticipate the sharper breaks of tonality in the *Cantos*. The narrative context that Browning provides is sloughed off in Pound's writing, which operates by juxtaposition without the inhibition of surface continuity. Early and late—say in the shift from the hell-scene to the sudden paradisal vision in Canto 16; and from the morass of associations, compounded of immediate prison-camp details, political outbursts, bits of rueful memory, and random items, to the sustained, increasingly exquisite Dionysian hymn that rises from the morass in Canto 79—Pound has made brilliant use of this method adapted from Browning. The leap, in the passage just quoted, to the images of "heaven and the terrible crystal" and of "the life to be lived / In the blue solitudes" is of the same order (in the intrinsic, rather than the overt poem) as Pound's leaps of focus and affect, or Yeats's sudden "The swan has leaped into the desolate heaven" in "Nineteen Hundred and Nineteen." In "The Englishman in Italy" itself it prepares us for another miraculous shift at the heights, again anticipating Pound's method, that brings us some lines later into the mythical realm of Ulysses and the Sirens.

It also, however, introduces between these passages a strange element of hubris, from which perspective the ordinary folk below seem contemptible. An unexpected arrogance, foreshadowing the poem's amused condescension toward the end, intrudes on the tone of high solitude so thrillingly reached. By contrast the plains below are now seen as "soft" and "cowering"—"a sensual and timorous beauty" that is "how fair! but a slave." Thus we are prepared for the descent, which, when its moment comes, is so charmingly managed in the 57-line fourth section that one forgets the withering tone in the interim. But it returns. The sirocco has spent itself, and the child is directed to all the activities people have resumed—just a matter of looking out the window and seeing the gypsy, "our tinker and smith," all set up for his work and keeping his eye on the mischievous urchins all around, and the whole village preparing for the Feast of the Rosary's Virgin, and the scaffold ready for the musicians—

> All the fiddlers and fifers and drummers
> And trumpeters bold,
> Not afraid of Bellini nor Auber—

and the fireworks laid out to be "religiously popped" when the "flaxen wigged Image" of the Virgin is carried in.—And so we have moved from protective concern to sheer life-zest to transported exaltation to responsive appreciation, at once exuberant and nastily snobbish, of the villagers' noisy bustle of religious celebration. Then, after a quick return to the world of cheerfully brutal earthy things ("a scorpion with wide angry nippers"), the poem ends in a seven-line section contrasting all this blazing life with England's repressive atmosphere. Specifically, the ending concerns the reactionary resistance to abolishing the Corn Laws—the politically black British "Scirocco," in Browning's metaphor.

The balance of affects—radiant tonal centers of specific qualities, and intensities, of emotionally and sensuously charged awareness—in Browning's poem provides the germ of how a sequence works. It precisely indicates the nature of lyrical structure, which is based on dynamics: the succession and interaction of units of affect. (We of course assume the work of a real poet, by which we mean someone superbly gifted in creating affects and building them into an organic structure.) The relevant questions are simple: What are the successive affects, and what poetic resources have been used to create them? is there a cumulative, psychologically satisfying curve of movement? is there sufficient variety to make for a rich and complex experience? Lyrical structure, incidentally, is by no means restricted to poems. It is a characteristic of all literary genres: plays, novels, short stories, sermons, speeches, even prose exposition. It is, precisely, the concrete aesthetic dimension of any piece of verbal expression. We are here deliberately oversimplifying what, in practice, makes for some subtle and demanding problems of evaluation, but the downright simplicities are essential to bring criticism in direct alignment with the literature it is concerned with.

Once the clarification of a work's progression of affects, its *dynamics* or curve of movement, has been accomplished, then all sorts of extrinsic considerations may be brought to bear, such as the pressures exerted on the lyrical structure by personal experience, historical events, poetic influence, myth and religion, psychoanalytical theory, or whatever. All such considerations are relevant if they can be shown to be shaping forces within the work's movement and not, under the guise of poetic criticism, pursued for their own sakes.

We have stressed an approach to poetry that is not only analogous to the study of music but almost identical with it. Aaron Copland, in his essay "The Composer in Industrial America," could be a poet speaking of his art:

> What, after all, do I put down when I put down notes? I put down a reflection of emotional states: feelings, perceptions, imaginings, intuitions. An emotional state, as I use the term, is compounded of everything we are: our background, our environment, our convictions.

The composer's "emotional state," defined thus broadly, is a quality of his music similar to what we mean by a poetic affect—although, putting the matter more rigorously, we would say that a passage *presents* the affect rather than *reflecting* it. Further, Copland says that whatever "meaning" music has resides precisely in the particularizing of these "fluent emotional states." Just because it "particularizes and makes actual," art opens onto these states—the subjective "meaning" of the human condition. The "worlds of feeling" that musical themes project make up the "expressive plane" in music, which Copland sharply distinguishes from the "sheerly sensuous" and "sheerly musical" planes. The expressive plane, he says in "How We Listen," is the heart of the musical experience and should be the first concern of performers. "Professional musicians . . . often fall into the error of becoming so engrossed with their arpeggios and staccatos that they forget the deeper aspects of the music they are performing."

Here Copland's professional musicians are very much like those trained readers of poetry—whether students, professors, or critics—who likewise ignore the heart of artistic experience. They too forget that the "very nature" of poetry, as of music, is "to give us the essence of experience transfused and heightened and expressed in such fashion that we may contemplate it at the same instant that we are swayed by it." In music that "essence of experience" lies in the progression of themes; in poetry it resides in the progression of affects within the work's lyrical structure.

Now *Song of Myself* can be considered the first modern poetic sequence because it is the first poetic work of considerable length whose ordering is overridingly lyrical. It is not bound by thematic, philosophical, or formal conventions in the way that so many earlier so-called "sequences" were—Elizabethan sonnet sequences in particular—or in the way that more varied works such as Sidney's *Astrophil and Stella* or Herbert's *The Temple* were. *Song of Myself* involves a more sustained curve of movement than these or than such more nearly contemporary works as Blake's *Songs of Innocence and of Experience*. Nor is it like any exactly contemporary British work. Thus Whitman may be passionately concerned with death and immortality ("themes" in an abstractly topical rather than a musical sense); but the whole character of his sequence is entirely different, say, from that of *In Memoriam*. Tennyson's poem is intellectually far more self-conscious and at times attempts to set philosophical discussion and resolution to rhyme and meter. Even in *Maud*, published the same year as *Song of Myself*, Tennyson felt the need to set up a narrative-dramatic frame to justify his real structure: the juxtaposition of states of mind so intense they went beyond what his society considered normal. But *Song of Myself* is structured—on the surface as well as in its innermost life—neither to resolve a problem nor to conclude an action but to evoke the keenest, most open realization possible. As far as we know, only Dickinson was working along the same lines.

It may strike some readers as curious or naïve to speak of the "genius" of modern poetry, and especially to link it with the terms "heroic" and "epic" and with Coleridge's phrase "organic unity." Very serious people are now calling literature obsolete and seeking to discharge its mysteries along with its traditions and its distinction as art. Indeed, there is a strong tendency to deny the validity of both "art" and "distinction." Yet we must insist—whatever the theoretical attractions of exalting intellectual systems and ennui-generated positions over living works—that the individual work of genius is our one touchstone of value in the arts. As for "the genius of modern poetry," we refer to the stripping-away process, or liberation of sensibility, that marks the evolution of modern art. We cannot attribute the process to any one person or movement; it is a cultural phenomenon, a reflex of deep historical change. Poetry has, in a famous phrase, wrung the neck of rhetoric—that is, of "poetic" attitudinizing, sentimentality, and religious or philosophical preaching. It has sloughed off any notion that it is primarily a medium for story-tellers and moralists. Ideas, dramatic situations, narrative suspense—these have by no means disappeared from poetry, any more than has the sonnet or iambic pentameter or any other mode developed in the past. All have their uses. But the real poem, its dynamics always active beneath the surface structure of poems in any age, has come into its own.

As a result, the poet is now free to let his or her poem present itself directly in its own right and to create a movement, reversible and always in flux, of vital immediacies. Lyrical structure as we have described it is improvisatory but not undisciplined. Its rigor lies in the play of tonal depths and shadings and shiftings, grounded in human intensities but not translatable into general ideas or empirical problems and solutions. Those human intensities are centered in moments of realization, shared in life and art, in which aesthetic conversion becomes possible. *Language*, with all its traditions and evocative possibilities, connects with the complex of human awareness in action and takes over from it. Poetic phrasing and structuring are a phase of our human lives (a reflex, as we have noted) that enables us to reorient, reassemble, and explore felt meanings and imagined states with but one responsibility: faithfulness to our most candid perceptions and to the implications of language. The most developed poets have clarified these simple principles in their practice and thus have revolutionized the art of poetry. The climactic development in the revolution is the modern sequence, still in process of formation.

3. Earlier Models: Shakespeare and Tennyson

We are convinced that this development was always implicit in the nature of poetic structure. The sonnet sequences of the distant past—say those of Petrarch or of Shakespeare—suggest a driving emotional pressure toward a balanced resolution; but they tend to be more successful in individual poems or very small clusters of poems than in their movement as a whole. For one thing, the formal redundancy—Shakespeare's sequence contains 154 sonnets—militates against either the poet's or the reader's seeing the whole work in terms of different major units of affect. Shakespeare's greatness as a poet reveals itself not in the unfolding of the sequence but in the isolated poems and clusters we have mentioned. There are magnificent lines and passages throughout, but scarcely twenty-five of the sonnets might be called outstanding as complete poems. Shakespearean scholarship has hardly distinguished itself in arguing this or that possible order of the poems. The chief attempts have been to arrange them thematically and in order of composition. J. Dover Wilson is characteristic, dividing them into eight basic groupings with a supposed autobiographical bearing: "The Marriage Sonnets," "The Coming of Love," "The Poet Goes a Journey," "Liaison Sonnets," "The Rival Poet," "Farewell Sonnets," "To the Dark Woman," and "Independent Sonnets" (the last heading being a desperate effort to deal with, but not account for, poems that cannot easily be categorized).

Looking at the order of the sonnets in Wilson's scheme and trying to conjecture what their actual dynamics might be, one does feel a modulation toward an actual lyric structure: from a light overture of involvement (the somewhat absurd campaign to persuade a young man to marry and beget children), to an unfolding of love (in whatever sense), to a vaguely introduced unhappiness at separation, to bitter misery at betrayal and neglect, and to renunciation—and then follows a parallel but briefer curve of poems revolving about an unfaithful mistress. But the undifferentiated form of *The Sonnets*, its pointless gaps in development, the heavy presence of its more boring poems, and its general unevenness make of it a very weak *sequence* no matter how the ordering is shuffled. The artistic problem is compounded by the way Shakespeare often throws away the endings of sonnets that begin superbly, just as he often throws away entire poems. C. S. Lewis's claim that *The Sonnets* as a whole constitutes "the supreme love-poetry of the world" is heartwarming but artistically almost meaningless. E. K. Chambers's assertion that the order of the poems is "an autobiographical one, following the ups and downs of an emotional relationship," is not much better either as description or evaluation. Few modern poets have equaled the power of the best of Shakespeare's sonnets, but many have created more successful sequences. (Incidentally, although we have no space to develop the point here, Shakespeare's plays are far truer forerunners of the modern sequence than *The Sonnets* or any other traditional sonnet-sequence. If we read *King Lear*, for instance, in this perspective—as a succession of variously charged affective units—the dynamics of a sequence is implicit in its movement.)

Among British poems of power in the last century, Tennyson's *Maud* may come the closest to the modern sequence of any of the works that do not quite break free of subservience to a plot line. It is at the opposite extreme from the Elizabethan sonnet-sequence, which was too redundant and theme-ridden to provide a genuine dynamics. The music of each sonnet was an isolated affair; at best one might find a few poems, more or less in the same key, driven by a unifying impulse and energy, and perhaps a few related shifts of tonal nuance and contrast as well. *Maud*'s structure, though reined-in by its character as a sustained dramatic monologue in several major movements—"the history," Tennyson said, "of a morbid poetic soul under the blighting influence of a recklessly speculative age"—is beautifully articulated in its dynamics.

Maud is at the very meeting point of long poem and sequence. True, its narrative continuity prevents release into the freest play of relationship among centers of feeling and awareness. It presents an unfolding love story, shot through with melodrama and fused with a tale of social ruthlessness, hatred, and violence. As a sort of verse-novel (one would have to qualify this term considerably, however), it has a touch of *The Sorrows of Werther* but in some ways approaches the profound and ironic complexities of *The Sentimental Education* and *The Magic Mountain*. Yet Tennyson attempted something very like a modern sequence in the succession of tonalities and degrees of intensity, and the prosodic variety, of the twenty-eight separate poems making up his "monodrama" (twenty-two in Part I, five in Part II, and one in Part III: a slow accumulation of manifold affects, then a plunge into the world of guilt, remorse, and madness, and then a final leap of self-sacrificial patriotic idealism that serves as a desperately makeshift effort at self-redemption). The work's ultimate emphasis, bringing it close to the great modern sequences despite its clankingly restrictive armor of plot, is on the psychological pressures within its cyclothymic and perhaps paranoid protagonist. He is *Maud*'s single overt voice, and his "morbid" state provides a rationale for its volatile leaps of affect.

For Tennyson's contemporaries such a rationale would have been indispensable. As it is, *Maud* still disturbs many readers in spite of the façade of a speaking character overloaded with carefully spelled-out motives. It is obvious that the fevered pitch of feeling and unpredictable mood-shifts are hardly mere fictional contrivance, and even recent critics instinctively shy away from the exposure of raw private intensities and pressures bursting through the fictional mask.[3] Even as late as 1963 a British critic and poet as astute and gifted as John Heath-Stubbs could say that *Maud* is "marred by a certain hysterical lack of control" (a position also taken by T. S. Eliot—last of the true Victorians—in his essay on *In Memoriam*). This, 108 years after both *Maud* and *Leaves of Grass* and forty-one years after *The Waste Land*! Thus does critical theory lag behind poetic practice, even in our all-knowing day and even when the critic is someone who really knows better. The suggestion that "hysterical lack of control" is no fit ground for poetry is objectionable here on several important grounds. Apart from the faintly moralistic implication that the artist must always keep a stiff upper lip, the phrase confuses the protagonist's condition in *Maud* with the character of the work itself. Whatever its faults, *Maud*'s original force derives from its artistic isolation of the emotional energies supposedly driving the protagonist, energies that have welled up in Tennyson's own psyche in such a way as to demand conversion into poetic process and structure. And *Maud* deploys these converted energies into a carefully articulated structure, by means of highly sophisticated skills. Thus Tennyson met the challenge of pressures demanding enough and intractable enough to be worthy of great artistic effort.

The major section of *Maud* is Part I, whose twenty-two poems make up almost four-fifths of the whole. Taken by itself this part, heavy though it is with fictional elements, would almost be a true sequence. Nothing could be more beautifully controlled than the way it moves, from the

3. Tennyson's own use of "morbid" to explain his protagonist's thoughts and behavior is a reflex of this circumstance. But he *was* a master of "morbid" poetry—in the sense of ability to cope with depressive awareness—and as such has provided a model for moderns like Yeats, Eliot, Bunting, and Kinnell.

bitterly alienated mood at the start to the anticipatory ecstasy at the end. Nor is this movement a simple one. There are counternotes all along the way—for instance, the warm ray of possible delight introduced near the end of the first poem and the darker notes amid the Keatsian joy of the closing one. The movement involves shifts of *proportion*. The ratio of harsher and more joyous affects (as well as, say, of more violent and gentler or more fearful ones, and of degrees of intensity), marked by many subtle modulations and lyric turns, is never quite the same at any two points along the way. The exploratory emotional search, with its dynamic juxtapositions, does constitute a potential sequence within the "monodrama."

We must be as clear as possible about the aesthetic conversion involved when we speak of "bitterness" or "joy" or "emotional search" in a poem. This is not a matter of rejecting the presence of the author's own empirical feelings and ideas but of seeing that they are initial pressures on the poem without determining its character as an artistic construct in language. The distinction explains itself as soon as we concentrate on a poem's particular language and form instead of speaking more generally. Take the opening quatrain of *Maud:*

> I hate the dreadful hollow behind the little wood;
> Its lips in the field above are dabbled with blood-red heath,
> The red-ribb'd ledges drip with a silent horror of blood,
> And Echo there, whatever is ask'd her, answers "Death."

The succession of images in the context of speech-tone and rhythmic arrangement defines itself intrinsically: an affect of horror and fear cumulative in force. This affect, naturally, is moving in a space of time, between "I hate" and "Death," that is an echo-chamber of patterned elements. After the abrupt personal assertion of the first line, in which the words "hate" and "dreadful," like a burst of offstage music, prepare us for what is to come, an equally abrupt visual metaphor personifies the "dreadful hollow" as a head whose lips drip blood and out of which the word "Death" oracularly echoes "whatever" question might be asked. The second and third lines present a figure with no exact referent save its own suggestions of a continuous revulsion and "horror," an endlessly bleeding wound of sensibility. The fourth line, personifying the inescapable death-enmeshment of the scene, is again psychologically rather than literally indicative. So we can respond to the quatrain's gathering movement, from its initial private outcry through its fierce surrealist images to its vision of a primeval scene of blood and personified doom.

One reason the quatrain is so powerful is its succession of hexameter lines, unusually long lines for English verse and made all the more so in the first three because each is sustained without an internal break. These are followed by the climactically arranged fourth line with its two caesuras. The second and third lines, crammed with imagery of grisly bleeding that recalls scenes of Dantean agony, are relentless in their increasingly appalled vision. The fourth line presents an aural reflex of this vision, a hideously reverberating monotone. Intellectually, it is as if the pondering mind had buckled under pressure and could only repeat the one death-message—brokenly, as the two caesuras indicate. Even the inexact rhymes sustain the affect of horror and fear. The whole impact of the quatrain falls heavily on the words "blood" and "Death," without any *facile* echo of sounds.

Compare the quatrain with this stanza that opens the final poem in Part I:

> Come into the garden, Maud,
>   For the black bat, night, has flown,
> Come into the garden, Maud,
>   I am here at the gate alone;
> And the woodbine spices are wafted abroad,
>   And the musk of the rose is blown.

The tone of the final poem (deservedly one of our famous love-poems) is well represented in this stanza. It mingles passion and sensuous, joyous arousal with faintly morbid notes, matching the intensity and the authority of the opening quatrain of *Maud* though very different in affect. The same extremely heightened state of sensibility is present, but not the horror. The short lines, the refrain-like repetition in the first and third lines, the emphatically musical exact rhymes, the graceful expansion and then contraction in the two closing lines, and the associations of dawn and garden fragrances and a lover's insistent calling—all these elements make for an atmosphere that is the exquisite opposite of horror. In isolating these opposed yet related affects of the two passages, we see that we can become alert to them and characterize them but cannot *translate* them. Like the echo in the quatrain, they reflect meanings but have their own reality.

Because it has so many such centers of lyrical self-containment and naked self-awareness, Tennyson's *Maud* comes nearer to the modern sequence than his extended philosophical elegy, *In Memoriam,* which one would ordinarily think of first as a possible forerunner. It is certainly evident from these two brilliant improvisations that Tennyson was on his way toward the open road that Whitman discovered. But *In Memoriam* is more like an Elizabethan sequence than *Maud.* Despite many lyric virtues, it is weighed down by its endless succession of uniform stanzas (tetrameter quatrains rhyming *abba*) and dominantly meditative tone, and by the redundancies one stumbles over in the long course of its 131 poems written during sixteen years (1833–49). It is less a sequence than it might have been, too, because of the steadily developing discursive argument and continuous account of a spiritual struggle.

This is to oversimplify. We are speaking of nothing less than a very moving work brought to birth because of the shock of an intimate friend's early death, the acute sense of life's bleakness when devoid of their finely affectionate relationship, and the consequent challenge to idealism and religious faith. Not a single poem is without its touch of quickened or deepened music, often startlingly poignant, in the midst of the more usual gravely somber intonations. Typical of such moments are the lines "O Sorrow, cruel fellowship" (III), "And, thy dark freight, a vanish'd life" (X), "How fares it with the happy dead?" (XLIV), and "Never morning wore / To evening, but some heart did break" (VI). In a few piercing instances, too, a whole poem breaks out of the meditative mode and stands as a touchstone of the deepest feelings animating the whole of *In Memoriam*. One thinks at once of certain poems in particular—notably, Poems VII ("Dark house, by which once more I stand"), XI ("Calm is the morn without a sound"), CVI ("Ring out, wild bells, to the wild sky"), and CXIX ("Doors, where my heart was used to beat"). Other poems, more quietly or delicately alive in texture, contribute to the rich lyrical matrix. They provide an essential field of emotive reference that wards off the danger of tedious and lugubrious semi-profound philosophizing which accompanies an undertaking of this sort, much as some of Pound's more purely lyrical cantos counteract the rhetorically driven ones that surround them.

There is no question of *In Memoriam*'s place in the pantheon of English verse, although T. S. Eliot's praise ("never monotony or repetition") is subject to further discussion—as is his dispraise of *Maud* for its "effect of feeble violence . . . the result of a fundamental error of form." Eliot saw the problem of *Maud* as the poet's failure to commit himself to a clear choice between dramatic and lyric form. We would suggest that the issue was different: a struggle to recover the dramatic immediacy underlying lyrical structure—just the one thing lacking in the earlier *In Memoriam*, which Eliot describes with winning sympathy but not altogether accurately:

> It is unique: it is a long poem made by putting together lyrics, which have only the unity and continuity of a diary, the concentrated diary of a man confessing himself. It is a diary of which we have to read every word.

Eliot wrote his essay "In Memoriam" (originally an introduction to an edition of Tennyson's poems) in 1936, long after he had left *The Waste Land* behind him. That poem directly confronts the same artistic challenge Tennyson deals with in *Maud:* to project and relate different internal states with the greatest possible immediacy. The "peculiarity" of *Maud,* Tennyson observed, "is that different phases of passion in one person take the place of different characters." The methods of the two poets are certainly very unalike; yet it is reasonable to see a direct continuity in their work, with one great distinction. Eliot has shed narrative, discursive, and dramatic continuity and is free to use every possible mode presentatively. That was in 1922. But by 1936 Eliot's writing was far closer than it had once been to the spirit of *In Memoriam* and growing ever more distant from the nerve-end exacerbations of his own earlier work, let alone *Maud.* The year of the Tennyson essay was the year that Eliot's *Collected Poems 1909–35* appeared, containing "Burnt Norton" and the flatly discursive and liturgical "Choruses from 'The Rock.' " Compared with the "Choruses," even the lyrically most intractable passages of *In Memoriam* are all quicksilver and fire.

Be all that as it may, with Tennyson we are at the borders of the modern sequence. As for *In Memoriam,* Eliot's comment has enough validity to suggest its genuine artistic significance. In it, Tennyson's tremendous virtuosity blends with the whole past poetic tradition from Shakespeare to Landor and Keats and many another. We hear *their* magic, assimilated to Tennyson's own. It is also striking how we hear the music of future poetry as well, from characteristic notes of Fitzgerald's *Rubaiyât* (which would appear just a few years later) to Yeats and Eliot. However overloaded, conventional, repetitive, and occasionally pontifical, *In Memoriam* prepared the way for the more stripped-down experiment of *Maud.* By its plastic inventiveness within a constricting form, it helped clear the ground for the lyric sequence. But for all this it was Walt Whitman's *Song of Myself* and Emily Dickinson's fascicles (although their importance has only lately begun to be recognized) that first broke the new ground of the modern sequence.

Chapter Two

American Originals I: Walt Whitman's *Song of Myself*

1. "A Graduated Kinship of Moods": The Dynamics of *Song of Myself*

Walt Whitman's *Song of Myself* and several of Emily Dickinson's fascicles, Numbers 15 and 16 in particular, are the great nineteenth-century exemplars of the dominant form of serious modern poetry in English. Critics and scholars have studied Whitman's major sequences—*Song of Myself, Calamus,* and *Drum-Taps*—extensively. Few have explored the arrangement of Dickinson's fascicles, or even entertained the promise they hold of being something more than miscellaneous collections of her poems stitched together in housewifely tidiness. With both poets, anyhow, the search has been for a conventionally conceived unity of theme, argument, or dramatic (or autobiographical) situation. Poetically, such "unity" would make little difference to the lyrical structure, and we propose, rather, to look at these sequences—and primarily at *Song of Myself* and Fascicles 15 and 16—in the terms developed in our preceding chapter.

When it first appeared in 1855, *Song of Myself* looked physically like a traditional long poem. Its division into 52 sections came only later; in one quasi-Biblical manifestation in the course of its evolution, its 372 stanzas were separately numbered. Naturally, it was no less a sequence in 1855 than in its final redaction in 1892; but the original form suggests that from the start Whitman felt the work as a whole, no matter what parts he completed first. The sheer variety of tonalities he found himself containing within it fragmented it enough so that it became our first great modern sequence. *Calamus, Drum-Taps,* and Dickinson's fascicles are more obviously assemblages of individual lyrics. The fascicles do not always cohere; but when they do they are true sequences like Whitman's. That is, their characteristic dynamics makes for an integrated effect quite unlike that of a loosely ordered grouping.

Since we are about to plunge into *Song of Myself,* a momentary digression may be in order here concerning the way sequences define themselves. It is important to remember that they exist within a continuum. At one extreme they present themselves as long but highly fragmented structures whose progression towards encompassing awareness is at once implicit and a matter of shifts in perspective—presented states at different intensities and levels of consciousness: *Song of Myself,* Stevens's "The Auroras of Autumn," Pound's *Pisan Cantos.* At the other extreme we find sequences clearly assembled from separate poems and fragments that have proved to cohere as a system of tensions, modulations, and reciprocal tonal forces: Eliot's *The Waste Land* and *Ash Wednesday,* Dickinson's fascicles, Yeats's *The Tower, Words for Music Perhaps,* and the lyric section in his *Last Poems and Two Plays.* In between the extremes are the myriad other works making up the genre.

And just beyond the tip, as it were, of each extreme fall the works that approach the sequence but are not, finally, sequences (although, precisely where questions of organic form are concerned, the issues raised are far more to the point than any overly rigorous answers might be). We have seen that Tennyson's *Maud* and Browning's "The Englishman in Italy" remain long poems in the traditional sense, despite being fragmented in their quite different ways. They verge on becoming sequences and no doubt embody the pressure to create such a form. Blake's *Songs of Innocence and of Experience* has many of the ingredients of the sequences at the other end of the spectrum, if we consider the two sets of poems as reciprocal. Yet for all their powerful attractiveness and tendency to interact magnetically, the "songs" do not click together structurally in a decisive way. Blake's constant rearrangements of them reflected this predicament. This is not to say that his poems do not surpass the quality of many sequences, but merely that they make for a work of another kind—very likely because of their involvement with the engravings and because the slightly schematic moral intention interfered with purely lyrical organization.

It was not until the first quarter of the twentieth century that the problem was explicitly addressed and by some of the finest poets writing in English. For example, Thomas Hardy—to anticipate our discussion in Chapter Four—felt the need to create a sequence-like structure but could not, except in his *Poems of 1912–13,* solve the problem. Thus he referred apologetically to the fact that his volumes were "miscellanies." In the preface to *Late Lyrics and Earlier* (and in that *annus mirabilis*

1922) he speaks of their unstructured character with regret, while at the same time chiding his critics for finding fault with tonal shifts between unrelated poems placed side by side in the same collection. These "journalists," he complains, are "deaf to the sudden change of key" when a satirical or humorous poem follows "verse in graver voice." Then, wistfully, and with an implication that it may be impossible to manage, Hardy suggests a method of ordering volumes that sorts exactly with the nature of lyrical structure.

> I admit that I did not foresee such contingencies as I ought to have done, and that people might not perceive when the tone altered. But the difficulties of arranging the themes in a graduated kinship of moods would have been so great that irrelation was almost unavoidable with efforts so diverse.

This is much to the point. If we allow the idea of a "graduated kinship of moods" full play, and subordinate the more mechanical one of "arranging the themes," then Hardy is proposing an artistic ideal close to what we have called the progression of specific qualities and intensities of emotionally and sensuously charged awareness. It is most interesting that not only was Hardy rueful at having missed creating such a progression but his critics too felt that way—in other words, something was in the air without which no one felt quite happy, although they could not say exactly what it might be: like the human race groping for expression the day before language was invented. But unlike his critics, and apparently without realizing it himself, he had found his way in the relatively compact elegiac grouping *Poems of 1912–13*.

But we are looking ahead.[1] To return now to *Song of Myself,* we shall try to proceed more as responders than as translators. The aim is to be open to its progression of tonalities, watching especially for peak moments when the language is most highly charged. We can then consider the relationship of such passages to the overall curve of movement: how the sequence gets from beginning to end. *Song of Myself* starts exuberantly and yet ceremonially, its speaking sensibility that of a man bursting with health and self-regard who yet is a bard chanting formulaically. "I celebrate myself, and sing myself"—the mixture of private and epic celebration is striking. The sequence ends with the dissolution of that sensibility, but in an ecstatic embracing of death and a vision of immortal communion with the reader. Its language is of delighted choice rather than of mournful brooding. "I effuse my flesh in eddies, and drift it in lacy jags"—one would think the prospect was sexual. And "I stop somewhere waiting for you"—a rendezvous, not a farewell.

1. Specifically, to pages 82–95.

So the movement involves a progression from one mode of sensuous immediacy to another, with both states deepened by their contexts: the sense of ritual occasion, the anticipation—assertive, consoling, gay, touching—of continuing identity beyond death. Especially if we glance at the language of physically aroused transport in Poem 2, we realize how much has been ventured on sheer sensuous openness in varied contexts in this sequence—how much self-exposure risked, how much glorious self-realization and transcendence sought, in making this openness the essential fabric of the work. The closing vision awakens even more glittering possibilities for sensuous as well as spiritual ecstasy. But within this curve are poems of desolation and excruciating sensitivity that resonate forcefully yet very subtly with the opening and closing affects. The opening poem, despite its celebratory tone, its attitude of nonchalance, and its proclamation of the speaker's "perfect health," does not forget his mortality and the risks and negative sides of life:

> I, now thirty-seven years old in perfect health begin,
> Hoping to cease not till death.
>
> Creeds and schools in abeyance,
> Retiring back a while sufficed at what they are, but never forgotten,
> I harbor for good or bad, I permit to speak at every hazard,
> Nature without check with original energy.[2]

And the final poem is flung out, as it were, in the face of death. The poem promises everything, yet there is a shadow of wistfulness over the language:

> The spotted hawk sweeps by and accuses me, he complains of my gab
> and my loitering.
>
> I too am not a bit tamed, I too am untranslatable,
> I sound my barbaric yawp over the roofs of the world.
>
> The last scud of day holds back for me,
> It flings my likeness after the rest and true as any on the shadow'd
> wilds,
> It coaxes me to the vapor and the dusk.
>
> I depart as air, I shake my white locks at the runaway sun,
> I effuse my flesh in eddies, and drift it in lacy jags.
>
> I bequeath myself to the dirt to grow from the grass I love,
> If you want me again look for me under your boot-soles.

2. Walt Whitman, *Leaves of Grass*, eds. Sculley Bradley and Harold W. Blodgett (New York: W. W. Norton, 1973).

> You will hardly know who I am or what I mean,
> But I shall be good health to you nevertheless,
> And filter and fibre your blood.
>
> Failing to fetch me at first keep encouraged,
> Missing me one place search another,
> I stop somewhere waiting for you.

Oh, how the speaker wants what he affirms to be true! The whole power of the sequence has depended on this yearning, expressed not as desire but as certainty. The proud affirmations and the courageous facing-up, at once generous and bitter, to the degradations of existence are both emboldened by a faith in the transforming power of intensity in itself. That faith is Whitman's great throw of the dice, the "hazard" admitted in the opening poem, the essential link between modern aesthetic and the secular mysticism of so much modern thought (both in process of evolving when *Song of Myself* was composed). *Song of Myself* projects the interaction among its major fields of affect (affirmation, confrontation of the negative, ecstatic experience) in a series of dynamically related states, in the context of a recurring struggle for transcendence. The transcendence we speak of here is literally poetic: a mobilization of units of charged language that, taken together and in the succession of their appearance, encompass many opposed motives and self-contradictory feelings. In the next-to-last poem (number 51) this very point is made, in a tone at once insolent, defiant, and defensive:

> Do I contradict myself?
> Very well then I contradict myself,
> (I am large, I contain multitudes.)

Because of its many radiating centers the sequence *is* "large," *does* "contain multitudes." To strike an idiosyncratic balance among competing moods and sensations is the overriding aim of lyrical structure. In the volatile realm of the subjective life no such balance can be reached unless the keen sharpness of each tonal energy the poem "contains" is actively present. *Song of Myself* is engaged throughout in realizing and "containing" these energies. Its final balancing point, in Poem 52, between the realms of the living and the dead bears a certain resemblance to the situation of Yeats's "All Souls' Night," the poem that completes *The Tower*. Both poets reach a pitch of pure volatility that is one great proof of their art.

The graduated kinship of moods (in Hardy's phrase) in *Song of Myself* proceeds through seven major groupings of poems, accumulating a dense context of emotional cross-reference, sometimes repeating an earlier tonal motif or reorienting it, sometimes leaping abruptly into a new level of thought and feeling. In its unfolding the process is not unlike that of a group of Pound's cantos; Pound's method is indeed a proliferation of Whitman's. The seven groupings we propose define the overall curve of movement of *Song of Myself*—its dynamics or lyrical structure broadly considered.

I. *Poems 1–7:* varied centers of emotional reference: key notations of sensibility
II. *Poems 8–17:* varied projections of identity: objects of love
III. *Poems 18–25:* negative extensions of II: the defeated, the forbidden; passion for elemental realities
IV. *Poems 26–29:* the sensitized responder; "touch" poems at the heart of the sequence
V. *Poems 30–36:* credo-poems converted into particular moments fixed in historical memory
VI. *Poems 37–43:* the prophetic, divine, *crucified* self
VII. *Poems 44–52:* mystical and cosmic extensions of the self; the open road of reaching into the unknown, including death; recapitulations

Put more concretely in terms of lyrical structure, the first group provides a brilliant scattering of tones and intensities locating the work's nerve-centers of significant awareness. In the next two groupings, the sensibility attaches itself to a variety of human types, both idealized and misery-laden or corrupt; the movement drops into realms of depressive recognition, but elemental union with the night and the earth (expressed in lovers' language) converts the depressive energy into a transcendent state of hovering intensity. (Nature—air, sea, sunlight, night, animal life—plays this role throughout the work.) In the fourth grouping all that has gone before is placed in one unifying perspective: an unbearable sensitivity to, first, music and, second, "touch"—rapidly and powerfully developed as sexual response of an uncontrollable yet miraculously ordering intensity. This small, crucially generative unit, Poems 26–29, retrospectively illuminates and reinforces the notes of high ecstasy in the first group—in Poem 2, which has the speaker mad for contact with the air; and in Poem 5, where body and soul unite sexually. Its shocks of excitement echo the female erotic fantasy of Poem 11 (Group II) and the voluptuous love song to earth, night, and sea in Poem 21 (Group III). Poems 28 and 29, especially, ground *Song of Myself* in the nerve-centers of a sensibility that is prey to all the pressures acting on an extraordinarily awakened responder.

After the extreme pitch of these two poems, the fifth grouping retreats—or tries to—into less frenzied, broadly philosophical utterance.

Yet each of these poems seeks confirmation of its pronouncements in touchstones of intense experience. This time the touchstones are reported incidents: dramatic moments of heroism, terror, and violence recalled by war veterans. The moments hang in the air forever, eternal tableaux of passionate action fixed in the psyche like mythical events in Homer. Outlasting their empirical source, vivid as the sensual arousal of the responder confessing his vulnerability in Poems 28 and 29, they are—like his instants of confusion and bliss—transcendent flashes of felt meaning. Such memories, nourishing but also tormenting the mind they haunt, make it unable to forget the terrible instant, say, of a shipboard amputation or the death of a valiant officer. They wreak a sort of havoc on the imagination that can see, hear, smell, touch them. (Whitman is very close to Pound in this respect.) Hence, in Group VI, the sensibility's adoption of the role of Christ crucified and the various thrusts of aggressive prophecy and ironic, self-mocking humor. Renewed, but with a darker, more painful awareness to wrestle with than before, the sensibility in the poems of Group VII launches a series of mystical extensions of itself. The emphasis shifts to exploration of an unknown, dangerous world that is nevertheless endlessly promising. We face the mystery of "the road," which is also the call of Death and whatever lies beyond it:

> This day before dawn I ascended a hill and look'd at the crowded
> heaven,
> And I said to my spirit *When we become the enfolders of those orbs,*
> *and the pleasure and knowledge of every thing in them, shall we*
> *be fill'd and satisfied then?*
> And my spirit said *No, we but level that lift to pass and continue be-*
> *yond.*
>
> (Poem 46)

The seven groupings of the sequence we have proposed are of course not indicated by Whitman explicitly, nor do we mean to suggest that they should be considered absolutely essential to a grasp of the work. The surges of feeling overlap, and often a new set of perspectives is introduced even while the preceding movement is completing itself. Thus, Poem 7 is both an afterbeat of Poem 6, which rounded off the first grouping beautifully with its reassuring and compassionate meditation on grass and death ("A child said *What is the grass?* fetching it to me with full hands"), and an introduction to the succeeding group that concerns itself with multiple empathy—the enormous leap of sensibility into the communality of experience. And Groups II and III might readily be considered together, as the bright and the dark sides of that communality. Whitman moves furthest into another being's sensibility in "twenty-eight young men bathe by the shore" (Poem 11), when he sinks into a young woman's secret daydream:

> The beards of the young men glisten'd with wet, it ran from their long
>   hair,
> Little streams pass'd all over their bodies.
>
> An unseen hand also pass'd over their bodies,
> It descended tremblingly from their temples and ribs.
>
> The young men float on their backs, their white bellies bulge to the
>   sun, they do not ask who seizes fast to them,
> They do not know who puffs and declines with pendant and bending
>   arch,
> They do not think whom they souse with spray.

Again, we might think of Poems 21–29 as the climactic movement of the sequence, overlapping with what we have called Group III (Poems 18–25). This is primarily because of Poem 21, which introduces the deep-going surge of realized love that finds its climax in Poem 29. It is the moment of richest convergence of opposites in the sequence, rising up out of the identification with the depressive realm of loss and defeat and leading to the poems of self-acceptance in the midst of despair (Poems 23–25). Poem 21 is the night complement to the poems of Group II. Instead of the earlier daytime smells and sounds, the "play of shine and shade on the trees," the "full-noon trill," the poet "walks with the tender and growing night."

> Press close bare-bosom'd night—press close magnetic nourishing night!
> Night of south winds—night of the large few stars!
> Still nodding night—mad naked summer night.
>
> Smile O voluptuous cool-breath'd earth!
> Earth of the slumbering and liquid trees!
> Earth of departed sunset—earth of the mountains misty-topt!
> Earth of the vitreous pour of the full moon just tinged with blue!
> Earth of shine and dark mottling the tide of the river!
> Earth of the limpid gray of clouds brighter and clearer for my sake!
> Far-swooping elbow'd earth—rich apple-blossom'd earth!
> Smile, for your lover comes.
>
> Prodigal, you have given me love—therefore I to you give love!
> O unspeakable passionate love.

This love song is not addressed to any individual but to ecstasy-producing nature. To turn to worship of one's own body as a holy seat of sensation is natural, then, and in Poem 24 Whitman offers a paean to the "luscious" all of him, in exclamations that parallel those of the paean to nature in Poem 21. A shorter excerpt will suffice:

> Root of wash'd sweet-flag! timorous pond-snipe! nest of guarded duplicate eggs! it shall be you!
> Mix'd tussled hay of head, beard, brawn, it shall be you!
> Trickling sap of maple, fibre of manly wheat, it shall be you!
> Sun so generous it shall be you!
> Vapors lighting and shading my face it shall be you!

Paradoxically, the extreme delight here, isolating the glories of one's own body, also breaks down the distinction between the private self and the outside world entirely. The images for the body are all metaphors from external nature. "Divine am I inside and out, and I make holy whatever I touch or am touch'd from." The sexual exuberance threatens to get out of control: "Something I cannot see puts upward libidinous prongs, / Seas of bright juice suffuse heaven." The sensual overloading of sensibility almost breaks the sequence down at the end of Poem 26: "I lose my breath, / Steep'd amid honey'd morphine, my windpipe throttled in fakes of death." And in Poem 28 the structure of control does break down:

> I am given up by traitors,
> I talk wildly, I have lost my wits, I and nobody else am the greatest traitor,
> I went myself first to the headland, my own hands carried me there.
>
> You villain touch! what are you doing? my breath is tight in its throat,
> Unclench your floodgates, you are too much for me.

But with the masterly Poem 29—"Blind loving wrestling touch, sheath'd hooded sharp-tooth'd touch"—all the sensual elements fall into place. Now a new calm begins to pervade the sequence, until the historical memories in Poems 33–36 renew the violence of emotion in a new context. Thereafter the moments of high intensity come far less frequently. The body and context of the sequence have been established, and it would be possible to regard Poems 37–52 as the final, overlong grouping, with too much explanation and recapitulation. But this would be to ignore the leap into association of agonizingly hard-pressed sensibility with the world of the "outlaw'd and lost" that justifies the assumption of the role of crucified Christ in Poem 38 and gives rise to entirely new tones of self-irony in Poems 41 and 42. These poems, in Group VI, are farther-reaching afterbeats of Group III. And in Group

VII, the remarkable poem of the open road (Poem 46, whose compassion echoes that of "A child said *What is the grass?*" but whose tone also contains a cold insistence on the loneliness and rigors of the journey) and the closing poems on the "bitter hug of mortality" and the challenge of death add a new unsentimental rigor to the vision of the sequence.—Still, an open structure of affects can never be pinned down to a scheme. The individual poems are discoveries reaching in many directions. What we *can* assert is that we have a series of interacting surges of affect whose general limits and directions can be discerned and therefore repossessed by the responsive reader.

Now, given the passionate search by other critics (Strauch, Miller, Cowley, *et al.*, each with his own five or seven or nine divisions) for a logical or thematic structure, perhaps we should reiterate our own rationale here for breaking *Song of Myself* down into several groupings. Our divisions are not based on stages of discursive reasoning or on correlation with some process extrinsic to the poem such as mystical experience or Oriental religious disciplines or even Whitman's psychological history. No matter how incidentally useful and illuminating such correlations or logical components may be, our prime attention is to the aesthetic of the work: *its* dynamics of organic structure. We see, then, how its most powerful poems and passages gather others around them so that certain tentative groupings or surges of affect reveal themselves. These overlap and modulate or feed into one another. But they are also isolable, pragmatically speaking, and define the work's lyrical structure in the large.

2. Peak Moments and Renewal of Momentum in Altered Contexts

It should be clear, we hope, that approaching sequences through their lyrical structure forbids translation into abstractions and general principles, for every formulation—however tentative—must be tested, and inevitably corrected, by reference to the fluid life of poems. We are not allowed to forget that, although an overview is a grand thing for helping to relate poems to one another, it will not be worth much unless it results from engagement with one poem at a time. And not to grow too metaphysical, we must also say that immersion in any poem of substance will for the moment change the proportions of the whole sequence and make everything center on the one chosen center. This is a crucial consideration in getting in touch with a sequence; the object is never to *schematize* its movement, but rather to experience it. The successions of groupings we have proposed are not meant to diminish the centrality of each key poem in its turn. Description of a process can never be more than indicative.

Thus, if we take Poem 1 ("I celebrate myself, and sing myself") in this spirit, we shall understand that despite its key position at the start of the sequence it does not exist only for the scheme of the whole. The bold, flaunting, debonair, yet completely democratic affect of this poem is a world unto itself, and therefore a complex of self-contained associations. For one thing, it presents a serious parody of conventional epic openings, echoing Vergil's "Arms and the man I sing" at the beginning of the *Aeneid.* To recall Williams's phrase, it is "a reply to Greek and Latin with the bare hands," the emphasis displaced from epic heroes to our common humanity and common corporeal being. The third line ("For every atom belonging to me as good belongs to you") is, however, more than a courtesy or democratic commonplace. Very early on, Whitman is telegraphing what later turns out to be an almost desperate absorption in physical process and its promise of a kind of immortality. That promise will be movingly asserted at the very end, in Poem 52. We may be forgiven for repeating just two lines from that poem here:

> I bequeath myself to the dirt to grow from the grass I love,
> If you want me again look for me under your boot-soles.

The image of "the grass I love"—bearing the speaker's own life-continuity—has appeared as early as the second strophe of the first poem. There the orotund expansiveness of the opening lines is suddenly dropped in favor of an entirely different tone, at once colloquial and revelatory. The speaker, a sort of American Rimbaud, gives us the key images of the whole sequence:

> I loafe and invite my soul,
> I lean and loafe at my ease observing a spear of summer grass.

Again, more is involved than a surface vignette. An early identification is being made between the speaker's soul and a whole state of alertness to the minute particulars of the natural world. The superbly casual posture of indolence is as of one of the elect in a pastoral paradise.

In its early versions, *Song of Myself* went directly from the first two strophes—with some modification—into what became the second poem, starting "Houses and rooms are full of perfumes." Originally the opening line was simply "I celebrate myself"; the addition of "and sing myself" reinforced the reference to earlier epic poems. But the most telling change was the addition of two strophes from "Proto-Leaf":

My tongue, every atom of my blood, form'd from this soil, this air,
Born here of parents born here from parents the same, and their parents the same,
I, now thirty-seven years old in perfect health begin,
Hoping to cease not till death.

Creeds and schools in abeyance,
Retiring back a while sufficed at what they are, but never forgotten,
I harbor for good or bad, I permit to speak at every hazard,
Nature without check with original energy.

Again, Whitman is doing more than appears at first glance. Both the essential Americanness of this work and Whitman's plunge into the natural worlds are affirmed more strongly here than in the first two strophes. Whitman is singing not only himself, and by implication everyone else—especially those involved in the expansive American experience—but is carrying us into the realm of the spirit and of cosmic experience: "I permit to speak at every hazard, / Nature without check with original energy." His interrelationship with the natural world is affirmed in concrete terms like those of Poem 52, but here the process is one of formation ("My tongue, every atom of my blood, form'd from this soil, this air") rather than dissolution ("I effuse my flesh in eddies, and drift it in lacy jags / I bequeath myself to the dirt to grow from the grass I love"). Yet this very dissolution, at the end, is countered by the image of new growth and by Whitman's vision of infusing air and grass so powerfully that he will be able to "filter and fibre" the blood of his future readers. His impact, the imagery suggests, is not only spiritual but actually physical, as physical as the grass and as enduring as all natural process:

What do you think has become of the young and old men?
And what do you think has become of the women and children?

They are alive and well somewhere,
The smallest sprout shows there is really no death . . .

(from Poem 6)

The "proof" of this immortality is given again and again in *Song of Myself* in images that seem, in themselves, to be charged with natural energy. This is a result of Whitman's intimate tones of expression, together with his gift for simple, immediate diction that makes a figure of speech seem a direct projection of experience. Poem 6 provides a brilliant instance:

Tenderly will I use you curling grass,
It may be you transpire from the breasts of young men,

It may be if I had known them I would have loved them,
It may be you are from old people, or from offspring taken soon out of
their mothers' laps,
And here you are the mothers' laps.

This grass is very dark to be from the white heads of old mothers,
Darker than the colorless beards of old men,
Dark to come from under the faint red roofs of mouths.

O I perceive after all so many uttering tongues,
And I perceive they do not come from the roofs of mouths for nothing.

The repeated word "dark," like the whole drift of the imagery, betrays the morbid side of hypersensitive awareness. It is the inevitable price of the volatile imagination that projects such satyr-like joy elsewhere in the sequence:

The atmosphere is not a perfume, it has no taste of the distillation, it
is odorless,
It is for my mouth forever, I am in love with it,
I will go to the bank by the wood and become undisguised and naked,
I am mad for it to be in contact with me.

The smoke of my own breath,
Echoes, ripples, buzz'd whispers, love-root, silk-thread, crotch and vine,
My respiration and inspiration, the beating of my heart, the passing of
blood and air through my lungs,
The sniff of green leaves and dry leaves, and of the shore and dark-
color'd sea-rocks, and of hay in the barn . . .

(from Poem 2)

This is a language of almost unbearable sensation. The state of arousal it presents is felt not as something shameful but as an absolute good. These lines are a touchstone for the most intensely alive moments of *Song of Myself*, found in Poems 28 and 29. "Mine is no callous shell," says the poet in the outstanding understatement in American poetry—in Poem 27—just before he absolutely lets go in Poem 28:

Is this then a touch? quivering me to a new identity,
Flames and ether making a rush for my veins,
Treacherous tip of me reaching and crowding to help them,
My flesh and blood playing out lightning to strike what is hardly differ-
ent from myself,
On all sides prurient provokers stiffening my limbs . . .

Poem 29, on the other hand, despite the pitch of erotic feeling, is as close to classical balance and control as anything in the sequence and is worth pausing over:

> Blind loving wrestling touch, sheath'd hooded sharp-tooth'd touch!
> Did it make you ache so, leaving me?
>
> Parting track'd by arriving, perpetual payment of perpetual loan,
> Rich showering rain, and recompense richer afterward.
>
> Sprouts take and accumulate, stand by the curb prolific and vital,
> Landscapes projected masculine, full-sized and golden.

The first line here projects some of the intoxicated confusion of the preceding poem, where the touch was "villain" in its mastery of the speaker. But the line's overall effect is one of power, weight, and deliberation, carried rhythmically by the overwhelming presence of nine stresses in a twelve-syllable line. The stresses, moreover, all fall on long vowels and are further strengthened by consonant clusters. The line's two balanced outcries to "touch" are followed by a gentle, loving question addressed apparently to a sexual partner. The language suggests a feminine as well as masculine consciousness: "blind loving wrestling touch" is an excellent image for a man's sensual experience of a woman, and "sheath'd hooded sharp-tooth'd touch" for the woman's of the man. Questions of Whitman's hetero- or homosexuality are irrelevant (as is the manuscript "evidence" that "sharp-tooth'd" had its source in a fairly explicit description of fellatio—"Must you bite with your teeth with the worst spasms at parting?"). What the finished poem offers is both sides of love-experience. And either partner could ask wistfully, tenderly, and proudly: "Did it make you ache so, leaving me?"

This dual perspective, the give and take of all creative process, is retained in the next stanzas as the speaker expands the sexual imagery to embrace all fecundity. And he turns again to the image of the grass (here suggesting life-sustaining wheat as well) as emblematic of all creative energy, of whatever we mean by the life force. More explicitly visionary experience has the same fundamentally sensuous base for Whitman. Thus he opens Poem 5 with a statement of the equality of body and soul—neither should be "abased to the other"—but when he actually launches into his description of enraptured transport, the primacy of the body, of the senses—especially touch—is established:

> I mind how once we lay such a transparent summer morning,
> How you settled your head athwart my hips and gently turn'd over
>   upon me,
> And parted the shirt from my bosom-bone, and plunged your tongue
>   to my bare-stript heart,
> And reach'd till you felt my beard, and reach'd till you held my feet.

Here, as in Poems 1, 29, 52, and so frequently elsewhere, grass, sensation, vision, and sensually responsive communion combine in a complex image for Whitman's passionate desire to penetrate and change the being of others. If Whitman actually succeeds in allowing "Nature without check with original energy" to speak through his words, the effect on his readers will be equally transporting, and it is significant that the poet's apostrophe to his soul is also applicable to the kind of fundamental power he wishes his poetry to have:

> Loafe with me on the grass, loose the stop from your throat,
> Not words, not music or rhyme I want, not custom or lecture, not even
> the best,
> Only the lull I like, the hum of your valvèd voice.

The dominant vision in Poem 5 is of sensuous existence penetrated at all points with spiritual existence; distinctions between God and man are unreal; love binds all beings together; nature's recurrence is an emblem of immortality:

> And limitless are leaves stiff or drooping in the fields,
> And brown ants in the little wells beneath them,
> And mossy scabs of the worm fence, heap'd stones, elder, mullein and
> poke-weed.

But *Song of Myself* is not limited to ecstatic affirmation. As we have seen, its affirmations are inseparable from its awareness of death and decay. This awareness is so keen that it must be converted into its opposite. Indeed, most of the more powerful passages in the second half of the sequence are the ones dealing with war. Here the poet takes on all suffering—"Agonies are one of my changes of garments"—and one thinks of Wallace Stevens almost a century later in "Auroras of Autumn" postulating a Being meditating on "The full of fortune and the full of fate." Whitman, exploring the "large hearts of heroes, / The courage of present times, and all times," brings out the blackness as well as the glory—the massacre at the Alamo as well as the old artillerist recalling a successful defense, a hellishly hounded slave as well as a dying fireman "exhausted but not so unhappy." The climactic poems of this sort are 35 and 36.

The brisk narrative of Poem 35 centers on the bravery and humanity of both sides, English and American, and on the charismatic leadership of the "little captain" who paraphrases John Paul Jones at a desperate moment (or who is himself Jones)—"*We have not struck,* he composedly cries, *we have just begun our part of the fighting,*" and the understated victory despite seemingly impossible conditions:

> Not a moment's cease,
> The leaks gain fast on the pumps, the fire eats toward the powder-
> magazine.

One of the pumps has been shot away, it is generally thought we are
sinking.

Serene stands the little captain,
He is not hurried, his voice is neither high nor low,
His eyes give more light to us than our battle-lanterns.

Toward twelve there in the beams of the moon they surrender to us.

And then comes the break to a new mode—"Stretch'd and still lies the midnight, / Two great hulls motionless on the breast of the darkness." Now the horrendous cost of victory can be reckoned: the sinking ship, the serene captain now "white as a sheet, / Near by the corpse of the child that serv'd in the cabin," the many dead—sailors and officers—leading up to the terrific closing lines:

Formless stacks of bodies and bodies by themselves, dabs of flesh upon
the masts and spars,
Cut of cordage, dangle of rigging, slight shock of the soothe of waves,
Black and impassive guns, litter of powder-parcels, strong scent,
A few large stars overhead, silent and mournful shining,
Delicate sniffs of sea-breeze, smells of sedgy grass and fields by the
shore, death-messages given in charge to survivors,
The hiss of the surgeon's knife, the gnawing teeth of his saw,
Wheeze, cluck, swash of falling blood, short wild scream, and long,
dull, tapering groan,
These so, these irretrievable.

Here are the barbarity and pathos of war set off by poignant reminders of all that the dying must give up. As so often in Whitman, his sensuous response to the sea evokes some of his finest passages. There is tremendous impact from the juxtaposition of images of carnage with the "soothe of waves" and the "delicate sniffs of sea-breeze." There is nothing in *Drum-Taps,* despite Whitman's actual experiences in the Civil War, that quite measures up to this extraordinary poem whose scenes are creations of imagination. The poem moves from the breath-taking image of the "Two great hulls motionless on the breast of the darkness," to the captain agonizing over the dead and wounded, to the unharmed survivors, and then to the passage just quoted, which ends with one of the simplest and most powerful indictments of war: "These so, these irretrievable." There is an absolute agony of war that nothing can expiate, and all of Whitman's affirmations of the positive meaning of death must be measured against this passage and the accumulated power of the last line.

We have been touching on peak moments in *Song of Myself* and thinking a bit about their interrelationships. We mean by "peak" those passages that provide keys to the essential dynamics of the sequence. It is the charge of language, rather than doctrinal content, that makes such passages decisive. We have not by any means discussed them all. Rather we have selected them with a double aim: to point up their relationship, as deep centers of emotive reference, to the work's curve of movement as a whole; and to focus on them in their own right, as units of affect whose quality irradiates the structure—not schematically but in the float and energy of the process that is the sequence.

Renewal of momentum in altered contexts—the uncoiling of unexpected movements of realization that nevertheless sustain the process—is essential to the life of a sequence. When this occurs, the individual poem involved becomes a new epitome of the way the sequence works. Poem 49, for example, begins with engaging bravado, in a tone not unfamiliar and yet startling because of the stinging metaphor at the center of its first line: "And as to you Death, and you bitter hug of mortality, it is idle to try to alarm me." The line stands on its own as a separate stanza: a wistful manifesto. One thinks, "Yes, I know that feeling. I know the pathos of that contradiction between the words 'bitter' and 'idle.' " And then the poem proceeds through certain recognizable phases, starting with a counter-imagery of birth in the next four lines. These lines, very gravely handled, have startling directness and force. "I recline by the sills of the exquisite flexible doors," says courageous Walt, imagining himself watching the accoucheur at work. And who is the accoucheur? We are not told in so many words that he is Death, and yet much in the poem—the opening; the use of the words "outlet," "relief," and "escape" (admittedly ambiguous here); the return to the subject of mortality in the next stanza, this time in direct address to a "Corpse," and the description further on of Life as "the leavings of many deaths" —sustain the implied conceit.

If this were all, particularly at this late point in the sequence, one might admire much in the poem and yet feel a certain weariness with its echoing of earlier tones and attitudes and its weathered air of ingenuity. (On the whole, Donne did this sort of thing far better.) But then, suddenly, we have one of those unexpected uncoilings and pure leaps of realization with which *Song of Myself* teems. When Whitman wrote, in the line addressed to the Corpse, that "I think you are good manure, but that does not offend me," he did not seem to realize that *he* might be the one giving offense. But now, in the final section of the poem, he reaches through to what he was preparing for—the realm of continued existence beyond death, a rebirth of a kind in the cold and lonely cosmos. Nothing here—although the passage does affirm—is cheerful or sentimental.

I hear you whispering there O stars of heaven,
O suns—O grass of graves—O perpetual transfers and promotions,
If you do not say any thing how can I say any thing?

Of the turbid pool that lies in the autumn forest,
Of the moon that descends the steeps of the soughing twilight,
Toss, sparkles of day and dusk—toss on the black stems that decay in
the muck,
Toss to the moaning gibberish of the dry limbs.

I ascend from the moon, I ascend from the night,
I perceive that the ghastly glimmer is noonday sunbeams reflected,
And debouch to the steady and central from the offspring great or small.

The reality of the decay out of which the speaker must ascend is presented starkly enough to counter the prophetic posturing of much of Groups VI and VII. Poem 49, in fact, may be one of those quietly definitive poems that make their impression almost subliminally. One can imagine T. S. Eliot's imagination being literally infected with the whispering, sinister, moon-saddened tones and images of this passage, which is extremely close in affect to *The Hollow Men* and other poems of Eliot's. Poem 50, too, is quietly self-contained and leaves its stamp on the sequence. It adds a note of openness and hesitancy about the whole *Song of Myself* enterprise:

There is that in me—I do not know what it is—but I know it is in me.

Wrench'd and sweaty—calm and cool then my body becomes,
I sleep—I sleep long.

I do not know it—it is without name—it is a word unsaid,
It is not in any dictionary, utterance, symbol.

Whitman of course goes on to define the "word unsaid" (one wishes he had let well enough alone for the moment): "It is not chaos or death—it is form, union, plan—it is eternal life—it is Happiness," but the essentially problematical nature of his answers does come through here. The same tentative attitude appears in Poem 51 in the slightly more grandiloquent "Do I contradict myself? / Very well then I contradict myself," and in the last lines of the sequence:

You will hardly know who I am or what I mean,
But I shall be good health to you nevertheless,
And filter and fibre your blood.

Failing to fetch me at first keep encouraged,
Missing me one place search another,
I stop somewhere waiting for you.

Song of Myself is great in part because Whitman—to alter Matthew Arnold's dictum—presents life steadily and presents it whole. We suggested in our first chapter that the purpose, or formal end, of lyric structure is the most open possible realization of experience. Despite his willed assertions of being at "peace about God and about death," Whitman does not gloss over the darker aspects of existence. The sequence is a satisfying whole, with a kind of humble, vulnerable, opening oneself to experience counterpointing easy modes of encompassment. It is also the most gloriously sensuous of poems, and its moments of highest intensity present ideal centers for the structure of a sequence in our modern sense. And despite the length, despite the catalogues, despite the exhortations, it remains fresh and alive. Much of its vitality depends on the quick shifts, variety, and brilliant improvisations made possible by the sequence form. Coming to the end of one poem, the poet is not constrained by the demands of logic or narrative to take his work in any particular direction. Nor does Whitman apologize for relying on his intuitive sense of the architectural demands of the sequence. As we have noted, the defiant "Very well then I contradict myself" is a properly impatient assertion of the new poetics that Schlegel via Coleridge and Emerson was hypothesizing.

Neither *Calamus* nor *Drum-Taps* is as successful as *Song of Myself*—no doubt unfair comparisons, since *Song of Myself* and the long poem "When Lilacs Last in the Dooryard Bloom'd" are high points of Whitman's art. Whitman tinkered considerably with the two later sequences—to the detriment especially of *Calamus*, for the 1860 version is obviously superior to the later one. Clearly *Song of Myself* came into being in a single surge of feeling, while the other two sequences involved more intractable materials. As the external facts of the Civil War and postwar periods, and Whitman's immediate experience of them, changed, affective discontinuities among the *Drum-Tap* poems became more extreme. Whitman faced a more difficult problem, compared with the ordering of *Song of Myself*, of balancing and rearranging the war-poems for maximum effect. With *Calamus*, he lost his nerve to some extent, excising the more confessional poems revealing his devastating vulnerability in love relationships. Yet it was precisely these poems, along with the more flagrantly erotic ones, that had formed his initial impetus for the group. In the interests of a public stance, he took the heart out of *Calamus:* its dark crisis of jealousy, despair, and devastating loss. In their essential directions both *Calamus* and *Drum-Taps* parallel the movement of *Song of Myself*—from affirmation into despair and then to darkened reaffirmation. But this is true primarily for the first version of *Calamus*, as we have suggested, and *Drum-Taps*, altogether honorably, betrays its fullest possibilities. The focus settles on compassion and love for the wounded, more than on their suffering and the terrible pity of war. The public stance (the necessity of the war to preserve the Union) to some extent displaces and distorts the subjective and confessional genius of the sequence.

Chapter Three

American Originals II: Emily Dickinson's Fascicles

1. What Are the Fascicles?

With both Whitman and Dickinson, the opportunity to improvise sequences had something to do with the publishing situation. They could both suit themselves in matters of arrangement, Whitman because he paid his own publication costs and Dickinson because she never put a whole book together. Both could have benefited from sympathetic and intelligent editors, but doubtless this deprivation was a boon to the new genre.

Emily Dickinson professed a reluctance to publish, and certainly she felt no public call in Whitman's sense. She never had, and perhaps never envisoned, a contemporary audience outside the half-appreciative few and her own sensibility. She realized that her thoughts might shock a genteel reader, but she had no *program,* even in fantasy, for doing so. Nor did she feel herself a Cassandra warning deaf ears. What she did share with Whitman, apart from exquisite lyrical gifts and sensuous alertness, was the passion that drives their work and the need to be discreet. The need was hopeless in both cases, because their most intense work gives the game away. Misunderstood, Whitman's self-exposure seemed unmanly and suspect to many readers. Dickinson's would have been thought unbecoming in a lady had she been published and understood.

But Whitman's discretion was a practical expedient, a minimal self-protection in a hostile environment. Dickinson's was a matter of her essential idiom of style and personality. As we shall see, much of her greatest writing evokes boldly clear states of awareness within an indeterminate context. It is both reticent and revealing: "pure" as Whitman's "touch" poems are pure. The principle at work in Dickinson, the pressure to get an extreme emotional complex into focus while repressing its private source, may account for the unevenness of her work, given her prolificacy and the absence of reaction from peers. It also helps account for her greatness, whenever a poem springs a concealed trap upon some wrenching perception rooted in passionate feeling. The great poems give overpowering direction to their sequences. Sometimes embarrassing and unpleasantly committed to carrying their perceptions through—and then reaching a sort of empyrean of pure states of emotional realization—they carry the lesser pieces along in triumph. However "unbecoming" the desire or fear or obsession they unfold and expose on the way to this triumph, they stamp their authority on the sequences they inhabit.

After Emily Dickinson's death in 1886, her poems appeared in a number of small selected volumes, and with a good many editorial alterations calculated to make them smoother rhythmically and clearer grammatically. Then, in 1958, Thomas H. Johnson published his three-volume edition, *The Poems of Emily Dickinson,* based on original manuscripts or the best available texts.[1] This was, of course, an enormous step forward. At the same time, the reader of Dickinson was suddenly swamped with 1775 poems (and variants), eccentrically punctuated and capitalized, and arranged chronologically rather than in the groupings the poet had often indicated for them. Seen in the mass, the poems were hard to assimilate—there are so many short pieces, with innumerable cries of the heart, vivid little impressions of birds and sunsets and moments in nature, compressed elegies for unnamed persons, ambiguously oriented expressions of grief and love and deprivation, and wildly moving comments, at once abstract and deeply personal, packed with evocative imagery. The artistry is so precise and idiosyncratic, the tonalities so brilliantly moving, that one knew there was no question of mere redundancy. Yet in a sense the range of attention *is* limited. Certain emotional predicaments and resolutions and stimuli recur again and again, as they would from day to day in a given life. It seemed that all one could do was to select as best one could from the flood of entries by the world's most articulate emotional diarist.

Yet Dickinson herself had given mute guidance. Of the 1775 poems in Johnson's collected edition, almost half had originally been arranged by the author in little booklets (generally called "fascicles") of from eleven

1. Thomas H. Johnson, ed., *The Poems of Emily Dickinson* (Cambridge: The Belknap Press of Harvard University Press, 1958), 3 vols. Our references to numbers of individual poems are all based on this text.

to twenty-nine poems. The fascicles, gatherings of four to seven folded sheets sewn together, have been reconstituted in R. W. Franklin's *The Manuscript Books of Emily Dickinson.*[2] It is the fate, and therefore the nightmare, of editors to ignore the obvious, and it seems not to have occurred to her early editors that the poet must have had an artistic reason for making the fascicles.[3] Although Dickinson's first editor, Mabel Loomis Todd, appears to have ignored the fascicles except as sources of completely independent poems, she at least is probably guiltless of the unfortunate mutilation and dismantling of the booklets.[4] Fortunately, later scholars (not least of them Johnson, despite the arrangement he decided on for his collection) have been able to reassemble them with reasonable confidence in the results. With the publication of Franklin's facsimile edition, we can expect general agreement about which poems go into which fascicles, and in what order, despite some inevitable problems.

That Emily Dickinson had something like sequences in mind is the most natural of conjectures. After all, we do have the evidence of her own groupings. Enough of her work, too, is both superbly accomplished and imaginatively far-reaching to make her a peer of other major poets who have put whole poems into larger structures of intrinsic relationship. Her mind was subtle and original, and the reciprocities among her poems show her grasp (perhaps the best among poets writing in English in the last century) of the way separate poems within the same surge and float of feeling reinforce one another and grow into an encompassing, unified body. Had she prepared the fascicle groupings for *publication,* she would doubtless have worked with them further—for reasons we shall suggest in discussing sequence-formation in relation to Yeats's *Last Poems and Two Plays,* in Chapter Six. But that the fascicles are, by their very nature, either sequences as they stand or sequences in the making seems as obvious as such things can be.

Sensing this reality without letting themselves be guided by the obvious implications—namely, a poem is not a treatise; its phrasing moves organically or not at all—students of Dickinson have proposed thematic schemes to account for the fascicles. Ruth Miller, for instance, argues that each one follows the same basic pattern: "Each is a narrative structure designed to recreate the experience of the woman as she strives for acceptance or knowledge, is rebuffed or fails because of her limitations,

2. R. W. Franklin, ed., *The Manuscript Books of Emily Dickinson* (Cambridge and London: The Belknap Press of Harvard University Press, 1981), two volumes.
3. See the apt comments on this situation in Richard B. Sewall, *The Life of Emily Dickinson* (New York: Farrar, Straus and Giroux, 1974), pp. 537–38.
4. See the lucid discussion of Mrs. Todd's editorial role in R. W. Franklin, *The Editing of Emily Dickinson* (Madison: University of Wisconsin Press, 1967), especially pp. 31–34.

but then by an act of will, forces herself to be patient in order to survive, fixes her hopes on another world where Jesus and God await her, and remains content meanwhile with herself alone."[5] But apart from the overriding fact that one is hard put to find this scheme in the actual poems, the work of a Dickinson can hardly be reduced to a monotonous formula. She wrote, in a storm of creation over a relatively short space of time, a vast number of poems of high intensity. She arranged them into physically linked structures that enabled her to give tentative order to the chaos of emotions by which the writing was seized. This was not a matter of repeating the same exemplary tale over and over. The poems penetrate a life of secret turmoil, each striking a certain held pitch of awareness; and the fascicles mobilize these little systems of subjective energy into larger ones, permitting a more complex equilibrium among affects. Why translate all this orchestrated sensibility into gray commonplaces?

2. Towards a Multiple Sequence

Study of the fascicles as sequences has hardly begun, although Franklin's edition of the manuscript books is an enormous aid. Already, however, we can see that a number of fascicles are full-fledged ones and a fair number seem at least modulations toward the form. We have chosen only two for concentrated discussion: Fascicles 15 and 16.[6] Both come out of the violent psychological and artistic upheaval (perhaps not fortuitously coincidental with the outbreak of the Civil War) during which Dickinson found her level as a great poet. They include such poems of tragic force as "The first Day's Night had come" (poem 410), "The Color of the Grave is Green" (411), " 'Twas like a Maelstrom, with a notch" (414), "I read my sentence—steadily" (412), "Before I got my eye put out" (327), "I felt a Funeral, in my Brain" (280), and " 'Tis so appalling—it exhilirates" (281). Poems like these are revelatory, magnetic centers in themselves. Their relation to other poems in their respective fascicles is like that within a planetary system: a process of tensions and counter-

5. Ruth Miller, *The Poetry of Emily Dickinson* (Middletown, Conn.: Wesleyan University Press, 1968), p. 249

6. We are following Franklin's essentially chronological fascicle-numbering—see *The Manuscript Books*, Vol. II, pp. 1391–1408 especially—rather than Mrs. Todd's random but hitherto standard numbering according to which 15 and 16 were originally 26 and 32. Also, the former originally ended with the nine poems Franklin places at the end of 14. (We had some years ago begun thinking of 26 and 32 as a double sequence, and so Franklin's purely scholarly reordering seems a welcome mutual corroboration by two disciplines.)

tensions in motion, self-contained yet not rigid. At the same time there is an accumulative forward motion from poem to poem. The whole system of relationships and balances, implicit from the start, reveals itself in the course of the sequence.

Franklin dates both fascicles "about 1862," and their inner reciprocities suggest that most of 16 was written after 15. But the literal order of composition of the individual poems is scarcely a vital question here; as we shall see with Yeats's civil-war sequences, the original order is often changed and even reversed during sequence-formation. What *is* important is that both fascicles issued forth in the same general creative period, under the same sustained impulse of feeling and at white-hot intensity. They seem to form a reciprocal or double sequence: a progression from the shock of devastating experience evoked in "The first Day's Night had come" at the start of 15 to the distanced, qualified, just-held affirmation of the final group of poems in 16—especially in "When we stand on the tops of Things" (242) and "He showed me Hights I never saw" (446). But each fascicle has its own independent curve of movement as well. For the reader wishing to reconstruct 15 and 16 on the basis of Johnson's text and numbering of poems, the order is as follows:

Fascicle 15: 410, 411, 414, 580, 415, 419, 420, 421, 577, 412, 416, 417, 418, 581, 413, 578, 579 (*17 poems*)
Fascicle 16: 327, 607, 279, 241, 280, 281, 282, 242, 445, 608, 446 (*11 poems*)

A certain affinity with Whitman is seriously present no matter what their differences were. Emily Dickinson might well have seen the 1855 and 1860 editions of *Leaves of Grass,* despite her humorously suspect, prim denial. Replying to an inquiry of T. W. Higginson's, she wrote: "You speak of Mr. Whitman—I never read his Book—but was told that he was disgraceful." Yet the two poets are kindred sensibilities, a fact that becomes more visible when their reciprocities are seen in the magnifying lenses of their sequences. Whitman, to be sure, was more direct and explicit about sex, but no more sensitized to it if we read Dickinson aright. He plunged more ardently, with strongly sexual overtones, into his intoxicating commerce with nature, and was relatively untroubled by any sense of division between himself and nature or deity. In general, he provides external detail and circumstance plentifully, addresses himself to the reader, and relatively seldom keeps his literal subject hidden from us. Dickinson, more complex and starkly conscious of the betrayal of manifold hopes, finds many more shadings and levels and degrees of half-alienated relationship among self and nature and God, just as she does in her relations with other persons and her sense of herself. She pushes further, with more psychological precision, than Whitman does into the subjective realm of self-doubt and of confusions and agonies of spirit.

This difference will be clear enough if we read the first poem of Fascicle 16 (poem 327) and compare it with the opening poems of *Song of Myself.* Whitman lays claim to the universe, nonchalantly, grandly, and of course assertively:

> Stop this day and night with me and you shall possess the origin of all poems,
> You shall possess the good of the earth and sun, (there are millions of suns left,)

In the Dickinson poem an introspective sensibility, disheartened by some debilitating personal experience, cannot face possessing and being possessed by the universe. Her dominant image, shocking when we visualize it literally, is flung at us in the first line: "Before I got my eye put out."

And yet both works place the central sensibility on a vast stage: the cosmos. Dickinson's opening poem in Fascicle 16 reads as if, after all, she *had* read Whitman's "disgraceful" book and was replying: "I was like you, Walt, until disaster struck. Myself *I* celebrated too, and I looked into the sun as its equal and was blinded. I was blinded because the joy was fuller than I could endure." Poem 327 is on Whitman's scale in its reach into infinity; at the same time it closely matches those passages of *Song of Myself* that confess extreme vulnerability and personal defeat.

> Before I got my eye put out—
> I liked as well to see
> As other creatures, that have eyes—
> And know no other way—
>
> But were it told to me, Today,
> That I might have the Sky
> For mine, I tell you that my Heart
> Would split, for size of me—
>
> The Meadows—mine—
> The Mountains—mine—
> All Forests—Stintless Stars—
> As much of noon, as I could take—
> Between my finite eyes—
>
> The Motions of the Dipping Birds—
> The Lightning's jointed Road—

For mine—to look at when I liked—
The news would strike me dead—

So safer—guess—with just my soul
Upon the window pane
Where other creatures put their eyes—
Incautious—of the Sun—

(327)[7]

So the two poets share extreme volatility and a vision of the cosmos as the theater of their feelings and imagination. One is tempted to make a major effort to account for the fact that they are soul-mates; it would not have taken much of *Leaves of Grass* to open Dickinson to her true possibilities and to the language of every kind of intensity. Stranger things have happened than that Dickinson should have read Whitman secretly. Whether she found him irresistible or disgusting or even mostly tiresome would not have mattered; the fertilization would have taken place. Compare, for instance, her poem 327 with poem 25 in *Song of Myself:*

Dazzling and tremendous how quick the sun-rise would kill me,
If I could not now and always send sun-rise out of me.

We also ascend dazzling and tremendous as the sun,
We found our own O my soul in the calm and cool of the daybreak.

My voice goes after what my eyes cannot reach,
With the twirl of my tongue I encompass worlds and volumes of worlds.

Both poets find the natural world, especially the sun, murderous in its glory. Whitman, magnificently, summons up a rival inner glory as his only salvation. The process of converting terror into triumphant imagination is presented swiftly by Whitman, its stages collapsed into an assertion unsullied by self-examination. Dickinson's "Before I got my eye put out" is altogether subtler and humbler, but the predicament and pressure are much the same in both poems. She was as capable as he of writing in great high spirits. (One of her poems, in fact—poem 308 in Fascicle 27—begins "I send Two Sunsets," thus going Whitman's creation of his own sunrise one better.) If in poem 327 she gives us a developed imagery of fear and withdrawal, an internal stocktaking that confesses failure to cope adequately with blazing life, even here she flaunts her imagination as flamboyantly as Whitman does.

The poem sets a sensational scene at once. The opening, "Before I got my eye put out," is grotesquely drastic. Two ways of seeing—with

7. We are using the fascicle version of 327 without the suggested change in line 15, "The Morning's Amber Road."

the eyes, or senses, and with the soul, or imagining sensibility—are contrasted in an atmosphere of extreme emotional tension. In a succession of startling, paradoxical intensities, the middle stanzas present the pain of being forced to retreat, because of the brutally blinding effect of natural beauty, into the soul's possibly "safer" enclave. The only leavening elements are the occasional colloquialisms and the pun of "I" and "eye" in the first line. But the pun slams us into the gross physical image of "eye put out," and the colloquialisms (e.g., "got my eye put out," "as much of noon, as I could take," and "strike me dead") all emphasize the destructive effect of too keen awareness and fear.

The tension here, between lust for more experience and crushed withdrawal, is at the heart of the poem's volatility. It is an intimate ingredient of the structuring energy in Fascicles 15 and 16. After the riddling of the first stanza—its unexplained allusion to a wound, and its half-promise to reveal some "other way" of vision unknown to ordinary beings—the poem breaks into a paean to the natural world despite what has happened. But the paean is punctuated with notes of fear, and we see that the speaker is locked into a complex depressive state. Absolutely intoxicated with life, she can no longer face it except in imagination, for she cannot risk opening herself up at close range to powerful feelings once again. The final stanza, as riddling as the first, concentrates the whole psychic situation into a new image: of the soul facing the sun as the eyes, before they were blinded by too fierce exposure, had used to do:

> So safer—guess—with just my soul
> Upon the window pane
> Where other creatures put their eyes—
> Incautious—of the sun—

As so often with Dickinson, we have an emotional state whose basis is a mystery. The ambiguous word "guess" is another riddle. Probably the suggestion is that it is safer to imagine the world than to experience it nakedly; but another possible reading is that this may not be so—I only *guess* it will be safer—and that the soul too is dangerously exposed. Either way, what is the situation—a literal onset or threat of blindness? a crippling disappointment or humiliation? a miserable passage-at-arms in love? All the *poem* gives us is its affect of passionate recollection in a context of wounded withdrawal. The stanzas present so many centers of evocation that we need no "story." And still, they have the *form* of narrative and refer to unspecified crucial events. The poem presents the aftermath of a heart's voyage into lovely, unexpectedly deadly places, and it offers something like a resolution—all in the elusive context of envisioned moments of choice and crisis, each with its own storm center of language. For these reasons, poem 327 epitomizes the process of the fascicles we are considering.

We have focused at some length on the one poem, and also on Dickinson's affinities with Whitman, to emphasize and re-emphasize the scope, variety, and shifting depths of which she is capable. Together, Fascicles 15 and 16 make an epic of the subjective life, as far-reaching and heroically brave as *Song of Myself* and far more pressing in its pursuit of the inmost depths of feeling—with no attempts to glamorize those depths. The drastic states with which these fascicles wrestle demand, and receive, large-minded, universally expressive presentation, and none of her successors—not even Eliot at his best—has surpassed her work.

Before going more fully into the two sequences and their reciprocities, we should note that Dickinson's proliferation of relationships among the fascicles goes beyond the connection of any pair of them. Fascicle 14, for instance (although we shall not pursue the matter here), seems clearly a testing ground, preparing the way for Fascicles 15 and 16. It has an opening group of poems that are in the main bitingly wry or bitter, then a single poem, "The feet of people marching home" (7), that is a masterpiece of restrained melancholy and longing, and then another group that reaches powerfully toward the kind of grief and shock with which Fascicle 15 begins. Similarly, Fascicle 17, after 16 has striven so bravely to contain and come to terms with that grief and shock, makes another turn on the struggle for equilibrium. In itself, it is an exquisite sequence that balances life's sources of ecstasy against its horrors—the first line of its opening poem (348) is perfectly indicative: "I dreaded that first Robin, so." And its antepenultimate poem, "The Soul has Bandaged Moments" (512), moves between these two states as Homer moves between a sense of divine prowess and the pathos of Hector's defeat in the *Iliad.* It seems likely that Emily Dickinson was heading toward a multiple sequence, compelled by the nature of her sensibility and her extraordinary copiousness, very much as certain successors—notably Yeats, Pound, and Williams—were to do later.

3. Fascicle 15: The Poetry of Psychic Trauma

The entire movement of Fascicle 15 is like that of poem 327 writ large. Its psychological and poetic coherence is extraordinarily powerful, and its dynamics are characteristic of the modern sequence. The opening poem, "The first Day's Night had come" (410), presents a protagonist—a sensibility, rather—unspecified as to sex or other external characteristics, staving off madness and chaotic self-disintegration in the wake of some unnamed catastrophe. The closing poem, "I had been hungry, all the Years" (579), presents a psyche stunned into numbness after devastation. We have been carried from terrified whirling to a quietened desolation. There is much more to the sequence, however, than this simple curve, so reminiscent of poem 327. In seventeen poems the poet gave herself room to create many more dimensions and to explore drifts of related tonality more richly. It might not be amiss to call Fascicle 15 a kind of purer *Maud,* without the contrived support of plot, named characters, and exposition.

The opening movement of three poems is unrivaled as a powerfully lyrical grouping, although only the third is at all well known. They are "The first Day's Night had come" (410), "The Color of the Grave is Green" (411), and " 'Twas like a Maelstrom, with a notch" (414). In these three poems the fascicle starts at the height of intensity: extreme feeling at once, together with an atmosphere of morbid horror, and then the ambiguous agony of the third poem. Yet all three have an iron discipline of sustained and subtly compressed association, as though a rigorously self-analytical Ophelia were speaking. This first movement culminates in a poem of anguished moral dilemma, recalled in a series of nightmare images beginning with the opening line, " 'Twas like a Maelstrom, with a notch," and rendered doubly painful by the rueful confessional question at the end: "Which Anguish was the utterest—then— / To perish, or to live?"

It would be dogmatic, and therefore pointless, to argue that ideally a sequence should begin with a poem of great power, immediately creating a decisive affective center. The rest of the sequence would then define itself in relation to this initial field of force. Obviously, the position is too limiting; for many reasons, the location of a work's magnetic centers must vary according to its idiosyncratic character. Nevertheless, there are special advantages to beginning, as it were, climactically. It is the sequence's way of beginning *in medias res.*

Fascicle 15 does precisely this. It copes with an encompassing poem whose several rapidly developed phases present psychic trauma that changes but does not cease. The "grateful" feeling after enduring the first shock proves deceptive. The condition of extreme suffering persists from that shock onwards. It goes underground into shattered numbness that seems like relief for a brief time, then returns in a monstrous tidal wave of obliterating "horror," and remains as hysterical "giggling" and something like schizophrenic self-division within the sensibility at the moment of the poem's utterance. The succession of tenses—past perfect to past to present—accumulates the whole movement in a single time-compressed curve.

The first Day's Night had come—
And grateful that a thing
So terrible—had been endured—
I told my Soul to sing—

She said her Strings were snapt—
Her Bow—to Atoms blown—
And so to mend her—gave me work
Until another Morn—

And then—a Day as huge
As Yesterdays in pairs,
Unrolled it's horror in my face—
Until it blocked my eyes—

My Brain—begun to laugh—
I mumbled—like a fool—
And tho' 'tis Years ago—that Day—
My Brain keeps giggling—still.

And Something's odd—within—
That person that I was—
And this One—do not feel the same—
Could it be Madness—this?

(410)

The poem is a grim mystery, and the sequence as a whole an extension of it. We are never told what the "thing so terrible" was. Later poems in the sequence imply that the shattering event was a lover's death or, alternatively, a symbolic death: the end of an overwhelming love relationship through separation or rejection or abrupt disillusionment. But other readings are conceivable; the sequence does not pin itself down to one limited, autobiographical event. The poems focus not on the event itself but on the way it is received—the sensibility's efforts to cope with hideous personal disaster. The "thing" is a secret, but the resulting hysteria and disorientation, the whole turmoil of feelings and introspection, must erupt. This double pressure, of what must be and what must not be expressed, fills the whole sequence with passionate paradox.

The first poem, just quoted, is a complex of riddles. Even the opening line, "the first Day's Night had come," may be read in two ways, both obvious: The "first Day"—first in the sense that it initiates a whole new chapter of the poet's life—was so dreadful it *was* "Night"; and, more likely, the evil day had at last come to an end. But this is all a language of feeling. We would be wrong to take the words altogether literally. In the poem language for states of feeling and language for points in time run together inseparably. Years have passed, yet the "Day as huge / As Yesterdays in pairs"—the day when madness or overwhelming horror akin to madness took hold once and for all—goes on forever.

The lines on the Soul hold some mystery as well. Like the lines on days and nights, they do not spell out logical meanings or easily paraphrasable associations. Yet they do project emotional and subjective states, very purely. The speaking "I" is split off from her personified Soul, another "she," whose specific nature is not altogether consistently presented. They give each other instructions. "I" tells the Soul to sing. The Soul replies that she cannot; her "strings" and "bow" (her music-making ability, no doubt standing in for her poetry-making one) are ruined now and the "I" must set to work to mend them. But the poem says mend *her;* the Soul is its ("her") own instrument, then. Confusing, and utterly clear. The task of makeshift therapy is short-lived, however. When the full, sickening second wave of reality-recognition strikes, the enterprise is forgotten. The "Brain," its alienation from the Soul completed, is reduced to blind and "giggling" helplessness. All sense of the self's integrity is lost; it is now "that person that I was." And the question at the end ("Could it be Madness—this?") raises a final riddle, since it has apparently already been answered, in the affirmative, in the preceding stanza.

Two-thirds of a century later, in *The Tower* and *The Winding Stair,* Yeats too employed misery-born riddling, sometimes in images strikingly like Dickinson's. Rich and vital as his poems are, Dickinson anticipates them and goes more directly to her crucial affects. The protagonist of "Sailing to Byzantium" seeks the very solution that "The first Day's Night" turns to at once: the Soul's "singing school." Yeats's poem loses itself in this dream, while the predicament of being "sick with desire / And fastened to a dying animal" persists. Dickinson's poem presses further, unremittingly. Again, she has anticipated, but with no self-indulgence, the condition of apparent madness envisioned in Yeats's "All Souls' Night" as the saving state that can put the living in touch with the dead and beyond intrusive reason. In "A Dialogue of Self and Soul," too, there is a similarity to the Dickinson poem (apart from the bitter pressures underlying the poems, and the fact that self and soul speak to one another so demandingly in both cases) in the use of innocent ecstasy to escape from the unbearable. But there is a vast difference in Dickinson's presentation of the condition as sheer breakdown rather than transcendence—although her question at the end may be introducing the latter possibility. It might be argued that her work presents those perspectives she in some sense shares with Yeats more cogently and, if not more bravely, less swaggeringly than he does. The important fact, though, is that she is in her way ushering in the poetry of psychological pressure through her drastically stripped down method and absolute directness of feeling. She shares the world of Yeats and the other great moderns and, in fact, points some directions they never had the opportunity to observe because so much of her work has been unavailable until fairly recently.

Sequences, ordinarily, tend to reach a state of equilibrium among centers of feeling that is elegiac in effect—a distanced perspective that marks the condition of "all passion spent." This too is anticipated in Fascicle 15, but at the same time too much power, and too much refusal simply to accept, manifest themselves for the sequence really to settle for such a balance. The succession of explosions of massive, painful insight in the first three poems would be enough to prevent it. But in any case there is another such explosion at the center of the sequence, in "If I may have it, when it's dead" (577). And the exhausted inability to accept newly proffered riches of life and love after all in the closing poem, "I had been hungry, all the Years" (579), is an afterbeat of the earlier poems of power, hardly a quietly passive resignation to fatality.

But to go on with the progression after the opening poem: The second poem, "The Color of the Grave is Green" (411), deserves to be known far better than it is:

The Color of the Grave is Green—
The Outer Grave—I mean—
You would not know it from the Field—
Except it own a Stone—

To help the fond—to find it—
Too infinite asleep
To stop and tell them where it is—
But just a Daisy—deep—

The Color of the Grave is white—
The outer Grave—I mean—
You would not know it from the Drifts—
In Winter—till the Sun—

Has furrowed out the Aisles—
Then—higher than the Land
The little Dwelling Houses rise
Where each—has left a friend—

The Color of the Grave within—
The Duplicate—I mean—
Not all the Snows c'd make it white—
Not all the Summers—Green—

You've seen the Color—maybe—
Upon a Bonnet bound—
When that you met it with before—
The Ferret—Cannot find—

(411)

The images here are even more immediate and wrenching than in the first poem, in which the speaker is describing a psychological event at least objectively enough to pin a name—"Madness"—to it and to present it, partly, in terms of a dialogue between self and soul that involves a project for healing the condition. In the second poem the dialogue form is dropped in favor of direct presentation of an inner desolation so black that the self—"That person that I was"—seems irretrievably lost. Horror is pushed to further extremes of presentation as the poem advances inexorably to its final, rank image of a ferret trying to unearth a corpse. This last image is telegraphed in the second line by the emphasis on the "outer" grave, which makes us think immediately of the blackness of the grave itself. But the opening four stanzas stress the gentle exterior—the grave as a natural part of the landscape—and the cozy companionableness of the dead. The shift comes with the fifth stanza, when Dickinson goes contrary to our expectations of the outer grave's complement. She is not talking of a physical grave at all; her concern is with the "Duplicate" grave within the human psyche. And whereas nature can beautify an actual grave and make it less terrifying, nothing can alleviate the inner blackness. There is so deep a nothingness that not even the corpse of the speaker's former self can be found. Spirit itself has shrunk into that same realm of concentrated, heavy non-being described long ago by Donne in "A Nocturnall upon St. Lucie's Day"—that touchstone of depressive lyric poetry of loss. But Dickinson has added the dimension of compressed hysteria close to madness.

Then, with a mighty torque-effect, the third poem, " 'Twas like a Maelstrom, with a notch" (414), appears in the sequence. It strikes full force: a poem in which the murderous, forced renunciation suggested in the two previous poems is weighed against the crushing guilt of letting go and yielding to desire:

'Twas like a Maelstrom, with a notch,
That nearer, every Day,
Kept narrowing it's boiling Wheel
Until the Agony

Toyed coolly with the final inch
Of your delirious Hem—
And you dropt, lost,

When something broke—
And let you from a Dream—

As if a Goblin with a Gauge—
Kept measuring the Hours—
Until you felt your Second
Weigh, helpless, in his Paws—

And not a Sinew—stirred—could help,
And sense was setting numb—
When God—remembered—and the Fiend
Let go, then, Overcome—

As if your Sentence stood—pronounced—
And you were frozen led
From Dungeon's luxury of Doubt
To Gibbets, and the Dead—

And when the Film had stitched your eyes
A Creature gasped "Repreive"!
Which Anguish was the utterest—then—
To perish, or to live?

(414)

To leap ahead for just a moment, this poem provides a significant link with the two most powerful poems of Fascicle 16: "I felt a Funeral, in my Brain" (280) and " 'Tis so appalling—it exhilirates" (281). It is the poem *par excellence* of agonized suspense and equally agonized release. Its complex of feeling is echoed in the whole of 280 and in the speaker's piercing thought in 281—after "bleak dreaded" woe has finally arrived—that "Suspense kept sawing so." Also, all three poems display drastic modes of imagery that serve as a kind of refrain in the two sequences: images of psychic numbness, terror, and dropping downward into bottomless disaster. But 414 is unique in the escalating power of its almost symmetrically balanced affects. It objectifies, with absolute authority, the quality of the predicament afflicting the central sensibility of both sequences.

Clearly, each of the poem's three major units of affect is an extended simile for an unspecified, suspense-ridden crisis of spirit. The crisis itself, in its dual nature of terror and release, is represented only by the word "it," in the contraction (" 'Twas") that starts off the poem. What "it" was like is presented to us in a long sentence—the first twenty-two of the poem's twenty-four lines—containing the three major affects. "It" was like a maelstrom sucking "you" into its center and dropping you downward (stanzas 1–2); like being in the grip of a monstrous being, "Goblin" and "Fiend," about to crush you in "his Paws" (stanzas 3–4); and like being on the verge of being executed for some crime after your

"Sentence stood—pronounced" (stanzas 5–6, except for the last two lines).

In each hideous image-complex, suspense builds up to an unbearable point, then collapses into release from the nightmare climactic moment. The escalation of power derives not from greater intensity of phrasing in each stanza-cluster but from new perspectives on essentially the same affect—the combination of helplessness in the grip of superior force and release by another superior force. Merely physical peril (Dickinson's maelstrom image is in some ways an intensification of the whirlpool with the coffin life-buoy at its center in the epilogue to *Moby-Dick*) is replaced by being helplessly in the grip of an evil supernatural force, and then by a death sentence because one has committed a crime. At the same time, the movement is accumulative, each stage of the poem providing a context for the next, so that the larger affect of the whole poem (through line 23) is of extreme physical and moral danger complicated by profound guilt. The question with which the poem ends then turns the poem right around in the opposite direction. All those last-moment escapes, always brought about from the outside—"something" breaking the maelstrom's force, God remembering, the "Creature" that gasps "Repreive"—are not blessed relief after all but a rival agony. One then recalls the sexually suggestive phrases in the opening stanza: "Maelstrom, with a notch," "boiling Wheel," and "delirious Hem." Hence, the struggle between God and fiend, and the sense of being condemned for a crime, take on the dimension of an inner moral struggle in which conscience is triumphant at the expense of fulfillment.

The poem thus puts into a certain focus the two preceding poems, which hove into view with such devastating impact at the start of the sequence. We do not mean that the poem is literally "about" the emotional and moral agony of a woman, stung by passion and almost drawn into yielding herself, who is "saved" from sinning but not from the poison of her loss thereafter. Yet the fascicle's contexts of association invite such a reading. They include the desperate extremes of feeling ("Could it be Madness—this?") of the first poem (410); the sexually suggestive admission that "I gave myself to Him" in a hopelessly uneven "contract" of the fourth poem (580); and the dimension of morbidity ("If I may have it, when it's dead") in the ninth poem (577). That is, the psychological float of the sequence includes drifts of guilt, sexual need, moral struggle, and dreams of gratification through imagined communion beyond death.

" 'Twas like a Maelstrom, with a notch," then, is the critical unifying poem of its fascicle. In it the speaker has been terribly buffeted and shamed, and a deceptive reprieve has been granted that only forecasts the disappointed and sickened equilibrium of the closing poem. But many of the intervening poems, as we shall see, suggest a great and vulnerable love, complicated by moral scruples, that has wrenched the speaker out of her simpler delight in the world and has been physically defeated either by a lover's death or an absolute termination of a relationship. Putting the matter this way oversimplifies the situations of the poems, which are always open to a number of possible interpretations. In " 'Twas like a Maelstrom," for instance, many signs indicate a state of sexual need, guilt, and frustration. At the same time, the condition *might* be one of threatened loss of religious faith, or of total loss of self-control, or indeed of a mixture of these and other predicaments. The poem consists of a series of extended similes telling us what "it"—the unnamed crisis around which the poem is spun in an effort to characterize its quality—was "*like*" but not what it *was*.

The three succeeding poems then swirl around this central complex of experience. The first of them, "I gave myself to Him—" (580), even lets go into an atmosphere of sweetly precarious risk-taking, in fine contrast to the painful affects surrounding it. It gently foreshadows certain poems further along, such as the fancifully joyful eleventh poem, "A Murmur in the Trees—to note" (416); and also the thirteenth and fourteenth poems, "Not in this World to see his face" (418) and "I found the words to every thought" (581), which recall an indefinable ecstasy, never to be surrendered, from the passionate past. Poem 580 is more than a beam of charming sunlight, however. It makes room for visions of rich delight and love—commits itself to them—and leavens the general tone, anticipating similar later effects in the sequence. Still, it has its darker underside—its sense of making do within disillusionment. It seems to hark back to a moment before the trauma of the opening poems and remind itself of a time of hope and danger that preceded it, when there was a "contract" with a lover (or life, or God—but the language suggests a love-relationship) that was "solemn" and was "ratified" yet still uncertain:

> I gave myself to Him—
> And took Himself, for Pay,
> The solemn contract of a Life
> Was ratified, this way—
>
> The Wealth might disappoint—
> Myself a poorer prove
> Than this great Purchaser suspect,
> The Daily Own—of Love
>
> Depreciate the Vision—
> But till the Merchant buy—

Still Fable—in the Isles of Spice—
The subtle Cargoes—lie—

At least—'tis Mutual—Risk—
Some—found it—Mutual Gain—
Sweet Debt of Life—Each Night to owe—
Insolvent—every Noon—

(580)

If we have been at pains to suggest the large vision, and even epic scope, of this very subjective and private poet's art, we should also make bold to note the sophistication of a poem like this one. The imagery is worldly, yet might be read as religious. The sense of love as a contract, entered into with one whose expectations may be disappointed while the dreams behind them somehow can still prevail, is equally ambivalent. Ambivalent or not, however, there are fear and distrust of love, and marvelous, hovering dreams, and some oppression by the sense of dependency on a male figure. The poem's view of love and of existence generally is a bittersweet, gambler's view, almost gaily pessimistic.

It is this darker, subtler side of 580 that connects it with the starker poems preceding it and the two poems of disequilibrium immediately following it: "Sunset at Night—is natural" (415) and "We grow accustomed to the Dark" (419). The first of these swiftly recapitulates the opening poem's sense of total, dreadful disorder. ("Midnight's—due—at Noon"; "Jehovah's Watch—is wrong.") The second plays on the imagery of light and darkness, which recurs in Dickinson's work in many contexts, once more. Its nearly innocuous first line, "We grow accustomed to the Dark," telegraphs a shift from normal, commonsense talk (about the adjustments of vision to literal darkness) to the "larger—Darknesses" of depression close to madness: "those Evenings of the Brain." Here too there is adjustment of a kind, but the brooding irony of its character is obvious:

Either the Darkness alters—
Or something in the sight
Adjusts itself to Midnight—
And Life steps almost straight.

The sequence maintains its wobbling balance, with its inability to lift itself out of the negative even when, as in the next poem, "You'll know it—as you know 'tis Noon" (420), it strikes a cheerful note of faith. There is an immediate lapse, in "A charm invests a face" (421), into the precise mood of 580: wary, minimally hopeful, refusing to shut possibility out somehow. We would almost be persuaded that the wounded sensibility of the sequence has kept itself in order pretty well after all, and has at least not lost all resiliency and capacity for self-renewal. But then a single strange poem, "If I may have it, when it's dead" (577), disabuses us and we are back in the world of the first three poems of the fascicle, albeit in a different key.

As with the ferret image in the second poem, so also the corpse-imagery in this ninth poem is rankly physical. At the very heart of the sequence, poem 577 is grotesquely obsessed with a lover's corpse. Its opening line, "If I may have it, when it's dead," is matched in necrophiliac ardor by its ending:

> Forgive me, if to stroke thy frost
> Outvisions Paradise!

This is at once one of the most riddling and most interesting poems in the sequence. In form it is a welter, a mélange of quatrains—some in ballad meter, both rhymed and unrhymed, and some made up of rhyming tetrameter couplets—together with an irregular cinquain ($aa \times^4 \times a^2$). The primary end-rhyme pair is *Thee-me* ("Thee" is used in terminal position four times, "me" three times), and the constant repetition helps considerably in binding the stanzas together. And it should be noted that the first and last stanzas, although separately unrhymed, together constitute somewhat of a rhyme scheme: *xaxb abxx*. The poem may have been conceived of as a "mad" poem, deliberately disoriented in various ways, that follows through on the state of "Madness" set up and then questioned in the first poem. Similarly, its morbidity may be an afterbeat of the second poem, whose attention is so riveted on the grave and what lies hidden within it (and on the self as its own grave once life has shattered its most cherished meanings). The initial affect has a certain literal grossness that counteracts the conventional sentimentality with which the passage is tinged:

> If I may have it, when it's dead,
> I'll be contented—so—
> If just as soon as Breath is out
> It shall belong to me—
>
> Until they lock it in the Grave,
> 'Tis Bliss I cannot weigh—
> For tho' they lock Thee in the Grave,
> Myself—can own the key—

The repeated "it" in three of these lines is so suggestive of an inert body—a cadaver (with traces of the head Isabella buried in her pot of basil)—that we must take the language at face value. The literal situation envisioned is that the "I" of the poem will have sole access to the body until it is buried, and that this contact will allow the speaker to bridge the distance between herself and her lover after death. The rapture at the prospect, combined with the macabre circumstances and hints elsewhere in the poem, indicates that this face-to-face meeting will be the first ever. That is, at last the poem's protagonist will be able to put a face on the warm presence, conjured up by her love and need, "buried" inside her and yet a living spiritual companion.

> Think of it Lover! I and Thee
> Permitted—face to face to be—
> After a Life—a Death—We'll say—
> For Death was That—
> And This—is Thee—
>
> I'll tell Thee All—how Bald it grew—
> How Midnight felt, at first—to me—
> How all the Clocks stopped in the World—
> And Sunshine pinched me—'Twas so cold—
>
> Then how the Grief got sleepy—some—
> As if my Soul were deaf and dumb—
> Just making signs—across—to Thee—
> That this way—thou could'st notice me—

These three middle stanzas seem to recapitulate the experience of suffering in the first two poems of the sequence, but now the imagined scene is one of trusting, self-revealing conversation by lovers (perhaps in one another's arms). We still do not know whether an actual lover has actually died, or whether the speaker is addressing a desired lover with whom there will never be a relationship, or whether there was a wrenching parting with a man still living—either of the last two possibilities would be a kind of death where such vulnerably passionate commitment has been present. But the deprivation, the sense of lonely abandonment, suggests that the poem's deepest motivation is the need for communion at all costs. The need is pathetic—felt to be desperately hopeless—and accompanied by the conviction that the desired communion will never be permitted, whether by God, or by society, or by the longed-for lover, or even by the speaker herself. The final stanza says:

> Forgive me, if the Grave come slow—
> For Coveting to look at Thee—
> Forgive me, if to stroke thy frost
> Outvisions Paradise!

Many tonalities in this complex poem come together in this stanza. "Thy frost," most literally, refers to a cold, dead body. The phrase also recalls the earlier conceit used to suggest an ambience of utter coldness: "And Sunshine pinched me—'Twas so cold." (We are probably also in the realm of poem 281, in Fascicle 16, where "Truth" is revealed as "Bald, and Cold.") The implication that it was sexual coldness, despite the intensity, which blocked off the relationship and prevented deeper communion, is an obvious possibility. The coldness may have been on either partner's side, or a matter of timidity or inhibition, or just the frozen-out isolation of the speaker from the person whose love she dreamed of but never experienced in any way. Even the reference to delaying Paradise in favor of a lesser but more immediate and perverse satisfaction ("stroke thy frost") may conceivably refer to a psychological problem related to fear of consummation. Literally, of course, in this near-conceit, the speaker is saying that she needs to cherish the dream of warm-cold communion even at the expense of putting off a quick death that would enable her to join her soul-mate in Heaven immediately—one of several instances of the simultaneous use and circumvention of sentimental clichés in this poem. However one reads it, the poem reveals enormous courage in facing and exposing psychic humiliation and embarrassment, whatever shocking imagery and morbid imaginings may be revealed in the process. In this respect, 577 more than matches the poems that precede it.

The quietly composed poem that follows, "I read my sentence—steadily" (412), has the taut understatement of "The Color of the Grave is Green." Also, it carries forward, in a different key, the humiliation or embarrassment (here called "the shame") that contributes to the emotional mixture of "If I may have it, when it's dead." It presents an excruciating experience of rejection, couched in courtroom terminology. Once more, the "I" does not spell out whoever or whatever has shamed her so deeply that she cannot feel death will be "a novel Agony." In the dynamics of the sequence, this poem is a magnificent moment of control. Its sudden, deceptive final calm ("And there, the Matter ends—") provides an essential conversion of affect. To attain a deathlike tranquility and so reduce the shock of being "sentenced" to eternal anguish, the speaker has trained her Soul to become "familiar—with her extremity" and has succeeded all too well. She has made her Soul so tranquilly friendly with "Death" that the dreadful abnegation of life's hopes has come upon her impersonally and irrevocably: "silently, invisibly," as Blake would have put it. (The subtle reciprocities with "If I may have it, when it's dead" are established simply by juxtaposition.)

Six poems in a relatively muted key follow in the wake of "I read my sentence—steadily," as if to hold its Lethe-tending balance steady for a prolonged moment before the fascicle's final poem. A fey whimsy marks "A Murmur in the Trees—to note—" (416), whose charm reminds us of the more popularly known poems by Dickinson but which, in context, may suggest a soul on its way to the unknown and thereby isolated from the known world. "It is dead—Find it" (417) reinforces this latter impression about 416, which might otherwise be dismissed as forced, by its riddling questions and answers about the state of being dead. "Not in this World to see his face" (418) archly puts off the opportunity to experience that state; and "I found the words to every thought" (581) cheerfully reports a "thought" so blazing it cannot be expressed in language—although the speaker has never had the problem before. The rhetorical questions ending this tiny pair of quatrains make for a miniature climax of morale-sustaining within the set of seven poems that began with "I read my sentence—steadily." The questions suggest uncommunicable inspiration:

> Can Blaze be shown in Cochineal—
> Or Noon—in Mazarin?

The two final poems of the set following "I read my sentence—steadily," however, lean more toward that poem's mood than the emotional buffer-poems we have been noting. They remain lighter in spirit than 412, but nevertheless are tinged with its dark humiliation. "I never felt at Home—Below" (413) is playful in its distrust of "Paradise," echoing "If I may have it, when it's dead" in a folk-comic key, and in picturing God as an all-seeing telescope. But it is serious in presenting a wistful disorientation, together with unease at God's constant vigil and at the thought of a Judgment Day. "The Body grows without" (578) begins brightly and wittily ("The body grows without— / The more convenient way"), yet turns at once to an image of the harried spirit seeking shelter in the flesh. But the second of the two quatrains loses its edge. It is at once slightly complacent in its moralizing (as though the speaker were saying, "I've had these tempests of passion but nevertheless I'm virtuous through and through") and weakly derivative. Wordsworth tells us, with stirring tenderness in its context in "Tintern Abbey," that "Nature never did betray / The heart that loved her!" Dickinson says of the body that

> Ajar—secure—inviting—
> It never did betray
> The Soul that asked its shelter
> In solemn honesty

The first line of this quatrain is daring and beautiful, a marvelous image of feminine self-regard in seductiveness. But its glory hovers alone in the stanza, the rest of which hardly matches its style and volatility.

But the final poem of the fascicle, "I had been hungry, all the Years" (579), precisely and humanly places the state of the sensibility that has been riding the entire sequence. Essentially, it narrates a moment of triumph that has come too late and is fraught with sick and weary disillusion. A distant echo of the heart-wrung third poem, " 'Twas like a Maelstrom, with a notch," it cannot rise to an eccentric fate that seeks to reverse the despair with which the sequence has seemed to come to terms, and that would stir one, again, to enjoy exuberant love or simply the available delights of being alive. Perhaps the affect here is a correction of the promising yet unsatisfying "The Body grows without." If that poem presumes self-satisfaction at resisting unworthy temptation (or self-reassurance at having, "in solemn honesty," yielded to another equally sturdy and sincere soul), the closing poem brings an end to such whistling in the dark. There can be no recovery from the blows evoked in the opening poems, and the trauma inflicted, and the morbid probings. The corruption of hope spelled out in "I gave myself to Him" is permanent:

I had been hungry, all the Years—
My Noon had Come—to dine—
I trembling drew the Table near—
And touched the Curious Wine—

'Twas this on Tables I had seen—
When turning, hungry, Home
I looked in Windows, for the Wealth
I could not hope—for Mine—

I did not know the ample Bread—
'Twas so unlike the Crumb
The Birds and I, had often shared
In Nature's—Dining Room—

The Plenty hurt me—'twas so new—
Myself felt ill—and odd—
As Berry—of a Mountain Bush—
Transplanted—to the Road—

Nor was I hungry—so I found
That Hunger—was a way
Of Persons outside Windows—
The Entering—takes away—

(579)

We noted, at the start of our discussion of Fascicle 15, that the poem that opens Fascicle 16, "Before I got my eye put out" (327), might be regarded as epitomizing the movement of the seventeen poems we were about to deal with. To turn from "I had been hungry, all the Years" to Poem 327 is to recognize an embodiment, in altered imagery that is also a comment on the poet's art, of a process of realization that has reached one level of transcendence. Even the first line tells us the condition in which the new fascicle starts and its reciprocal relationship to Fascicle 15 within an enfolding multiple sequence.

4. Fascicle 16: A New Start

It is in the aftermath of the cruel struggle for balance between two levels of very highly charged affect (the horror concentrated near the start of Fascicle 15 and the incomplete repossession and final weakening and lost hope at the end, with intervening poems contributing to the balance) that Fascicle 16 acts in such finely attuned reciprocity with the longer sequence. The two fascicles share an obsession with death, literal and symbolic, and are charged with the same overwhelming memory of irrevocable loss. The link, and the difference, is beautifully clear in the first poem of Fascicle 16, discussed in some detail in the second section of this chapter. It presents a wounded, life-fearing condition paralleling that at the end of Fascicle 15. The self, at a minimal level of vitality, is just barely holding itself intact; but also, it makes a new start in recalling the full energies, however destructive, of the beautiful natural world the sunlight reveals. That is the ambivalent state presented in "Before I got my eye put out" (327), which conjures up the splendors of the world that have reduced it to its present misery: splendors it still perhaps longs for but that would kill again, at once, through their sheer generous brilliance could the speaker's vision be restored.

The second poem, "Of nearness to her sundered Things" (607), stresses memory—a force so powerful in this poem that images out of the safer past eclipse the present scene. The movement of the five stanzas is like that in the first poem. This time, though, the central three stanzas are not devoted to natural phenomena; instead, they center on a "Mouldering Playmate" and other "Shapes we buried."

The relationship between these opening poems, 327 and 607, is a hovering one within a narrow psychological border area between renunciation of the vital world and cherishing of what is irrevocably past. To recapitulate, the speaker in 327 has suffered a great disaster because of the full impact of "the Sun." Its appearance to her blinded her to lesser delights; now she has lost that Sun and, reduced to naked sensibility and imagination, could not bear the shock of new sensuous and emotional arousal. Yet the imagery shows how keen her response has been and would be again. The poem stands as an expression of most intense love for the world of seen things amidst a condition of darkness and fear. It has its "positive" side in the implication that the world's bright life is so beautiful it would, if restored to the speaker, break her heart with joy and make her mistress of more than she had realized before. But that would be a freedom treasonable to the love that has destroyed her. Poem 607 "therefore" (we are speaking of the logic of emotional displacement) converts the mood and context of the opening poem. In it the power of sensation becomes the power of memory; the dazzling Sun and all it reveals is replaced by loved persons, "Things," and moments that return from death's realm: "Bright Knots of Apparitions." Each loss of something dear to me was a death of myself—that is the implication, and it throws an intense light on the predicament of the first poem:

> The Grave yields back her Robberies—
> The Years, our pilfered Things—
> Bright Knots of Apparitions
> Salute us, with their wings—
>
> As we—it were—that perished—
> Themself—had just remained till we rejoin them—
> And 'twas they, and not ourself
> That mourned.

These are the closing stanzas of 607. In them an extraordinary evocation of the persistence of the past in memory takes a brilliant turn. For here the past mourns the future, and in a personal sense the child mourns its perishing into the adult. (The "Bright Knots of Apparitions" must be our former selves as well as other memories, with all the ways of seeing and loving we have formerly known.) Again—as in poem 577 of Fascicle 15—literal corpse imagery enters, in the figure of the "Mouldering Playmate" who returns. Then we move from the concentrated perception, couched hypothetically, in the line "As we—it were—that perished" in 607 to the desperately exhilarated "Tie the Strings to my Life, My Lord" (279), and from there to the remarkable triad formed by the fourth, fifth, and sixth poems: "I like a look of Agony" (241), "I felt a Funeral, in my Brain" (280), and " 'Tis so appalling—it exhilirates" (281). In these poems we have a fling into an acid celebration of the end of doubts and ambiguities in death. The first is brief and fiercely unexpected—a remarkable foreshadowing of major affective modes in twentieth-century poets as far apart as Mayakovsky and Plath:

I like a look of Agony,
Because I know it's true—
Men do not sham Convulsion,
Nor simulate, a Throe—

The Eyes glaze once—and that is Death—
Impossible to feign
The Beads upon the Forehead
By homely Anguish strung.

(241)

The magnificent opening line of this poem seems emotional hyperbole. It was masterly to use the word "like" in such a harsh context—a shocking word full of savage irony here. The line cuts passionately, in deep recognition of inescapable truth, as though uttered by a cosmic diagnostician obsessed by the horror of things. From the third line on, the poem fixes on those physical symptoms of dying it is "impossible to feign." The cosmic diagnostician is also, in an angrily ironic sense, a comically impatient doctor who doesn't "like" uncertain symptoms, let alone mere play-acting at mortal suffering, in patients. People who aren't really dying are hypocrites. Meanwhile, the deeper affective energy comes from something like a thrill of recognition of the reality of death. The release of sensibility comes, once again, at the expense of a suspicion of morbidity or worse.

This rigorous little poem leads directly toward the even more dire "I felt a Funeral, in my Brain," a poem that intensifies the sickeningly liberating hard recognition of the self's own fate:

Then Space—began to toll,

As all the Heavens were a Bell,
And Being, but an Ear,
And I, and Silence, some strange Race
Wrecked, solitary, here—

And then a Plank in Reason, broke,
And I dropped down, and down—
And hit a World, at every plunge,
And Finished knowing—then—

"And Finished knowing—then" is a phrase that cuts two ways. The variant, "And Got through—knowing—then," also carries both meanings: the end of knowing or the plunging into truth. The first ties in with the preceding poem and the finality of the glazed eyes; the second with the following poem. In either case, the soul is "secure" in its grasping of ultimate horror—" 'Tis so appalling—it exhilirates"—an insight pounded home in the third and fifth stanzas of the sixth poem (281):

> The Truth, is Bald, and Cold—
> But that will hold—
> If any are not sure—
> We show them—prayer—
> But we, who know,
> Stop hoping, now—

and:

> Others, Can wrestle—
> Your's, is done—
> And so of Wo, bleak dreaded—come,
> It sets the Fright at liberty—
> And Terror's free—
> Gay, Ghastly, Holiday!

This hard-won clarity of perception, related to poems in 15, allows the distancing in the seventh and eighth poems: "How noteless Men, and Pleiads, stand" (282) and "When we stand on the tops of Things" (242). These have something of the calm of "Of nearness to her sundered Things" but with the important shift to acceptance of distance—a kind of cosmic indifference—rather than the recapturing of closeness as an earnest of immortality. In the seventh poem both men and stars are taken for granted and then suddenly vanish—"O'ertakeless, as the Air"—and human "disappointment" (how calm a word after the "Wo, bleak dreaded" of the sixth poem) at their passing is, from the point of view of the smiling heavens, simply that—disappointment, not agony of spirit. It is perhaps no more worthy of an answer than an overly importunate child:

> Why did'nt we detain Them?
> The Heavens with a smile,
> Sweep by our disappointed Heads
> Without a syllable—

The "Pleiads" become in the eighth poem the "Stars" that "dare shine occasionally / Upon a spotted World." Sorting out the somewhat confusing multiplicity of sources of light—mirrors, lightning, stars, suns—we recall the potent comment on the speaker's state at the opening of the sequence (327):

So safer—guess—with just my soul
Upon the window pane—
Where other creatures put their eyes—
Incautious—of the Sun—

This eighth poem, however (242), celebrates the opposite mode of behavior:

The Perfect, nowhere be afraid—
They bear their dauntless Heads,
Where others, dare not go at Noon,
Protected by their deeds—

Sun and lightning hold no terrors for the "Sound ones," who reflect heavenly illumination like "Mirrorrs on the scene— / Just laying light." These "Stars dare shine occasionally / Upon a spotted World," and provide a "proof" for the terrified soul that the "Suns" will endure. This poem brings the sequence to a fragile yet inevitable equilibrium, for its vision of an eternal order is based purely on the extension of human stability and strength: "And Suns, go surer, for their Proof, / As if an Axle, held."

Three more poems follow, however: " 'Twas just this time, last year, I died" (445), "Afraid! Of whom am I afraid" (608), and "He showed me Hights I never saw" (446). As with the preceding fascicle, they project an overpowering sense of having missed out and of sheer weight of sad awareness: the morbid envisioning of one's own death and the lonely waiting in 445; the flamboyant, nearly hysterical self-reassuring in the face of death in 608; and finally, in 446, the appalled acceptance, too late, of the cosmic offer first tended in the sequence's opening poem. *Now*, after all has been lost or renounced, the soul would venture into the great cosmos and realms of transcendence. But the opportunity has passed although the will has grown more urgent and decisive. So the fascicle ends in the most poignant sort of wish-fulfillment:

He showed me Hights I never saw—
"Would'st Climb"—He said?
I said, "Not so."
"With me"—He said—"With me?"

He showed me secrets—Morning's nest—
The Rope the Nights were put across—
"And now, Would'st have me for a Guest?"
I could not find my "Yes"—

And then—He brake His Life,
And lo
A light for me, did solemn glow—
The steadier, as my face withdrew—
And could I further "no"?

(446)

We have lingered over these two fascicles because of the special challenge Dickinson offers. She wrote with her nerve-ends, massing the results in poem-clusters whose internal coherence impresses itself on us only after sufficient absorption in the individual poems in relation to one another. No other poet of power demands so much internalization by the reader who wishes to get the precise emotional set of her poems. The least extroverted of poets, she *presents* her affects with a minimum of personal intrusion.

We might well have chosen other sequences for extended discussion. Fascicle 20,[8] for instance, is less demanding than 15 and 16 and unfolds exquisitely. It involves contemplation of nature's impersonal continuity and beauty (whatever our private woes) and a considerable music of courageous loneliness, love, sacrifice, and transport. A strong insistence on subjective vision as primary truth suffuses this sequence with the bright intensity of an indomitable will. It is supported by such poems as the inspired "Dare you see a Soul at the 'White Heat?' " (365), whose images embody the artistic process that produces pure affects from personal experience; the well-known "The Soul selects her own Society" (303); and the passionately felt love poem "How sick—to wait—in any place—but thine" (368).

The interaction between Fascicle 20 and the two fascicles we have been considering is a matter of obvious interest, as in fact is the aesthetic relationship among all the fascicles. This is a question very similar to the one that Pound's cantos present. We chose to begin with Fascicles 15 and 16 because they are representative but also because they contain so much work of sheer power and are extraordinarily adventurous psychologically—that is, in their deep associative diving. But a thorough study of all the forty fascicles still lies ahead. It promises to be one of the great voyages of discovery in modern criticism.

8. Fascicle 20: 1725, 1761, 364, 524, 525, 365, 526, 301, 527, 366, 367, 670, 302, 303, 368, 528, 369, 370.

Part Two
British Approaches

Chapter Four

W. E. Henley, A. E. Housman, and Thomas Hardy

1. Henley's *In Hospital* and Housman's *A Shropshire Lad*

Whether or not fully aware of what they were doing, Whitman and Dickinson came upon a new mode of lyrical discovery in the large. They grasped the principle that, under the pressure of a driving impulse, poems can interact with the same intimate reciprocity as units of phrasing do within a single poem. This natural extension of lyrical structure reclaimed for poetry the scope once reserved for epic and dramatic verse.

We need not be surprised that other poets did not learn at once from what Whitman and Dickinson had done. Dickinson's work had to await another century to be read in the versions now available to us: or, for that matter, in almost any version. And Whitman's poetry simply required time for the method to sink in while the superficial cheers and jeers subsided. Meanwhile, the search for what had already been revealed went on. Poets have always felt, strongly and uneasily, that a lyrical group should somehow have its own organic structure. Witness Hardy's apology for his arrangement of *Late Lyrics and Earlier*, cited in Chapter Two; and witness in general the care poets lavish on the ordering of their poems in any given book. As simple discursive or narrative continuity became ever less acceptable because imposed externally, two types of experiment with longer structures—the linked series and the "symphonic" poem—came into the foreground. Both were continuations of earlier modes, with a difference.

The linked series was simply the arrangement of a group of poems related to one another by their tonal range, or by a shared context, or by the repeated use of the same verse forms, as in the sonnet sequence. It is true that the sonnet sequence has endured and has engendered such offshoots as John Berryman's eighteen-line *Dream Songs* and Robert Lowell's irregular sonnets in *Notebook*. Old forms never die, nor do they quite fade away. But in many such works the set form, itself either improvised or tampered with, is played off against the unstable energies on which the poet is trying to ride herd. Indeed, the form is a kind of subject matter in the sense that the poet is testing its ability to control these energies. Poets of rich chaos like Lowell and Berryman are so skeptical in this matter of control that their use of a set form sometimes becomes an irony in its own right. On the other hand, the sonnets in W. H. Auden's *Journey to a War* are virtuoso pieces. The poet's casual display of his skill reinforces his account of the pathetic, self-defeating course of civilization: the tragic view as conveyed by a stand-up comedian who hardly cracks a smile during his patter. Like such characteristic late nineteenth-century groups of poems as W. E. Henley's *In Hospital* (1888) and A. E. Housman's *A Shropshire Lad* (1896), a series of Auden's type lacks the active dynamics of a developed sequence, whether in the interplay between its points of confessional inwardness and of special intensity, or in its organic handling of apparently lesser tonalities.

The second line of experiment is with the long associative poem whose continuity, as Conrad Aiken said of his *Senlin* (1918), is one of "symphonic form." The unity will lie in the speaking or imagining sensibility and its range of thought and of collision with life. Aiken's ideas—of ordering "emotion-masses" as "consecutive movements" in an "absolute music" of language—clearly approach the spirit of the modern sequence and take the discontinuities of subjective life into account. Neither in his prose comments nor in his poems, however, does Aiken feel any urgent pressure toward an encompassment of conflicting states. His work does not project powerful internal forces requiring equilibrium such as we find in *Song of Myself* and the Dickinson fascicles. Consequently, the dynamics of a system of intensities pressing upon a responding sensibility cannot get into action, or create the risk of unassimilability necessary for a true sequence. We shall return to Aiken later in connection with *The Waste Land*, for his review of Eliot's poem shows its subtle connection with what he conceived the key modern considerations of a longer structure to be.

At this point, however, we wish to pay some attention to a few outstanding British approaches to the sequence after *Maud*. First, it seems worth pausing for a moment over Henley's and Housman's linked series, specificially *In Hospital* and *A Shropshire Lad*, hovering as they are on the verge of the modern sequence. *In Hospital*—with its gloomy closeups of the hospital as seen upon admission and of the dreary, anxious wait before an operation, and then of going under ether and awakening afterwards; its portraits of doctors and nurses and patients; and its atmospheric moments and evocation of the relief of being discharged at last—is a finely human work. Despite its undistinguished versification it has touching moments, and one can see much in it that was later picked up by such poets as Siegfried Sassoon, Wilfred Owen, Edgar Lee Masters, William Carlos Williams, Hugh MacDiarmid, T. S. Eliot, Austin Clarke, and Ramon Guthrie. But the somewhat facile, journalistic impressionism of this verse-suite has little depth despite the genuine interest of such sections as these:

You are carried in a basket,
  Like a carcase from the shambles,
  To the theatre, a cockpit
  Where they stretch you on a table.

Then they bid you close your eyelids,
  And they mask you with a napkin,
  And the anaesthetic reaches
  Hot and subtle through your being.

And you gasp and reel and shudder
  In a rushing, swaying rapture,
  While the voices at your elbow
  Fade—receding—fainter—farther. . . .

(from "V. Operation")

Laughs the happy April morn
  Thro' my grimy, little window,
  And a shaft of sunlight pushes
  Thro' the shadows in the square.

Dogs are tracing thro' the grass,
  Crows are cawing round the chimneys,
  In and out among the washing
  Goes the West at hide-and-seek.

Loud and cheerful clangs the bell.
  Here the nurses troop to breakfast.
  Handsome, ugly, all are women . . .
  O, the Spring—the Spring—the Spring!

("XXVI. Anterotics")

In conveying an atmosphere and a psychological state, Henley's extroverted sensitivity is both successful and self-limiting. The trochees that dominate both these sections are almost too pat an embodiment of these characteristics. In both cases they contribute an impetus of controlled energy matching the general character of the presentation: that of a highly articulate dinner guest holding forth about an experience he has just been through, displaying his gifts of observation and zest for noting his own feelings as well. The effects and resonances here are about as subtle as Henley becomes anywhere in *In Hospital;* and about the only "advance" the work presents over Browning or Tennyson is the almost inevitable, involuntary metropolitan flavor: a contribution, in its way, to the accelerated secularization of poetry that has been one obvious direction of the art. Henley is hearty, very alert, capable of dark moods although certainly not prone to self-analysis or desperate psychic states and powerful imaginative play to deal with them. He is in touch with an important shift in sensibility—reflected in the urban, secularized tonalities we have mentioned, and in a certain intimacy of tone: developments from the work of much greater masters with richer technique and profounder vision. The linked series gave complete scope to his sort of talent—sufficiently competent, humane and vigorous in inspiration, and certainly not heaven-storming.

A Shropshire Lad, though lacking Henley's quality of being "one of us," a voice of the modern city, comes closer to the mark as a modulation toward the sequence. No more than *In Hospital* does it have the searching energies, inward reciprocities, and mobilized dynamics necessary, but its alternation of emotionally concentrated groups of poems with groups of more relaxed and varied character is compensatory. That is, the mixture *suggests* a true dynamics while concealing the redundancy and lack of real movement of the structure as a whole. Its organization is more a matter of careful strategy than dynamics. On the other hand, few readers would deny the poignant force of its more serious poems and the charm and pathos of others. Moreover, the whole book is deepened by its occasional confessional suggestiveness, however careful Housman was to surround the more vulnerable moments with wholesome pastoral notes and tones of pack-shouldering courage and patriotism. But one does have to recognize the redundancy—the recurring curse of all linked series of lyrical poems unless subjected to the psychologically athletic discipline that a lyrical *sequence* of quality by necessity imposes on itself. Too many of the poems in *A Shropshire Lad* labor the same life-ironies and dirges over mortality. We do not see a sensibility steadily interacting with the pressures that have forced its awareness open and compelled a search for a new equilibrium. That process (given Housman's gifts for classical brevity and powerful understatement) could easily have dispensed with the redundancy but might have made him show his emotional hand too openly. Housman's speaker sums up an artistic as well as a personal problem at the start of Poem XXX:

Others, I am not the first,
Have willed more mischief than they durst:
If in the breathless night I too
Shiver now, 'tis nothing new.

More than I, if truth were told,
Have stood and sweated hot and cold,
And through their reins in ice and fire
Fear contended with desire.[1]

The "mischief" might be any form of dangerous desire, and in these poems it is linked with suicidal feeling and death-obsession generally. Poem XLIV, especially, suggests that this associative complex is rooted in homosexual passion:

Shot? so quick, so clean an ending?
  Oh that was right, lad, that was brave:
Yours was not an ill for mending,
  'Twas best to take it to the grave.

The "lad" so sympathetically addressed here is one whose life would have led only to "long disgrace and scorn" as a "household traitor," a "soul that should not have been born":

Souls undone, undoing others,—
  Long time since the tale began.
You would not live to wrong your brothers:
  Oh lad, you died as fits a man.

Moved to honor this nobly motivated suicide, the speaker offers a "wreath" (the poem) in humble tribute—it will be immortal because of its wearer. Poem XLIV does not spell out its meaning although it seems to exclude all meanings but one, not only in itself but in the context of suggestions such as we have seen in Poem XXX and in elusive hints in many other poems—as, for instance, in the quick exchange of looks with a marching soldier made so much of in Poem XXII, or in the outcry to an unnamed "lad" in Poem XXIV where the speaker announces that a psychological moment has arrived:

Use me ere they lay me low
  Where a man's no use at all . . .

The homosexual intimations color the volume and, we infer, justify the darker moods of the poems without quite accounting for them, since

1. Our text of reference is A. E. Housman, *Complete Poems* (New York: Holt, Rinehart & Winston, 1959).

their source is implied in wisps of notation that amount to velleities. Not only is the central, driving anguish of *A Shropshire Lad* left undeveloped—an emotional motif constantly insisted on yet ultimately unearned—but one consequence of Housman's fear of openly realizing the essential feelings underlying the anguish is the blight of sentimentality. Even his highly accomplished verse technique may militate against a dangerous poetry of self-exploration. The simple patterns (mostly alternating rhymes, or couplets arranged in quatrains or longer stanzas, or comparable variations) are reminiscent of folk ballads and other folksongs. They encourage a defensive reliance on conventional situations and stock feelings: song effects that summon up a whole tradition in which deep subjectivity was unnecessary.

Admittedly, a poet can use such traditional simple forms very powerfully—Dickinson, Yeats, and MacDiarmid, for instance, provide splendid examples. Housman, however, rides with the forms he employs, going for uncomplex notes of emotional music that will emphasize a single tonality. That he often does so beautifully is a joy, and it is not denigration to suggest that his work falls short of its promise as a developed sequence. Very likely he would have done violence to his own dazzling but lesser genius had he tried to carry his effort, exhausting as he confessed it was—and beyond doubt threatening—any further.

2. Hardy's *Poems of 1912–13*

Despite his occasional notoriously bad passages, Hardy's intensity of vision takes him in *Poems of 1912–13* where the consistently polished Housman could never go. The laurels for the first developed British sequence, however uneven, belong to the grand old poet-novelist. His elegiac grouping of twenty-one poems, written when he was in his seventies, anticipated Yeats's civil-war sequences by a decade. The poems grapple with the complexities of personal disaster in a way typical of many sequences, by establishing equilibrium, in the manner we have discussed earlier, among various intense states of awareness without absolutely replacing one state by another. Hence the carping by some critics that Hardy should have preserved the contents of first publication (eighteen poems ending with the visionary "The Phantom Horsewoman" rather than with the more distanced and desolate "The Spell of the Rose," "St Launce's Revisited," and "Where the Picnic Was") is at best misleading. In fact, the opening poem, "The Going," telegraphs the final form of the sequence quite satisfactorily; in no sense are our expectations betrayed.

Essentially an elegy on the death of Hardy's first wife, the sequence is nevertheless not an avowal of simple grief. Nor is it an exercise in consoling oneself, eventually, by Tennysonian philosophical speculation or by the healing power of nature, time, or new love—whatever the biographical "facts" of Hardy's relationship to the one soon to become his second wife. The motivating pressure within the poems is the effect on the poet of the death of a woman once passionately loved but estranged for many years. The abrupt ending by her death of all chance for communication has generated a disturbing set of memories of earlier tenderness. It has also aroused a desperate need to fix the causes, quality, and ultimate meaning of a love that promised so much yet ended in appalling distance. Throughout the sequence certain basic tones are intermingled: vivid dismay, intolerable remorse, yearning to undo the past, loving compassion, intensely focused recollection of romantic excitement and joy, and that more distanced, desolate acquiescence in an essentially bleak reality we have already noted.

The first poem projects the affective complex of the whole work. Also, it sketches the motivating situation, central preoccupations, and psychological course of the sequence, with the concomitant changes of scene. Here and throughout, Hardy's gifts as a novelist for capturing the gist of scenes, characters, and incidents stand him in good stead. "The Going" is unusual among opening poems for the way it quickly suggests the nature of the whole sequence. It is therefore worth analyzing rather fully. Although not as finely successful as some of the other poems—notably "The Voice" (poem 9), "After a Journey" (poem 13), and "The Phantom Horsewoman" (poem 18)—it is a key poem in the sequence's structure, together with "At Castle Boterel" (poem 16) and "Where the Picnic Was" (poem 21).

"The Going" divides logically into three pairs of stanzas. The first pair treats the actual death; the second, the lost wife's haunting of the poet's imagination; the third, the estrangement between them that death has made irreversible. The motivating pressure in the sequence is to undo that irreversibility, to re-establish a communion lost since the couple's early days of love: logically a plain impossibility, psychologically an absolute necessity. The absurdity and the sheer emotional need combine to give the work its peculiar force, and Hardy's narrative skill aids him enormously in creating the illusion of a heartbroken ghost and her equally heartbroken survivor striving to re-create, together, the sources of their love that—once recovered in memory and repossessed in its original sensuous ambience—will free them both of their accumulated marital disappointment and sense of guilt. (It is interesting to compare *Poems of 1912–13* with Yeats's Crazy Jane poems some two decades later, and especially with the complex psychological realities implied in "Crazy Jane on the Day of Judgment" and developed in "Crazy Jane and Jack the

Journeyman." Hardy at his best is taking us into the same arena as Crazy Jane's, of the struggle toward self-transcendence through love and its commitments—no matter the inevitable contradictions and obstacles. Perhaps we can speak of the search for clarity in the most intimate aspects of life as one of the tremendous ultimate goals of modern art: certainly, such works as these are brilliantly exploratory towards this end. The affective tensions of the true sequence are essential to the effort, which also informs weaker structures like Henley's and Housman's without being able to do much more than simply indicating a direction or general state of feeling.)

Hardy's narrative instinct helps him give his sequence a grounding in literal places. His search for a renewal of love will take him back to where it began, and so the explicit narrative element in the sequence is a journey to Cornwall. It is sketched for the first time in the fourth stanza of "The Going" (the weakest stanza in the poem and yet resonant with adjectives, mostly participial, of strong psychic flavor):

> You were she who abode
> By those red-veined rocks far West,
> You were the swan-necked one who rode
> Along the beetling Beeny Crest,
> And, reining nigh me,
> Would muse and eye me,
> While Life unrolled us its very best.[2]

The stanza immediately following explicitly connects the hope for love's renewal to such a journey. In ruefully asking why the thought never occurred to "us" while there was yet time, the poem is at the same time projecting the future narrative course of the sequence:

> Why, then, latterly did we not speak,
> Did we not think of those days long dead,
> And ere your vanishing strive to seek
> That time's renewal? We might have said,
> "In this bright spring weather
> We'll visit together
> Those places that once we visited."

Did we claim, however parenthetically, that the *fourth* stanza is the poem's weakest? Surely the fifth too clamors for that distinction—and yet it has its redemptive surge of genuine remorse. Hardy is not squeamish when he feels he has to spell something out, particularly if it

2. Our text of reference for *Poems of 1912–13* is Thomas Hardy, *The Complete Poems*, ed. James Gibson (New York: Macmillan, 1978).

has an expository function. There is no point in defending—as opposed to just allowing for its inevitable presence—writing like this. We have been given, despite some violence to the ear, a projection not only of the work's narrative course but also of the characters of its two protagonists and the motives underlying the poet's imaginative undertaking. The other stanzas are considerably richer poetically although not without their awkward forcings of language and rhythm. For instance, the intimate, gently reproachful lines of the first stanza are addressed directly to the dead woman—not, "Why did she give no hint that night" but:

> Why did you give no hint that night
> That quickly after the morrow's dawn,
> And calmly, as if indifferent quite,
> You would close your term here, up and be gone
>   Where I could not follow
>   With wing of swallow
> To gain one glimpse of you ever anon!

The effect, of course, is of actual communion with the dead woman. Almost half the poems in the sequence are similar, with the poet addressing her, or her ghost speaking to him ("The Haunter," "His Visitor," "The Spell of the Rose"). The mode of direct address is thus absolutely reciprocal with the poet's need for communication, or at the least "To gain one glimpse of you ever anon!" There are also indications in this opening stanza of the state of their relationship at the end of her life. In retrospect, the reproach to the dead woman for quitting life so suddenly, without warning, does not quite conceal the bitter sense that her dying, in its mute finality, had been deliberate: her last, decisive expression of estrangement.

If in the first stanza the phrase "as if indifferent quite" has an ominous, reproachful ring, the second stanza presents the other side of the coin—self-reproach—as well:

>   Never to bid good-bye,
>   Or lip me the softest call,
> Or utter a wish for a word, while I
> Saw morning harden upon the wall,
>   Unmoved, unknowing
>   That your great going
> Had place that moment, and altered all.

Hardy achieves much in this superb stanza. First, he captures the atmosphere of sweet, loving intercourse in "lip me the softest call." The phrase serves as a transition between the memory of the wife's coldness ("as if indifferent quite") and that of the original dream of what the couple's life together would be. The dream is evoked in the fourth stanza's vision of her as "the swan-necked one" and of a superlative existence to come—"While life unrolled us its very best." At the same time, the second stanza's most powerful image—"I / Saw morning harden upon the wall"—suggests the indifference of inexorable fatality to human wishes. Finally, the husband's self-characterization implied in the line "Unmoved, unknowing" (the squinting modifiers here attach both to "I" and to "morning") is reciprocal with his wife's supposed indifference to him even at her death. In the context of the poem, especially its last stanza, and the sequence, words that in other elegiac settings would be absolutely trite—"your great going / Had place that moment, and altered all"—prove remarkably successful. For by the poem's end we find that, given the quality of the marriage, the wife's death might not have been expected to have such an impact. Moreover, the rest of the sequence is devoted to undoing her "great going."

The third stanza prepares us for the supernatural poems to come: "The Haunter," "The Voice," "His Visitor," and the reprise poem, "The Spell of the Rose." It also foreshadows the speaker's coming pursuit of the woman's phantom, which finds fruition in "After a Journey," "At Castle Boterel," "Places," and "The Phantom Horsewoman":

> Why do you make me leave the house
> And think for a breath it is you I see
> At the end of the alley of bending boughs
> Where so often at dusk you used to be;
>   Till in darkening dankness
>   The yawning blankness
> Of the perspective sickens me!

In its own terms, within a context of continued reproach, this stanza balances a moment of false elation with an even deeper depression. Complicating elements include the sense of compulsion—"*make* me leave the house"—the compelling need for the phantom truly to be there, the implied familiarity and domesticity of the scene—"where so often at dusk you used to be"—and the real, utter emptiness of a landscape in which she is no longer present.

The poem then turns to the landscape—the "red-veined rocks far West"—where she *was* once so dynamically present and to the tentative suggestion of seeking together "That time's renewal." Then, in the last stanza, there is a momentary attempt to dissipate all the pain of the ruefully questioning longing and disturbance and would-be communion through fatalistic musing, and the failure of that attempt:

> Well, well! All's past amend,
> Unchangeable. It must go.
> I seem but a dead man held on end
> To sink down soon. . . . O you could not know
> That such swift fleeing
> No soul foreseeing—
> Not even I—would undo me so!

Resolution of the knot of feelings will not come through philosophizing but through arrival at some altered state of realization. In the sequence this will happen much further on: in the thirteenth poem, "After a Journey," and the sixteenth, "At Castle Boterel." Meanwhile, the first poem ends in a welter of acute loss that is confronted helplessly.

Here we may pause to take preliminary stock. The modern reader will have to come to terms with Hardy's "old-fashioned" style. It is almost verbose, almost affectedly literary, and the phrasing can suddenly turn trite or banal or gauche. And yet the language is intimate and dramatic, a rush of active life-feeling that catches the speaker's turmoil directly and unmistakably. The varied stanza forms Hardy employs—another kind of "old-fashioned" contrivance—provide a sort of emotional control, channeling the torrent of words into a simple repetitive patterning that allows considerable metrical flexibility. The result is a paradoxically combined formality and immediacy of effect, appropriate to the profound seriousness of the poems and to their intensely private subjectivity. The implied "story" of the marriage that ended with the wife's death, and the progression in the sequence toward a psychological reversal that is almost a triumph over reality, demand such a combination.

It is characteristic that one of the climactic poems in the lyrical structure of *Poems of 1912–13* uses a stanzaic pattern that places awkward strains on Hardy's diction but whose unselfconsciously confiding simplicity of manner makes it infinitely graceful. "At Castle Boterel," which comes toward the end of *Poems of 1912–13*, is the point of resolution toward which the sequence strives. Here a moment of sweet companionability, which occurred early in the relationship, is recovered and placed in a perspective lacking in the intense turmoil of the magnificent thirteenth poem, "After a Journey":

> As I drive to the junction of lane and highway,
> And the drizzle bedrenches the waggonette,
> I look behind at the fading byway,
> And see on its slope, now glistening wet,
> Distinctly yet
>
> Myself and a girlish form benighted
> In dry March weather. We climb the road

Beside a chaise. We had just alighted
 To ease the sturdy pony's load
  When he sighed and slowed.

What we did as we climbed, and what we talked of
 Matters not much, nor to what it led,—
Something that life will not be balked of
 Without rude reason till hope is dead,
  And feeling fled.

It filled but a minute. But was there ever
 A time of such quality, since or before,
In that hill's story? To one mind never,
 Though it has been climbed, foot-swift, foot-sore,
  By thousands more.

Thus the first four stanzas of this quietly decisive poem. In itself the passage merely confides the remembered, hardly melodramatic moment from the past we have mentioned. Nothing could be plainer, or more unpretentiously and sweetly cherishing, than the language used to evoke this moment. It is so unspectacular that one may well ask: "Why should we consider the passage a special point in the sequence?" Unlike such other pieces as "The Voice" and "After a Journey"—both of which we shall take up shortly—"At Castle Boterel" is not an overwhelming poem. Yet in it the searching memory behind the sequence has found an epiphany that fulfills the wish uttered in the first poem, "to seek / That time's renewal" in "Those places that once we visited." The impossible wish has become a prophecy. The place of "At Castle Boterel" in the sequence has given the poem special power. Various preceding poems, in particular "The Going" and "I Found Her Out There," have prepared us for its denouement of realization by stressing the depth of the young bride's homesickness for her native Cornwall. In the latter poem the widower, who had taken her to his inland home after marriage, imagines that

Yet her shade, maybe,
Will creep underground
Till it catch the sound
Of that western sea . . .

Still further on in the sequence, three poems ("The Voice," "After a Journey," and "Beeny Cliff") show the speaker being lured by the illusion of his wife's voice and presence back to the Cornwall she had loved with a childlike, unshakable natural devotion. Those three poems, and others accompanying them, build a haunted atmosphere of psychological complexity—the speaker's compulsion to return to *her* places in Cornwall and somehow rejoin her there and make things right again—for it was taking her from those places that had spoiled her life. To him, at least, it seems so, and it seems also that he had inadvertently neglected her needs just as he had not realized how close she was to death at the time described in "The Going."

What happens in "At Castle Boterel," then, is that the quest to set things right succeeds in the repossession of their early joy together in the everyday world of *her* loved Cornwall. After the distraught intensity of remorse, grief, nostalgia, and desire to remake the past in preceding poems, the almost matter-of-fact tone of "At Castle Boterel" has enormous impact. One moment of remembered literal experience—here, a clear, pure moment of shared love—serves to fulfil the speaker's passion to recall the irrevocable and undo the ineluctable. He brings the woman back to life in a place most congenial to her; there is no failure of understanding between them; he is the fully sensitive husband and lover he should have become after that remembered moment.

The way that this transcendent moment emerges is a marvel of poetic modulation. It happens within the speaker's gloomy present life, well symbolized in the line "And the drizzle bedrenches the wagonette." The couple's suffering, at the center of so much of the sequence, is present as a context or frame for the happy moment he has dreamed his way back into, in that long-ago spring with its "dry March weather." It is not replaced but displaced; its pressure remains, giving the poem even more vibrancy than the happy memory alone would contain. But that memory is vibrant anyway, even without this counter-pressure. The whole atmosphere is charged with delight of a special, humanly satisfying kind: laboring up the hill together, the pony's gentleness, the earnest conversation, the sense of an enchanted minute of incomparable "quality."

The moment of triumph is a moment of respite from the anguish of "The Voice" and "After a Journey." The ninth poem, "The Voice," comes between two "ghost" poems in which the dead woman speaks: "The Haunter" and "His Visitor." The first of these is a poignant attempt to tell the grieving man that "If he but sigh since my loss befell him / Straight to his side I go"; it represents loving forgiveness of all the times he did not wish her with him:

> Tell him a faithful one is doing
>   All that love can do
> Still that his path may be worth pursuing,
>   And to bring peace thereto.

That this imaginative attempt is a failure is clear from "The Voice" and "His Visitor." As in "The Going," the woman "much missed" calls him forth from the house, but now he imagines her as

> Saying that now you are not as you were
> When you had changed from the one who was all to me,
> But as at first, when our day was fair.

Yet the element of delusion may be paramount; he knows really that she has been "dissolved to wan wistlessness," but the intense desire that it not be so draws him after her:

> Thus I; faltering forward,
> Leaves around me falling,
> Wind oozing thin through the thorn from norward,
> And the woman calling.

"His Visitor," a weaker poem, supposes the end of her haunting; there is no place for her in "this re-decked dwelling":

> I feel too uneasy at the contrasts I behold,
> And I make again for Mellstock to return here never,
> And rejoin the roomy silence, and the mute and manifold
> Souls of old.

At this point in the sequence, then, the stage is set for a journey; if the poet is to repossess the essence of her being it must be somewhere away from a house that no longer suggests her presence. It is, in fact, not at "Mellstock" (a reference to the cemetery at Stinsford) but in Cornwall, the locale of the twelfth poem, "A Dream or No," and in the past. "A Dream or No" has the narrative function of alerting us to the change of scene of the next six poems:

> Does there even a place like Saint-Juliot exist?
> Or a Vallency Valley
> With stream and leafed alley,
> Or Beeny, or Bos with its flounce flinging mist?

"After a Journey," the thirteenth poem, is placed at Pentargan Bay and opens with a vision of her as she was "forty years ago":

> Where you will next be there's no knowing,
> Facing round about me everywhere,
> With your nut-coloured hair,
> And gray eyes, and rose-flush coming and going.

Yet by the end the emphasis is on their mutual frailty and on her inevitable vanishing. The last lines are the heartbroken reciprocal to his hope in "The Voice" that she is telling him she is again "as at first, when our day was fair." Except for the effect of padding the phrase "though Life lours" gives, the last lines of "After a Journey" are masterful:

> Trust me, I mind not, though Life lours,
> The bringing me here; nay, bring me here again!
> I am just the same as when
> Our days were a joy, and our paths through flowers.

The anguish of the loss of that time cannot be undone; and even "At Castle Boterel" itself, like most of the rest of the sequence, ends in a tone of resigned fatalism. One more poem that follows it, however—"The Phantom Horsewoman"—insists on the speaker's unfading "vision" of his "ghost-girl-rider" as she was "when first eyed." In this poem the speaker deliberately divides his own nature. He speaks of himself in both the first and third persons, thus distancing himself from his own keenest state of openness and apologizing for it as a kind of madness. "Queer are the ways of a man I know," the poem begins, and goes on to describe the "man's" condition of "careworn craze." What Hardy does here is very much like what Tennyson does with his protagonist in *Maud*. He deliberately avoids suggesting that the affects of the poem reflect personal states of his own. His supposed speaker (in this one poem only—but the effort at detachment is crucial) cuts himself off from full responsibility for the obsessive states, whether misery-wracked or momentarily transcendent, that the sequence has hitherto summoned up. And indeed, all the poems reach beyond the speakers in them. Their real transcendence lies in the way they hold their contending emotional energies in a marvelously sensitized, continuously probing system of interacting memory, will, self-torment, and tentative distancing.

To recapitulate: The sequence begins elegiacally, with grief and remorse (complicated in ways we have suggested) dominating the first seven poems. The key poems in this first movement are "The Going" and "I Found Her Out There." These poems establish the empirical situation of the marriage gone wrong as well as the wife's unexpected death, which is felt as a last gesture of rebuke as well as a painful shock. The ensuing three poems constitute the heart of the five-poem second movement ("The Haunter" through "A Dream or No")—a sense of the wife replying, as it were, in a counter-thrust concentrated in "The Voice": "Woman much missed, how you call to me . . ." In this second group the wife's essential self, loving after all, returns to the living world to lure her husband into understanding. He is to be aroused to search back through the past for tokens of the way their life felt to her. "A Dream or No," following the relatively easy irony of "A Circular"—with its fashions forcibly reminding the "legal representative" of his dead wife's presence—introduces the motif of an actual journey back to her places. Images from the past surface of the "Fair-eyed and white-shouldered, broad-browed and brown-tressed" maiden, but in an atmosphere of considerable ambivalence about the accuracy of the memory and the usefulness of revisiting her Cornwall.

In the splendid third group, "After a Journey" (poem 13) through "Places" (poem 17) and the transitional eighteenth poem, "The Phantom Horsewoman," the search occurs and succeeds. It is marked by an eerie insistence, a supernatural imagery that objectifies the speaker's full agony of frustration as he is "led" to pursue "her" to the loved haunts of her youth. The third movement is a complex one. Within its headlong plunge of yearning despair it nevertheless holds intact the brilliantly conceived vision of "At Castle Boterel." That vision has itself been prepared for by the paradoxical "Beeny Cliff," which combines a bright memory with a bitter sense of the abyss of death. The movement is further complicated by the braking effects of the fourteenth and seventeenth poems, "A Death-Day Recalled" and "Places." These present an irony directly engaged with the main psychological effort of the sequence: to repossess the wife's world, her sense of life in the places she loved. Had Hardy written in the tradition of the pastoral elegy, he might have transmuted her into the indwelling divinity of these places or, at the least, conjured up a landscape in mourning for her. But this is a more tough-minded apprehension of death's finality, and the landscape remains unheeding—uttering no "dimmest note of dirge." The places care nothing for her any longer—what, and *who,* is one more lost young woman to this or that Cornish town or strip of beach or mountainside? Indeed, what importance have they to the speaker either, apart from her? It was she alone, in life and in her visionary reappearance, who animated them. Without the girl remembered in her "air-blue gown" ("The Voice"), with her "nut-coloured hair, / And gray eyes, and rose-flush coming and going" ("After a Journey"), they are nothing. "Places" sums it all up in its closing stanzas:

> Nobody calls to mind that here
> Upon Boterel Hill, where the waggoners skid,
> With cheeks whose airy flush outbid
> Fresh fruit in bloom, and free of fear,
> She cantered down, as if she must fall
>   (Though she never did),
>   To the charm of all.

Nay: one there is to whom these things,
That nobody else's mind calls back,
Have a savour that scenes in being lack,
And a presence more than the actual brings;
To whom to-day is beneaped and stale,
    And its urgent clack
    But a vapid tale.

There is one, then, who remembers and is able to set the young woman before us as she was; who is so haunted by the vision out of the past that external reality fades beside it. "The Phantom Horsewoman" expands on this inner vision, rounding off the third movement and initiating the closing one. As we have mentioned, the poem abruptly displaces the sequence's major continuity of voice, and a new first-person speaker here disclaims responsibility for the disordered state—"Queer are the ways of a man I know." The shift helps break carefully developed dramatic reciprocities into independent emotional and psychic energies. Their final balance is an open play among the tonalities and intensities indicated from the start and only partly controlled by the self-lacerating clarity of the intelligence assaulted by them at every turn.

The sequence does not end with "The Phantom Horsewoman." The change in speaker modulates into "The Spell of the Rose," in which the dead woman speaks again. Unlike her manifestations in the earlier haunting poems, however, she can only trace the growth of the estrangement and, poignantly, is ignorant

. . . whether, after I was called
to be a ghost, he, as of old,
  Gave me his heart anew!

In the twentieth poem, "St Launce's Revisited," and the last, "Where the Picnic Was," her spirit vanishes entirely. Both poems end with the stress on her distance from him; in the first, the poet balances a wistful hope that, although the inn at St Launce's where he stopped so many years ago is now staffed by strangers, perhaps all is still as it used to be in her home on the coast. The poem ends in utter desolation:

  Why waste thought,
When I knew them vanished
Under earth; yea, banished
  Ever into nought!

"St Launce's Revisited" serves a narrative function similar to "A Dream or No," only now the speaker is traveling away from "that place in the

West" ("Dream"), or at least no longer heading toward "the faces shoreward" ("St Launce's"). With "Where the Picnic Was" we are presumably back at Max Gate and on the Dorset, rather than Cornish, coast. The contrast with the Cornish scenes is as strong as that between the fourth and sixth stanzas of "The Going," between his vision of her as the "swan-necked one who rode / Along the beetling Beeny Crest" and the grim acquiescence and insight of the last stanza:

> Well, well! All's past amend,
> Unchangeable. It must go.
> I seem but a dead man held on end
> To sink down soon. . . .

So, too, "Where the Picnic Was" ends with a similar desolation (rather marred by the phrase "urban roar"):

> Yes, I am here
> Just as last year,
> And the sea breathes brine
> From its strange straight line
> Up hither, the same
> As when we four came.
> —But two have wandered far
> From this grassy rise
> Into urban roar
> Where no picnics are,
> And one—has shut her eyes
> For evermore

Poems of 1912–13, then, is very close to a modern sequence in important ways despite the strong narrative dimension. It is more than a mere series of poems linked tonally and by a shared context, although Hardy criticism tends to single out certain poems in it for discussion and ignore its character as a moving structure. The coherence of the work as a total structure lies in its powerful push toward self-discovery against the terrible odds—the self-discovery here consisting of the recovery of one's most generously empathic possibilities. To reconstruct a self that will meet the demands of a vision of what was once possible, to redirect fate by cultivating an obsession one fears may be at the pitch of madness, and to hold the new stasis fast while absolutely alert to the dead emptiness in which one is actually trapped—that is the effort in *Poems of 1912–13.* Memory and need supply the passionate language rooted in this effort; and the "ghost-girl-rider," whose voice and fleeting image appear in so many guises and circumstances throughout the sequence, draws the language through its dynamic movement in search of a point of perfect control.

No one would quarrel with the singling out of individual poems in a sequence for special attention. Certainly "After a Journey" (one of the poems any reader might well recall if asked to list the most beautiful, or most moving, lyric poems in the language) stands by itself in a quite real sense. But then there is another sense, also quite real, in which it surges in the context of the whole sequence. For instance, it echoes, enriches, and redirects the spirit of "The Going," addressing the lost wife with a grief and ardor beyond what the opening poem was in a position to do. We cannot recognize the reciprocities between these two poems without becoming aware of the affects building up from poem to poem in the sequence. As a clue to the organic structure of sequences, the relationship between "The Going" and "After a Journey"—each gaining fuller dimension from the other and from its specific placement in the whole work—is one important touchstone in *Poems of 1912–13.*

Chapter Five

The Evolution of William Butler Yeats's Sequences I (1913–29)

1. "Upon a Dying Lady" and the Civil-War Sequences

At the time Hardy was writing his elegiac masterpiece, Yeats was composing his first sequence, also elegiac. "Upon a Dying Lady" is, despite fine moments, the least successful of Yeats's ten or so sequences. Certainly it never approaches the heights of Hardy's *Poems of 1912–13.* But it does share, with all but one of Yeats's sequences, one major structural advantage of the genre: freedom from dependence on plot. (The one exception is the group of seven poems beginning with "The Three Bushes," a ballad that provides a clear narrative reference for the six "songs" following it.) As with most great sequence writers of our century, Yeats presents inner, associative voyages of discovery that rely on tonal dynamics rather than on surface continuity. In this respect he follows in the wake of Whitman and Dickinson, not of Tennyson and Hardy, despite the deceptive surface conventionality of his verse and its reliance on something like normal discursive syntax—aspects that have encouraged ideational rather than dynamic readings of his work by his critics.

With Yeats, as pre-eminently with Pound, we have an embarrassment of sequential riches. In order of composition, they are "Upon a Dying Lady" (primarily 1913),[1] "Nineteen Hundred and Nineteen" (begun in

1. See George Brandon Saul, *Prolegomena to the Study of Yeats's Poems* (Philadelphia: University of Pennsylvania Press, 1957), p. 111, for a discussion of the dating problem. We agree with him that the January 1912 date given for one of the sections by Richard Ellmann in *The Identity of Yeats* is too early. Poetic texts are taken from *The Collected Poems of W. B. Yeats* (New York: Macmillan, 1956).

1919, published in *The Dial* in 1921), "Meditations in Time of Civil War" (the first poem dating from 1921, the rest from 1922), "A Man Young and Old" (1926–27), "A Woman Young and Old" (1926–29), *Words for Music Perhaps* (1929–32), "Vacillation" (1931–32), "Supernatural Songs" (*c.* 1934), the cluster of seven poems starting with "The Three Bushes" (1936), and the nineteen poems of *Last Poems and Two Plays* (primarily 1938) in their original order—not immediately obvious in the incredible jumble of the "Last Poems" section of *Collected Poems.* (This last sequence was published posthumously in 1939 with the closely related plays *The Death of Cuchulain* and *Purgatory.*) *The Tower* (1928), in which the two Irish civil-war sequences and "A Man Young and Old" appear, can also be considered as a larger sequence. The volume as a whole, with special attention to the civil-war sequences, has been discussed at some length in Rosenthal's *Sailing into the Unknown: Yeats, Pound, and Eliot.*[2] In the course of exploring relationships among all the sequences, however, we shall be adding some further observations on those in *The Tower* as well.

In Yeats's sequences, we find a world of legendary, conventional, and stock or generic types. Not only do they mingle on the page with figures out of the poet's own life, and with the poet speaking in the first person; they also provide alternative personae, masks by which the floating sensibility can objectify this or that phase of itself. Figures like A Man Young and Old, A Woman Young and Old, and A Dying Lady may be given familiar human attributes though not even identified by name. In *Words for Music Perhaps,* identified figures (Crazy Jane, the Bishop, Jack the Journeyman) exist in a realm of archetypes burning with pure intensities suitable to their symbolic roles. Ribh in "Supernatural Songs" is a Blakean prophet of profane mysticism. Lady and Lover and Chambermaid in "The Three Bushes" are modern re-creations of types often found in folk ballads. Other legendary and mythical figures move through the directly personal sequences, often silently. The range of interaction is enormous in Yeats between the private sensibility and the impersonal forces of history and tradition and natural or cosmic process. It frees his sequences from dependence on narrative and dramatic plot for structural unity, and of course from merely thematic and discursive development. Indeed, as with Whitman and Dickinson, it changes our conception of the unity of *any* complex work, sequence or not. For this reason, the evolution of Yeats's sequences is particularly worth observing in some detail because of their evolving contribution to the form. (He wrote more sequences than anyone else among our major poets except Pound.)

2. New York: Oxford University Press, 1978, pp. 26–44, 116–55.

We are hardly proposing that Yeats's sequences are uniformly successful. The poems sometimes trip over their archetypal mysteries; and the poet's obsession with aristocratic values, inseparable from his dependence on heroic tradition and in any case to be understood idiosyncratically, can be an intrusive nuisance. Witness the fate Yeats wishes on the "dying lady" of his first sequence:

> When her soul flies to the predestined dancing-place
> (I have no speech but symbol, the pagan speech I made
> Amid the dreams of youth) let her come face to face,
> Amid that first astonishment, with Grania's shade . . .

To "come face to face . . . with Grania's shade" is a thrilling thought in a lovely phrase, but how far it is from any serious engagement with a real dying woman! From Yeats's letters and certain allusions in the sequence we know that the lady's character is based on that of the actress Mabel Beardsley, sister of Aubrey Beardsley. That is hardly the point, however; the sequence betrays its human concern by trifling with it. The dying lady, "our Beauty," is translated into a Nietzschean heroine, an aristocratic spirit at one with "Achilles, Timor, Babar, Barhaim, all / Who have lived in joy and laughed into the face of Death"—as if her demeanor and condition could be one whit ennobled by such rhetoric. In this early experiment with an extended structure fusing intimately private feelings with cosmically impersonal perspectives, Yeats's sense of proportion was unclear. He wished to write an adequate and moving elegy but also to make his dying lady a model of heroic transfiguration. That is, he wished to connect her crisis as a vital, suffering person with an ideal image of the artist as one whose aesthetic self-transformation could transcend her own bodily illness and death.

No doubt this aim is defensible, but the effort in "Upon a Dying Lady" is uneven and too often forced. The sequence moves falteringly despite its best moments: a sudden rush of compassion or admiration, an angry slash of resentment against a priest whose arrival in the hospital puts the lady's artist-friends and their gaily playful gifts in a demeaning light, a note of fine courtesy. Only the final poem of the sequence, "Her Friends Bring Her a Christmas Tree," finds the right balance of bravery and pathos:

> Pardon, great enemy,
> Without an angry thought
> We've carried in our tree,
> And here and there have bought
> Till all the boughs are gay,
> And she may look from the bed

On pretty things that may
Please a fantastic head.
Give her a little grace,
What if a laughing eye
Have looked into your face?
It is about to die.

This works superbly, in itself and because the center of interest has shifted from the earlier self-conscious and sometimes trite or pretentious descriptions of the sickroom scene, with their theological, mythopoeic, aristocratic, and other contrived implications. Instead, we have the poet confronting death directly, serving as knight-intercessor for the dauntless, uninhibited, childlike lady. Death, now cast in the role of *his* "great enemy" as well as hers, provides a strong personal interest that mobilizes the language. The poem is an early instance of Yeatsian interplay between the heroic and the vulnerable as the key to the tonalities informing a sequence. The technical virtuosity with which rhyme, meter, and enjambment are deployed in the increasingly concentrated and emotionally intense succession of sentences is both unobtrusive and climactically effective. Both the vision and the craft foreshadow the power of the later Yeats.

From the bold but awkward start represented by "Upon a Dying Lady," the poet moved on to the magnificent civil-war sequences of *The Tower,* paired in reverse order to that in which they were written: "Meditations in Time of Civil War" and "Nineteen Hundred and Nineteen." Here our special point of departure for catching the whole spirit of these sequences must surely be Yeats's complex obsession with aristocratic and heroic—and political—values. For this obsession, the civil-war sequences provide a far more congenial field of operation than does "Upon a Dying Lady." The feelings and attitudes involved are no mere sentimentally Miniver-Cheevyish nostalgia for a glamorous past, nor can they be reduced to the crude class-arrogance and eugenicism Yeats sometimes displays. He *can* be foolish and irritating in the latter regard, as in the fifth poem of "Upon a Dying Lady" when he actually writes that

She knows herself a woman,
No red and white of a face,
Or rank, raised from a common
Unreckonable race . . .

Even more disheartening, because calculated to demoralize his own children, is a passage in "My Descendants," the fourth poem of "Meditations in Time of Civil War":

And what if my descendants lose the flower
Through natural declension of the soul,
Through too much business with the passing hour,
Through too much play, or marriage with a fool?

Some of the thought here may be taken as merely cautionary. But the second line, and the eugenic implications of "marriage with a fool," are uncomfortable foretastes of the racist and fascist notes that crop up in the prose sections of *On the Boiler* (1939) and in certain of the last poems. This of course is the period of Nazism's swift rise and the coming on of World War II. Such thinking, however innocent of any lust to create death-camps, took on sinister implications despite the facts that Yeats was not Cinna the conspirator but Cinna the poet, and that he belonged to a generation among whom his old-fashioned assumptions and prejudices were rather widespread. As Auden put it in his elegy "In Memory of W. B. Yeats," he was "silly like us." That is, he had preoccupations engendered under maddening circumstances of history—our "nightmare of the dark," Auden calls it in the same poem: the "intellectual disgrace" of our inability to unfreeze "the seas of pity" and create a humane world. In the civil-war sequences he brings to bear on the world's and Ireland's post-World War I crises a crisis in his own sensibility, drawing on all the resources of his political thought and feeling. And as Auden's elegy, again, suggests, Yeats's art—whatever his private prejudices—discovers "a rapture of distress." His image for the modern soul, and his own, in crisis issues forth dazzlingly in the third poem of "Nineteen Hundred and Nineteen":

The swan has leaped into the desolate heaven:
That image can bring wildness, bring a rage
To end all things . . .

The passage embodies Yeats's effort, in the civil-war sequences, to confront the murderous desolation he feels at the heart of modern existence and to find a stay against personal chaos. The sequences gave him scope to test and qualify his desire for heroic assertion in the face of desolation and chaos, and for correcting the relatively oversimplified "tragic joy" of "Upon a Dying Lady." That work did include qualifying notes—quiet dismay here and there, and the gallant pleading and pathos of the closing poem—that expand, in the later sequences, into open grief, confusion, and bitter self-analysis. Even so, it cannot be compared with the civil-war sequences, which developed around the poet's distress and excitement at the violence raging in Ireland and the rest of Europe during the first World War and immediately afterward. Viewed as a music of consciousness in the eye of the storm, they are at once prophetic, introspective, heroically visionary, and skeptically antiheroic—altogether beyond "Upon a Dying Lady" in range and depth.

In these qualities they parallel, despite all surface differences, the early sequences of Ezra Pound and T. S. Eliot. It is extremely interesting that Yeats was composing them during 1919–22, a period that encompasses the making of Eliot's *The Waste Land* (published in 1922) and the publication of Pound's Canto 4 (1919), Cantos 5–7 (1921), and *Hugh Selwyn Mauberley* (1920). The civil-war sequences, *The Waste Land,* and Cantos 5–7 appeared originally in *The Dial,* and Canto 4 was reprinted there in 1920, as were the first six poems of *Hugh Selwyn Mauberley.* Pound's closeness to Yeats and Eliot during this period is common knowledge, and he may well have been the catalyzing agent in the creation of prototypes of the twentieth-century lyric sequence. In fact, the early versions of Pound's Cantos 1 and 3 had appeared in 1917 (very different from the cantos so numbered in the final text). Read with an eye to comparable tones and interests in the Yeats sequences, they suggest a similar struggle to repossess and reorient values of the classical, medieval, and Renaissance past, when "greatness" is supposed to have lived up to itself rather than abdicating its responsibilities in the face of violence.

We have oversimplified this last point of comparison in the interest of avoiding a detailed case for the younger poet's literal "influence" on his Irish friend, a case that would involve arguing too emphatically in any event. And we are talking less about themes than about an athletic effort of morale. The important thing to note here is the seriousness of the *poetic* engagement. In all the major sequences we find one overwhelming pressure at work: an urgent need to cope with an intractable situation. This need becomes an energizing element in the aesthetic of any given work. It presents itself as a sense of being balked, or of being beset by an engulfing flood of circumstance and consciousness that can be stayed only momentarily. The counter-efforts of sensibility lead to clarifying an inner state and relating it somehow to the intractable principle—the irresistible torrent of fatality, the sheer mass of sensations and memories and feelings, the impersonal sweep of history, the uncontrollable forces of political life, the cycles of existence, the unpredictable working of the psyche, the erosion of cherished values—through poetic equilibrium.

The connection between these counter-efforts and the traditional epic task or tragic choice of older literature will be apparent. So also will be the tremendous changes in modern emphasis. As *Maud* and *Song of Myself* and the Dickinson fascicles revealed long ago (to say nothing of the obvious parallels in Symbolist poetry), it has devolved more and more upon private sensibility to carry the impossible yet necessary heroic burden. In the process, classical narrative and dramatic structure have given way to the subjective primacy of lyrical structure. The concepts of "protagonist" and "plot" and "argument" and even "voice" are still convenient for critical discourse, but these terms are not, finally, accurate. Instead, the succession and interaction of the larger units of affective language and the related streams of tonality running through a work have emerged as the key to the intrinsic movement and quality of poems, and to poetic art itself. They alone create whatever semblance of consciousness may be said to inhere in a charged structure of verbally evoked awareness.

Yeats's double civil-war sequence clearly reflects this development. Its immediacy of presentation—the distressed engagement with bloody political chaos and with the failure of personal dreams—makes it a work of crucial significance. It *places* the modern sequence in its artistic character as a reflex, in its human function, of the predicament and dreams of the creative spirit in our century. In this role it is matched by only two other sequences of the modern age. The first is *The Waste Land*, with its brilliant succession of portentous and satirical and lyrically yearning tonalities and its confessional self-placement, just before the end, in the line "These fragments I have shored against my ruins." The second is *The Pisan Cantos*, with its glow and sweat of humiliation and heroic failure. Both these latter works begin with challenging notes of disturbance or frustration: "April is the cruellest month"; "The enormous tragedy of the dream in the peasant's bent shoulders." Similarly, Yeats's whole opening poem, "Ancestral Houses," is obsessed with the intractability of the degenerative principle. The historically irreversible betrayal of nobility leaves the tradition-saturated sensibility high and dry:

> Homer had not sung
> Had he not found it certain beyond dreams
> That out of life's own self-delight had sprung
> The abounding glittering jet; though now it seems
> As if some marvellous empty sea-shell flung
> Out of the obscure dark of the rich streams,
> And not a fountain, were the symbol which
> Shadows the inherited glory of the rich.
>
> Some violent bitter man, some powerful man
> Called architect and artist in, that they,
> Bitter and violent men, might rear in stone
> The sweetness that all longed for night and day,
> The gentleness none there had ever known;
> But when the master's buried mice can play,

> And maybe the great-grandson of that house,
> For all its bronze and marble, 's but a mouse.

What can the inheritor of that ancient companionship of warrior-aristocrat and artist do? He must play the roles of both, creating an illusion of power to remake the world "out of life's own self-delight." Yeats ends "Ancestral Houses" with a question engendered by the sight of a characterless élite, their privileges now pointless:

> What if those things the greatest of mankind
> Consider most to magnify, or to bless,
> But take our greatness with our bitterness?

The question reaches further than the local situation. It touches on what civilization itself will do with its heritage, and on what the individual sensibility will do with the rich gifts of the entire past. The poem's tone is such that it presents a challenge and an ideal, both embedded in the passages just quoted. At the same time it projects an elegiac pathos, quietly devastating in its ironic evocation of effeteness and languor. The two closing stanzas, in fact, are each a periodically structured question whose essence the final three cited lines catch perfectly. These stanzas present a subtle catalogue of possessions in a context of exhausted passions:

> gardens where the peacock strays
> With delicate feet upon old terraces,
> Or else all Juno from an urn displays
> Before the indifferent garden deities . . .

Everything recalls power now grown obsolete: "the glory of escutcheoned doors," "buildings that a haughtier age designed," "famous portraits of our ancestors." The telling over of such possessions is nostalgic, sometimes witty, sometimes ironic. These mingled tonalities make for a stream of disillusionment, not without its bittersweet edge, throughout the poem. But the *vital* energy comes from elsewhere—from a paradise of pure vision out of which Homer's art sprang, and from the furious need that engenders a civilization in the first place.

The vision and the need enclose an intensity and a force of desire beyond the poem's overt thought. The language for them—such phrases as "the abounding glittering jet," "some violent bitter man," and "the sweetness that all longed for night and day"—quickens the whole reach of the double sequence. It is a language of urgent power, of haunting loss and the keenest will to repossess. The next poem. "My House," leaves behind the tones of meditation and of scornful telling over of empty possessions but picks up the stream of intensity and rich desire of "Ancestral Houses," giving it a new context. Now we are in the toiling poet's domain, an old farmhouse and tower on "an acre of stony ground." No "flowering lawns" and straying peacocks here, but "old ragged elms, old thorns," wild waterfowl, some cows. The Norman tower is a link with a harsh past that accords both with the pastoral yet rugged landscape and the poet's life of reverie and hard work:

> A winding stair, a chamber arched with stone,
> A grey stone fireplace with an open hearth,
> A candle and a written page.
> *Il Penseroso's* Platonist toiled on
> In some like chamber, shadowing forth
> How the daemonic rage
> Imagined everything.

Here the telling over of possessions lacks complacency and "slippered ease." The urgent power, the passion to take over the "violent bitter" drive that longs for a sweetness that is its opposite and therefore creates in vibrant pride, now resides in the poet and mystical thinker. His task is to summon up the "daemonic rage" as Milton, in "Il Penseroso," dreamed of doing

> in some high lonely tow'r,
> Where I may oft out-watch the Bear,
> With thrice great Hermes, or unsphere
> The spirit of Plato to unfold
> What worlds, or what vast regions hold
> The immortal mind that hath forsook
> Her mansion in this fleshly nook;
> And of those daemons that are found
> In fire, air, flood, or under ground,
> Whose power hath a true consent
> With planet, or with element.

Yeats's own tower, the literal one, is presented as more prosaic than Milton's imagined one. Its previous "founder" was an obscure "man-at-arms," a cavalry officer stationed with his men "in this tumultuous spot"—then as now "tumultuous" because exposed to wind and floods and to "long wars and sudden night alarms." In this place the poet, equally heroic in his own way, and equally vulnerable, isolated, and obscure, will assume his stern symbolic task of repossessing the spirit abdicated by the privileged classes.

> Two men have founded here. A man-at-arms
> Gathered a score of horse and spent his days
> In this tumultuous spot,
> Where through long wars and sudden night alarms
> His dwindling score and he seemed castaways
> Forgetting and forgot;
> And I, that after me
> My bodily heirs may find,
> To exalt a lonely mind,
> Befitting emblems of adversity.

"Ancestral Houses" and "My House" superbly illustrate what we have called Yeats's aristocratic obsession as transmuted in his poetry. The transmuting involves three major streams of tonality, the first of which we have been discussing: the language of urgent power. We have taken note, too, of the second such stream: the language of something like heroic staunchness, mingled with that of vulnerable isolation. The third stream, less insistent and yet decisive for the dimensions of intelligence and self-questioning it introduces, is that of sardonic irony. We find it most clearly in the opening of "Ancestral Houses" ("Surely among a rich man's flowering lawns"—*surely,* indeed!) and again in the closing couplet of the third stanza:

> And maybe the great-grandson of that house
> For all its bronze and marble, 's but a mouse.

In dwelling on "Ancestral Houses" and "My House," we have been concerned, precisely, with this "transmuting" process. This is the process whereby initial pressures on sensibility enter the dynamics of a work. The resultant streams of tonality constitute the active principle animating and relating the otherwise separate units of affect. At any one point along the way in a sequence, one or another of these streams is likely to be dominant, although it may soon disappear only to surface again further along. Or it may, of course, merge with one or more of the other streams; and in fact once it has appeared its presence will be felt thereafter. It may attach itself to highly varied objects and contexts, as when the haunting loss evoked in "Ancestral Houses" has added to it, in "My House," the sense of isolation and vulnerability of the stalwart man-at-arms and his retinue and then, more intimately, the whole complex of intermingled feeling that swirls about the poet's own person. To this variable of context we must add another variable: of intensity. Thus, the initial irony of "Ancestral Houses" is converted, in the second stanza, into fierce and imaginatively soaring disillusionment. And in "My House" the sympathetic evoking of the man-at-arms' courage and hardships is converted into a different sort of confrontation of adversity, artistic and quixotic and presented with a climactically *singing* concentration at the poem's close. These two variables, of the context of a continuing stream of tonality and of the degree of intensity (such as can convert somewhat distanced irony to darkly bitter exclamation) are essential in the structural and qualitative life of a long poetic work.

Following through very quickly, we can see that the third and fourth poems, "My Table" and "My Descendants," complete the diversion of the first tonal stream from the context of a degenerate aristocracy to that of the embattled speaking self. The shift is emphasized by the use of "my" or "I" in the titles of all the poems of "Meditations in Time of Civil War" except the first. "My Table" continues the tonality of tragically darkened exaltation reached at the end of "My House"; but it does so in an altered key, becoming a celebration of an ancient Japanese sword, a "changeless work of art" embodying great chivalric and mystical values the world now considers obsolete. This celebration is interrupted by the negative fears of "My Descendants," a poem whose low intensity (by contrast with the succession of ecstatic couplets in the single unbroken stanza of "My Table") makes it seem almost genial despite its arrogance and pessimism. The very familiar language toward the end of this poem, when the poet says he has chosen the house "for an old neighbour's friendship" and has "decked and altered it for a girl's love," contributes to this tonal relaxation. But then, in the two succeeding poems, "The Road at My Door" and "The Stare's Nest by My Window," the second major tonal stream, of vulnerability and isolation, re-enters the work mightily. We are in the midst of civil war. Fighting men from opposed factions appear at the poet's door and shock him into realizing that he has been living in "the cold snows of a dream" while brutality prevails all about him. The poems arrive as traumatic recognitions, the one in anecdotally personal terms, the other in its delineation of horror:

> Last night they trundled down the road
> That dead young soldier in his blood.

In both poems, meanwhile, natural life is shown going calmly on. The moor-hen guides her chicks, "those feathered balls of soot," on the stream. The bees build in the tower's "loosening masonry." "Mother birds bring grubs and flies" to their fledglings. The contrast is deadly ironic, picking up from the mockeries of "Ancestral Houses" but more bitterly personal. The bitterness wells up particularly in the refrain to "The Stare's Nest by My Window," with its invitation to honeybees (and perhaps unnamed others) to "come build in the empty house of the stare." The fullest retreat into self-depreciation, however, occurs in the closing poem of "Meditations in Time of Civil War." This is "I See Phantoms of Hatred and of the Heart's Fullness and of the Coming Emptiness." Here all the external terms of reference of the sequence so far—houses, possessions, civil war, the generations—are replaced by visionary figures. Yet the major streams of tonality reappear in this totally altered context. In the exquisite yet harshly self-knowing first stanza, the world of the preceding poems fades into a mist that recalls yet distorts it. In the following stanzas a succession of fantasies arise in reverie—a senselessly vengeful mob emerging from the recesses of the far-off past; elegant ladies from faerie, mounted on unicorns and lost in voluptuous dreams; and, finally, an "indifferent multitude" and "brazen hawks" with "innumerable clanging wings that have put out the moon." These are all "monstrous familiar images" arising from the previous desperate confrontations of the sequence. The poet, falling into extreme stylistic buffoonery by mimicking Wordsworth's solemnity of diction at the very end of the poem, confesses that these "daemonic images" are all that his "half-read wisdom" amounts to. They must (in Wordsworthese) "suffice the ageing man as once the growing boy."

"Nineteen Hundred and Nineteen," the second part of the double sequence, is in part a compressed reprise of the first part's tonalities. Its poems are numbered but untitled, so that the tonalities relate to one another with less overt thematic guidance from the poet. The opening poem, in the same *ottava rima* as "Ancestral Houses," to some extent presents similar concerns. Its intensity is of a more burning order, however; it has a more profoundly elegiac aspect, and its stunned horror at human savagery picks up from the phrasing in "The Road at My Door" and "The Stare's Nest by My Window." In short, it is more complex and driven than any single poem in "Meditations in Time of Civil War" and it raises "Nineteen Hundred and Nineteen" to a higher pitch of feeling from the start. Its special energy is confirmed in the extraordinary third poem, the climactic internalization of the double sequence's turbulent course. All the important tonal streams we have encountered converge in this, the psychological centerpiece of the civil-war poems:

> Some moralist or mythological poet
> Compares the solitary soul to a swan;
> I am satisfied with that,
> Satisfied if a troubled mirror show it,
> Before that brief gleam of its life be gone,
> An image of its state;
> The wings half spread for flight,
> The breast thrust out in pride
> Whether to play, or to ride
> Those winds that clamour of approaching night.

A man in his own secret meditation
Is lost amid the labyrinth that he has made
In art or politics;
Some Platonist affirms that in the station
Where we should cast off body and trade
The ancient habit sticks,
And that if our works could
But vanish with our breath
That were a lucky death,
For triumph can but mar our solitude.

The swan has leaped into the desolate heaven:
That image can bring wildness, bring a rage
To end all things, to end
What my laborious life imagined, even
The half-imagined, the half-written page;
O but we dreamed to mend
Whatever mischief seemed
To afflict mankind, but now
That winds of winter blow
Learn that we were crack-pated when we dreamed.

Such is the force of the proud image of the swan readying itself for flight and then leaping "into the desolate heaven," and so absolutely concentrated is the meditation on the worth of human accomplishment, that one might almost miss the "rage" toward suicide in this poem. The three main tonal streams of the double sequence—of urgent power, staunchness in isolation, and sardonic irony—are drawn into the double vortex of the swan's leap and the poet's "rage / To end all things." The power of the poem gives it independent life, yet its reciprocities with the poems before and after it in the two sequences give it a clarity of resonance it could not otherwise have. The swan has reciprocities with the "abounding glittering jet" and even the "marvellous empty sea-shell" of "Ancestral Houses"; the "Platonist" who "affirms" that it is "lucky" to have all one's works vanish has affinities with "*Il Penseroso's* Platonist" toiling in "My House"; and the poet's "rage" is like the "daemonic rage" whose imaginative creation of "everything" Milton "shadowed forth." These are but a few reciprocities of the sort we have mentioned. Another of importance is indicated by the fact that "My House" has the same intricate stanzaic form as the poem now under scrutiny. The aims set forth in "My House" are now seen to have been vain, yet the staunchness is of the same order—in the one instance the careful construction of a dream-architecture out of unpromising reality; in the other, its heroic deconstruction.

The tragic exulting of the third poem of "Nineteen Hundred and Nineteen" can be compared—another reciprocity—with that of "My Table," but its wild power and deeper inwardness make it at once the unquestionably climactic poem of the double sequence and the most decisively introspective as well. Similarly, the short closing poem at the end of "Nineteen Hundred and Nineteen" goes beyond the closing poem of "Meditations in Time of Civil War" ("I See Phantoms of Hatred and of the Heart's Fullness and of the Coming Emptiness") in its terrified recognitions. In the thirteen poems of the double sequence Yeats had discovered the extraordinary dynamic possibilities afforded by the genre. The compressed recapitulation enabled by the double-sequence form was an added discovery—conceivably influenced by Pound's double sequence *Hugh Selwyn Mauberley,* whose second part, "Mauberley (1920)," is a subjective, associative reprise of the more externally directed and active first section.

2. "A Man Young and Old" and "A Woman Young and Old"

Very noticeably, there are no poems dealing with love in the Yeats sequences we have so far discussed. Civil war and hard-fighting politics have made the heart grow "brutal," as "The Stare's Nest by My Window" puts it. The sexual emphasis, if that is the right term, is on possible racial degeneration in these poems. It is true that both parts of the double civil-war sequence end with poems that include figures of erotic fantasy: the ladies with "hearts . . . full / Of their own sweetness, bodies of their loveliness" in the one poem; and "Herodias' daughters" with their "amorous cries," "the love-lorn Lady Kyteler," and "that insolent fiend" who is her incubus in the other. But even in the latter instance there is a certain emphasis on racial degeneration—the violation of a lady by a loutish demon—in the riddling passage that closes "Nineteen Hundred and Nineteen":

> thereupon
> There lurches past, his great eyes without thought
> Under the shadow of stupid straw-pale locks,
> That insolent fiend Robert Artisson
> To whom the love-lorn Lady Kyteler brought
> Bronzed peacock feathers, red combs of her cocks.

In a very superficial sense, the double sequence might be considered a kind of love poem, since "My Descendants" speaks of having "decked and altered" the tower "for a girl's love." And it is true that the heights and depths of feeling—the chivalric dreaming, the deep longings and disillusionments, the sustained celebration of creation out of powerful need—through which the poems move could well be projected onto a love relationship. Nevertheless, the poet's momentary gesture in the one poem, however affectionate toward his wife, has no clear relationship to those heights and depths. If they are relevant to the deeper psychic meaning of his personal life—as they must be—then we are in the realm of unconscious character formation beginning with the early loss of any nourishing maternal presence and developing into the emotional landscape and sexual denial and loss that marks so much of Yeats's poetry. Artistically, his mastery of the intricate problems of structuring the civil-war sequences prepared him for the even more demanding, intimately committed tasks he faced in the later love sequences.

By the time of the first of these sequences, "A Man Young and Old"—published in 1927 and reprinted the next year in *The Tower*—Yeats had of course written many love poems. (We are using this term very broadly here, to mean not only poems about love or declaring love but work engaged with erotic and romantic feeling in a wide range of contexts, the volatile range of a highly developed and endlessly associative sensibility.) His earliest volumes present delicately noble, tremblingly fragile idealizations of womanly beauty and of love, although it would be easy to overstress their unworldly and pathos-ridden character. By the time of *The Wind Among the Reeds* (1899), for instance, one finds poems like "The Heart of the Woman," "He Hears the Cry of the Sedge," and "The Lover Pleads with His Friend for Old Friends"—the first a passionately direct forerunner of poems in "A Woman Young and Old" and in the Crazy Jane series, the second sounding notes of world-desolation and thwarted love not to be heard again so powerfully until many years later, and the third a mixture of worldly dismay and fear of love-abandonment that reveals a mind as unsentimentally aware as it is romantic.

The next four volumes each contain a group of love poems, mostly clustered together, that is at least a modulation toward a sequence. In the first three of these volumes—*In the Seven Woods* (1904), *The Green Helmet* (1910), and *Responsibilities* (1914)—the emphasis is all on a sense of loss, but one counteracted, often, by memories of shared elation and tenderness. This emphasis is colored by a painful conception of a kind of love that is out of phase with modern attitudes, and by rueful and yet resistant feelings engendered by the aging of a beloved woman who has never given herself wholly to the poet. But the group of love poems that comes closest to a sequence appears in *The Wild Swans at Coole* (1919). The group includes thirteen poems, beginning with "Men Improve with the Years" and ending with "Presences." These poems get past the pure dejection and assumption of having been overborne by fate that dominate the earlier volumes. They play many tones of love-engagement against one another: self-ironies, sexual exuberance, confessional notes, evocations of enduring passion and of its exaltation above personal happiness, and recognitions in varied tonal modes of the irrevocable ravages of age and of "that monstrous thing / Returned and yet unrequited love" ("Presences"). Then, in *Michael Robartes and the Dancer* (1921), two piercing poems of intimate self-revelation, "An Image from a Past Life" and "Towards Break of Day," bring Yeats's love poetry to a very high pitch of candor, openness, and mature self-knowledge. They show his complete readiness to undertake the explorations of his first true love-sequence: "A Man Young and Old."

It is characteristic of Yeats's developed artistry, and of his related willingness to take risks, that he placed this sequence in *The Tower* just as he did. Just as the thirteen poems of the civil-war sequences follow two powerful opening poems, "Sailing to Byzantium" and "The Tower," so the love-sequence precedes the volume's two closing poems: the slashingly political "The Three Monuments" and the maniacally perfect "All Souls' Night." "Sailing to Byzantium" and "The Tower" present an excruciating predicament, the frustration of passionate old age. Yet, by a series of contextual conversions, they transfer their initial feeling into transcendent states that, without denying the private anguish, convert its intensity into a wider sphere of balanced energies. Extremely self-revealing, these two poems prepare the way for the civil-war poems, which place a troubled, romantically dreaming, but sharply observant sensibility amid the confusions of revolutionary violence. In a reciprocal manner, the poems that follow the love-sequence recall *The Tower*'s wider orientations, which have been opened into by the poems that end "A Man Young and Old," particularly the very last poem: "From 'Oedipus at Colonus.' " "The Three Monuments" restores direct connections with the bitter political cast of the civil-war poems; and "All Souls' Night" (somewhat prepared for by the variations on the theme of madness throughout "A Man Young and Old") dives into the state of crazed openness needed to sustain the pressures of all experience and all loss. The volume as a whole presents a vast wash of suffering and fatality while summoning up the energies of mind and imagination needed to assert the desired primacy of man's creative spirit. Together with the poems of old age, the sequences near the beginning and near the end of the book contribute the major elements of its organic body.

We may, indeed, think of the whole of "A Man Young and Old" as a confirmation of that organic body. It has something like an autobiographical structure, in that it takes us through phases of the speaker's love-experience from first love to an old man's memories and regrets.

We do not suggest that anything in the sequence is Yeats's literal autobiography or confession. But in it the male speaker's misery and his means of coping with love's pressures—including the proliferating personae he adopts—do strongly suggest *some* autobiographical basis.

To review the progress of the sequence summarily: the first four poems are those of the Man Young, the next six are of the Man Old, and the eleventh is a choral comment from Yeats's translation of *Oedipus at Colonus.* The first three take up the murderous, maddening cosmic beauty and indifference of the young man's beloved. She is unearthly, existing in Poems I and II ("First Love" and "Human Dignity") as the moon and in Poem III ("The Mermaid") as an immortal creature. As the moon she maddens, literally making her hapless lover "lunatic." As mermaid she takes her pleasure, indifferent as she pulls him down to be drowned. Then the fourth poem, "The Death of the Hare," reverses field, introducing a new complex of feeling. The moon-mermaid figure of the first poems is replaced by another who is earthly and vulnerable. In the midst of gallant wooing, the speaker suddenly sees her as his victim, in the image of a hare he has hunted down. (Perhaps there is a suggestion that the tables have been turned in another phase of the first relationship.) Yeats had used this same hare-image, though in a very different context, is his earlier and better poem "Memory." There it suggests the undying force of a once perfect, reciprocal relationship—

the mountain grass
Cannot but keep the form
Where the mountain hare has lain.

In "The Death of the Hare" the case is different. The speaker realizes he has hunted a woman down, treated her as a quarry rather than been to her as the mountain grass to the mountain hare. At the same time, he is remorseful, and thus utterly unlike the moon-woman who, in Poem II, is indifferent to her lover's "heart's agony," viewing it as though it were "a scene / Upon a painted wall." Nor does he at all share the mermaid's "cruel happiness" of Poem III, since he suddenly becomes aware, as she does not, of what he is doing—that he has been heedlessly dominating another person just as the moon-mermaid dominated him. This change from victim to hunter, but with the added dimension of rueful self-knowledge, is the first alteration of persona in the sequence—a bit contrived, and just possibly more swaggering than rueful after all, but nevertheless making for a new tonal direction in the sequence.

The fifth poem brings about another shift, contrasting predicaments of youth and age. This pivotal poem, "The Empty Cup," juxtaposes a "crazy man"—the "lunatic" young man of the first poem—and the older speaker whom he has grown to be, maddened by his memory of opportunities lost through his "moon-accursed" infatuation. Once he was too young and timid to drink boldly from the full cup (even the lady's lunar impersonality was very likely an illusion); now that cup is empty and dry. So the persona has added to itself the embittered consciousness of an ungratified life. Then, in "His Memories" (Poem VI), comes another turn like that in Poem IV. Here the poignancy of old bodies "broken like a thorn / Whereon the bleak north blows" is heightened by comparison with a heroic youth, when the speaker had Helen of Troy herself as his mistress. In the next four poems other aged ex-lovers—Madge, Margery, Peter—extend the range of comparison between now and then. The past is still alive in them. They live in an unfolding field of memory full of unresolved griefs and passions and ambiguities: a compost of contradictory emotions suggested by the poems' titles: "The Friends of His Youth," "Summer and Spring," "The Secrets of the Old," and "His Wildness."

The last two of these are the richest in this small group. "The Secrets of the Old" (Poem IX), a most singable psychological ballad, has a perfect music of memory and strange distancing. It begins:

> I have old women's secrets now
> That had those of the young;
> Madge tells me what I dared not think
> When my blood was strong,
> And what had drowned a lover once
> Sounds like an old song.

"His Wildness" (Poem X) is at the opposite pole of feeling—not this nostalgic acceptance of a cooling spirit but a frenzy to escape time's trap, with a suicidal impatience like that of the swan-poem in "Nineteen Hundred and Nineteen." The poem's beginning sums it all up: the passing of joy, beauty, and heroism from the speaker's life and from all life, and the passion for release into another sphere:

> O bid me mount and sail up there
> Amid the cloudy wrack,
> For Peg and Meg and Paris' love
> That had so straight a back,
> Are gone away, and some that stay
> Have changed their silk for sack.

This plea for release from human life is also a plea to enter the transcendent realm of the inhuman beloved of the first poems. It is reinforced by the next poem, the chorus from *Oedipus at Colonus* ending the sequence. "Never to have lived is best," goes the famous message. The "second best" is to put all behind and welcome death gaily, like the "dying lady."

In the civil-war sequences the speaker's roles of a battered Don Quixote of the imagination and of an artist-philosopher heart-broken over a civilization gone wrong helped create the elegiac and visionary structuring. In "A Man Young and Old" these mantles are shed. Now the speaker appears first as a suffering Petrarchan lover, as he had been in Yeats's much earlier writing—for instance, in the love poems of *The Wind Among the Reeds* (1899). But there is an important difference from the earlier work: the starker, more drastic engagement with destructive love. Moreover, the protagonist is not consistently a suffering lover but splits off into alternative *personae*. The Petrarchan emphasis disappears, for instance, from the boastfully guilty "The Death of the Hare" and the even more self-glamorizing "His Memories," as though the protagonist were being shown in several perspectives simultaneously: a Picasso-view of the multiple self emerging over a lifetime. Thus, in "His Memories," the stunned lover of the first poems is replaced by an opposite self. The speaker has become great Paris, Helen's lover. He recalls how she "who had brought great Hector down / And put all Troy to wreck" once lay in his arms, fearing her love-ecstasy would make her "shriek" aloud. We may speculate as we will—that this is a gallant personal tribute by Yeats to a remembered past love, or that the triumph and delight remembered are merely imaginary, or that the supposed memory is really the invocation of another age, a phase of the human past when miracle and heroic emotion saturated life as now they cannot. The poem invites many readings, extending the range of both passion and loss with which the whole sequence is engaged.

As we have seen, "A Man Young and Old" provides extreme shifts of affect, with logically irreconcilable contrasts of experience and feeling. One of the most interesting things Yeats does is to add new characters as the sequence progresses who seem to be formed out of the original two in "First Love." The "I" is split off into a more successful yet more self-castigating lover, and later into an old man in different moods, into Paris (in memory or imagination), and into his own former rival, Peter, who is a kind of mirror-image enabling the speaker to objectify what he had thought was a relationship uniquely his own. The "she" is moon-woman, then mermaid, then hare-like victim and Madge and Margery and "Peg and Meg and Paris' love." Bitter and sweet are so mingled by the end, as are the degrees of frustration and gratification, that despite its unevenness the work is strikingly a music of complex masculine feeling.

The movement is from the strange blending, at the start, of young love's illusions with cold, inhuman magic ("beauty's murderous brood") to the equally strange blending, by the end, of humorous, half-resigned humane wisdom with sheer frenzy. The eighth and ninth poems, "Summer and Spring" and "The Secrets of the Old," charmingly juxtapose the intensities and jealousies marking the intimate conversation of the young and the mellower perspectives of former lovers recalling the past. Then, without warning, "His Wildness" springs into the foreground, with its plea for freedom from all pain and loss and its speaker's need to utter the screech of a peacock—harsh, eerily metallic, untranslatably out of the heart of the life-force. (We are in the same realm, here, as in the third poem of "Nineteen Hundred and Nineteen," with its swan-image counterpointed to the speaker's sublime despair.) The sequence casts a remarkably wide net of affects, based partly on the split selves and their interaction. Together these affects create an organic body of complex love-involvement within psychic time.

Not everything works equally well. Poems like "The Mermaid" and "The Empty Cup" are far below the general level. And there is a certain amount of forced diction and symbolism. We think particularly of all the "shrieks" and "stones" and the ninetyish effects of the final Sophoclean chorus in Yeats's translation. Yet the shrieks do embody the inarticulable, impersonal powers by which we are driven. And the stone-image, introduced in "First Love" when the young man has "laid a hand" on his beloved's body but "found a heart of stone" (for she is made of lunar, not fleshly substance), is at once of a mythical order and associated with love's frustration.

The sequence thus opens with a Petrarchan complaint that is also a serious parody and reversal of the myth of Endymion and Selene. Later poems then pick up the stone-image to evoke barrenness and loss. "Old Madge," in "The Friends of His Youth," holds a stone to her breast as though it were a nursing baby. "His Wildness" presents the male speaker doing the same thing, maddened by the loss of all cherished things into a kind of transport of sexual confusion. "His Wildness" connects the deep elegiac float of *The Tower* as a whole with the drive toward utter volatility—beyond ordinary grief or any other ordinary emotion—perfectly embodied in "All Souls' Night." This volatility, finally, is the *sole* means of transcendence available to human feeling and to art. Yeats may have seen this fact more clearly than any other writer of our age, and in one sense, therefore, may have been our ultimate realist despite thinking of himself as among "the last romantics."

The intractable condition, the pressure, at the heart of the sequence is the will to create a sense of the whole of love's possibilities and meanings out of an initial state of loss and deprivation. Yet we never have a poem of love-reciprocity anywhere in "A Man Young and Old." Like the rest of *The Tower,* this sequence creates its structure out of a sense of deep adversity, a sense that prevails in Yeats's later sequences as well. But in the other love-sequences the major persona is female and the poems imagine more deeply into "her" varied states and perhaps allow the poet to clarify masculine experience a little further as well. Having confronted sexuality and love in "A Man Young and Old," he could deal with them less haltingly in "A Woman Young and Old" and in *Words for Music Perhaps.* The female personae seemed, too, to enable him to present passion and reciprocal love (and their disappointments) unselfconsciously because direct confession is not even remotely implied.

Both these sequences appear at the end of *The Winding Stair and Other Poems* (1933). "A Woman Young and Old," which was written earlier, concludes the volume and is clearly a companion-group to "A Man Young and Old." It too consists of eleven poems, the final one a translation of a Sophoclean chorus, and employs astronomical symbolism. But the correspondence is not exact. "A Man Young and Old," for instance, introduces its astronomical symbolism at once, with the lover a maundering lout and the loved woman imaged as the moon, radiant, changing, and inhuman. "A Woman Young and Old" starts with a realistic, contemporary dramatic confrontation: "Father and Child." The comically exasperated father speaks, yet the point of view is the daughter's—an interesting device for suggesting the subtler and more complex perspectives of this second love-sequence.

> She hears me strike the board and say
> That she is under ban
> Of all good men and women,
> Being mentioned with a man
> That has the worst of all bad names;
> And thereupon replies
> That his hair is beautiful,
> Cold as the March wind his eyes.

In "Father and Child" the man has the almost supernatural beauty and indifference. But the "child," unlike her male counterpart of the other sequence, has all her wits about her. Her reply to her father's explosion of social conformity is unanswerable, for it assumes that love exists in an archetypal ecstatic realm beyond trivial conventions and mores. Poetically speaking, moreover, what is unanswerable is the perfect placement of the closing line. It presents the poem's one pure image in a brilliant culmination that sloughs off the dramatic framework, making questions of "point of view" finally irrelevant. Hence we experience something of a jolt when the next poem, "Before the World Was

Made," bounces into view. This poem, with its metaphysical argument set to the rhythm of a music-hall number, reverses the emphasis on physical magnetism just embodied so perfectly in "Father and Child." The flesh-and-blood woman gives way to her ideal Platonic self, of which her physical beauty (partly self-created) is but a remote image. Her language, that of a vainly self-centered and coldly flirtatious woman, gives character and verve to the poem. But her true role is not as a seductress but as a passionless instructress, who can guide her lover to the worship of ideal beauty—her Platonic self, as it were:

> I'd have him love the thing that was
> Before the world was made.

Here, if we may digress for a moment, Yeats foreshadows his surprisingly pragmatic position in such very late poems as "Under Ben Bulben," "The Statues," and "Long-Legged Fly," in which the artist's prime duty is seen as the creation of images of the ideal: to give "women dreams and dreams their looking-glass," as we are told in "The Statues." We shall see in our discussion of *Last Poems* how such dream-provoking images act as a stay against spiritual and physical degeneration. They "bring the soul of man to God"—in the words of "Under Ben Bulben"—and affect the future of the race by showing men and women how to "fill the cradles right." All this is at least potentially implied in "Before the World Was Made," whose speaker wishes to be desired for her original, ideal self rather than her bodily one.

After this poem, however, the Platonic dream is dropped for a while in "A Woman Young and Old." The next poem, "A First Confession," is an abrupt reversal toward the purely sexual as the speaker confesses she has concentrated all her being on satisfying "the craving in my bones." Yet the last stanza provides a further turn, elevating sexuality to a cosmic principle whose impersonality at least parallels that of Platonic absolutes:

> Brightness that I pull back
> From the Zodiac,
> Why those questioning eyes
> That are fixed upon me?
> What can they do but shun me
> If empty night replies?

Yeats's note to the sixth poem in the sequence, "Chosen," refers to this poem as well. He says that he has "symbolized a woman's love as the struggle of the darkness to keep the sun from rising from its earthly bed."[3] The simplest reading of the somewhat cryptic stanza just quoted, then, would be as a defense of flagrant sexuality. That is, it is as natural for a woman to keep her man from leaving her or turning his attention elsewhere as for the moon to reflect the sun, filling the night with brightness. The imagery is reciprocal with that in the first poem of "A Man Young and Old," where the bright moon emptying the heavens of stars is seen as a woman's beauty emptying her lover of thought. In these poems, as in "Before the World Was Made," the major pressure derives from sources far beyond individual choice or personality. The purposefulness demanded in *Last Poems* seems a development from the perception of the helplessness of human will to control these sources—a condition more tolerable to Yeats at the height of his powers, perhaps, than when he was less able to relish pure volatility and its emotional demands as ends in themselves (as he does in "All Souls' Night," at the close of *The Tower*).

In any case, the first three poems of "A Woman Young and Old" set up certain axes of reference for the following four love poems, which alternate meditation and song. "Her Triumph" is the young woman's grateful recollection of the change her lover has wrought in her. He is at once her Christian and her pagan rescuer, St. George *and* Perseus, and she is translated into a realm of ecstasy both spiritual and sensual: "And now we stare astonished at the sea, / And a miraculous strange bird shrieks at us." The next poem, "Consolation," continues to blend spiritual and physical in a witty, lyrical reply to the chorus that ends "A Man Young and Old." To the assertion that "Never to have lived is best" our young woman has a sexually knowledgeable retort—that as for "the crime of being born," we may be sure that "where the crime's committed / The crime can be forgot." The word "crime" is a moral, legal, and theological sledgehammer, swung on poor mankind as it helplessly conceives in mortal sin. Ironically, delightfully, Yeats's young woman reminds us that all distress and guilt are forgotten precisely at the source of our entry into the world.

With the sixth poem, "Chosen," we return to the Platonic and cosmic imagery of the second and third poems. Love's ecstasy, the woman asserts, makes up for its pain; sexual fulfilment has a strong spiritual component. Her "utmost pleasure" is realized through the senses but transcends them as the lovers rise to a still realm, beyond time,

3. This note appeared only in *The Winding Stair* (New York: The Fountain Press, 1929). It is reprinted in *The Variorum Edition of the Poems of W. B. Yeats*, ed. Peter Allt and Russell K. Alspach (New York: Macmillan, 1957), p. 830.

> Where his heart my heart did seem
> And both adrift on the miraculous stream
> Where—wrote a learned astrologer—
> The Zodiac is changed into a sphere.

Like "Consolation" this description of pure transport, glowing with erotic joy in its sense of miracle but also in its Aristophanic touch, is as playful as it is deeply serious. Playfulness and mystery and exploratory discovery run through these woman-centered poems far more richly than in the companion sequence. Thus the seventh poem, "Parting," an aubade in dialogue form, picks up something of the tonality of "Consolation" and brings us back to simplicities after the elaborations of "Chosen." It also counteracts the deeper seriousness of "Chosen" and its fear of the encroachments of empirical reality: the "horror of daybreak" that intrudes on the lovers' transport. Everything becomes more simply lyrical and sweetly amusing. From stanza to stanza the young man's resolve to leave his mistress with the dawn crumbles, as she challenges daybreak with her more exciting sexual dark: "I offer to love's play / My dark declivities."

From this lighthearted poem of sexual byplay, the sequence shifts to "Her Vision in the Wood," the first of the poems in which the speaker is an old woman. This is in part a Dionysiac counterpart to the second poem, with the man's symbolical role emphasized, and with an indication of the role of art in projecting archetypes. The procession of "stately women moving to a song," carrying the wounded man, could be "bodies from a picture or a coin." Music, dance, painting, sculpture, concretize the ideal embodied in religious ritual or dream-vision—and the ideal is not that of some remote moon figure but of the archetypal lover or fertility god:

> That thing all blood and mire, that beast-torn wreck,
> Half turned and fixed a glazing eye on mine,
> And, though love's bitter-sweet had all come back,
> Those bodies from a picture or a coin
> Nor saw my body fall nor heard it shriek,
> Nor knew, drunken with singing as with wine,
> That they had brought no fabulous symbol there
> But my heart's victim and its torturer.

The old woman who in the first stanza "stood in a rage / Imagining men" now sees through all her former lovers to the archetype they embodied, just as she too was an archetype for each "heart's victim." And she finds the truth not in some beautiful young body but in someone as foul and bloody as she—in "That thing all blood and mire, that beast-torn wreck." This poem stands with "Byzantium" and "The Circus Animals' Desertion" (also in the modified *ottava rima* of "Her Vision in the Wood") as a particularly powerful projection of the yoking together of the opposites of human reality—the mortal "blood and mire" and the wide-ranging human spirit that must imagine a transcendent opposite. Here the violent anguish of the speaker—caused by age, lust, and loss—gives birth to the image of other "grief-distraught" women. Yet these can shape their grief into a ritual with universal meaning and themselves belong to the transcendent realm of art. Unlike the speaker, and like the lovers on Keats's urn, they are "for ever young."

The ninth and tenth poems, "A Last Confession" and "Meeting," drop away from the intensity of "Her Vision in the Wood." "A Last Confession" anticipates the split in the "Three Bushes" sequence between sexual and spiritual love. As there, complete love must have both components (compare the unmoved speaker in "A First Confession"). For the old woman, this completion can take place only after death—when her soul "naked to naked goes." The language for the joining of the two lovers after death, as in "Supernatural Songs," is intensely sexual. It exquisitely recalls the little aubade "Parting," as the speaker's soul seeks out that of her true lover to

> Close and cling so tight,
> There's not a bird of day that dare
> Extinguish that delight.

In "Meeting" the enraged old woman of "Her Vision in the Wood" confronts her equally raging male counterpart, as savage as she at the old age that has masked their essential natures. The very intensity of their once young love now causes the intense hate of their old age, a reciprocity made clear in the closing lines:

> But such as he for such as me—
> Could we both discard
> This beggarly habiliment—
> Had found a sweeter word.

We have noted that, like "A Man Young and Old," this sequence closes with a chorus from Sophocles. This time Yeats chooses the *Antigone,* whose chorus sings not of death's attraction but against death and on behalf of life's prime intensity—the paradoxical "bitter sweetness" of love. (Recall "love's bitter-sweet" in "Her Vision in the Wood.") The chorus prays that through love—"that great glory"—the "soft cheek of a girl" will be enabled to hurl heaven and earth out of their places and bring "calamity" to all order and power, even to the gods. That is, the song is a deliberately reckless curse against the destiny that could crush the pure aspirations of such a woman as Antigone. The prayer will not succeed; destined injustice has cosmically triumphed. Death, since there is no love among the dead, is not to be celebrated as in the chorus from *Oedipus at Colonus* at the end of "A Man Young and Old." Instead, it is to be hated and grieved at:

> Pray I will and sing I must,
> And yet I weep—Oedipus' child
> Descends into the loveless dust.

The "gay goodnight" of the earlier chorus is not canceled. But the deeper insight this sequence achieves into the ecstatic reaches of love (perhaps reflecting Sophocles' insights into womanly devotion in the *Antigone*) carries with it a heightened understanding of the agony of its loss. These dual intensities become even greater in *Words for Music Perhaps,* especially in the Crazy Jane poems.

Chapter Six

The Evolution of William Butler Yeats's Sequences II (1929–38)

1. *Words for Music Perhaps*

The twenty-five poems of *Words for Music Perhaps*, with the exception of the opening seven Crazy Jane poems, were written in two separate batches: Poems VIII–XX in 1929 and perhaps early 1930, and XXI–XXV in the latter half of 1931. The Crazy Jane poems span these years, three dating from March 1929 (I, VII, II), one from October 1930 (III), one from July 1931 (V), and two from November 1931 (IV, VI). They contain the heart of the sequence and make up a smaller sequence of their own whose essential preoccupation—like that of the volume as a whole—is the true nature of love. Yeats's exploration, however, moves forward by juxtaposition of highly agitated or otherwise intense states of awareness, not by the more logical development of, say, Plato's *Symposium*. Yeats uses to the hilt the opportunities provided by a cast of characters and by various refrains to play off a number of passionate beliefs and experiences against each other. Crazy Jane, all by herself, serves as a mouthpiece for quite a number of moods and perceptions. Indeed she, of all Yeats's characters or personae, approaches most closely the complex sensibility that informs the more directly personal great poems like "Byzantium," "Among School Children," "The Circus Animals' Desertion," and "The Man and the Echo."

However, despite the variety, from the opening defiant imprecations of "Crazy Jane and the Bishop" to the exultant torment of "Crazy Jane Grown Old Looks at the Dancers" the shaping force is the same: a driven vision of enrapturing love that can fulfill both body and soul. Concomitant notes of pain, loss, and frustration—inescapable aspects of the mortal, human condition—punctuate the fundamentally ecstatic set of the group.

"Crazy Jane and the Bishop," the opening poem of the sequence, presents Jane's sensibility at its most exacerbated and sets certain key axes of thought and feeling that the rest of the sequence explores:

> Bring me to the blasted oak
> That I, midnight upon the stroke,
> (*All find safety in the tomb.*)
> May call down curses on his head
> Because of my dear Jack that's dead.
> Coxcomb was the least he said:
> *The solid man and the coxcomb.*

With typical economy, Yeats plunges us into the middle of an emotional storm. The speaker here (and throughout the sequence) is identified as Crazy Jane only by the title; no time is wasted introducing a cast of characters or describing a dramatic situation. But by the end of the second stanza we know all we need to of the external events that have brought Jane to such a pitch:

> Nor was he Bishop when his ban
> Banished Jack the Journeyman,
> (*All find safety in the tomb.*)
> Nor so much as parish priest,
> Yet he, an old book in his fist,
> Cried that we lived like beast and beast:
> *The solid man and the coxcomb.*

Here are the raw materials for a ballad or other melancholy narrative of lovers' separation and the death of the beloved wandering somewhere in foreign lands, but Yeats is after a different effect entirely. The plot is subordinated to a succession of lyric intensities; Jane is hardly a stock figure from romantic balladry; and the main drive of the sequence is towards an understanding, in emotional, intellectual, and sensuous terms, of the highest possibilities of love and its connection with artistic creation.

The emotional assault of the opening poem—dominated by rage—establishes immediately the enduring nature of the bond between Jack and Jane. Interference with their passion by the narrowly moralistic has been neither forgotten nor forgiven with the years; nor, seemingly, could the gap between two such antagonistic ways of life be bridged in any rational way. What can Jane do except "spit" at the distorted being who subverts love in the name of arid spirituality? And what could a man dominated by such a vision have to say to a woman who exemplifies, for him, sin and beastliness—except to continue railing? Yeats starts us off on Jane's side in this debate between orthodox religion and sexual passion—sexual passion that has endured beyond the grave.

With the second poem, "Crazy Jane Reproved," the hyper-aesthetic joins the moral-religious and sexual-pagan views on love:

> I care not what the sailors say:
> All those dreadful thunder-stones,
> All that storm that blots the day
> Can but show that Heaven yawns;
> Great Europa played the fool
> That changed a lover for a bull.
> *Fol de rol, fol de rol.*
>
> To round that shell's elaborate whorl,
> Adorning every secret track
> With the delicate mother-of-pearl,
> Made the joints of Heaven crack:
> So never hang your heart upon
> A roaring, ranting journeyman.
> *Fol de rol, fol de rol.*

The reproving speaker here is neither the Jane of the first poem, nor the Bishop, nor Yeats exactly. Rather, the voice is that of an artist who perceives that divinity manifests itself in the difficult creation of beautiful form divorced from human turbulence and sexual passion. This speaker would discard Zeus with the bull and would deny the violent energy associated with artistic transcendence. Jane and Yeats would reply "*Fol de rol, fol de rol,*" for this truth is only partial. (The artistically ritualized violence and brutality of the dance in the seventh poem are more to the point,—and poems I, II, and VII, remember, were the original starting points of the sequence.)

The next three poems, "Crazy Jane on the Day of Judgment," "Crazy Jane and Jack the Journeyman," and "Crazy Jane on God" explicitly introduce the supernatural. The Judgment Day in the first of these, if the "he" in that poem is God, is interestingly idiosyncratic. Jane does most of the talking, with God limited to sympathetic-sardonic agreement: " '*That's certainly the case,' said he.*" Also, the small *h* in "*he*" suggests that this is not the final Judgment of traditional Christianity, and obviously Jane is still caught in her obsession with profane rather than divine love. The preposition "on" in the title may of course mean "about" rather than "at," in which case one must read the poem as an intimately relaxed, half-bitter exchange between Jane and her lover—and the "judgment day" of the title simply indicates some ultimate point in time from whose perspective all the implications of love (and life itself) will at last become clear. In any case, this remarkably vivid poem centers on the evocation of extreme states of feeling hardly containable within any boundaries of rational thematic statement. It is clear, though, that Jane's concept of human love involves soul as well as body, and that the first stanza is close to a key perspective already seen in "A Woman Young and Old" and developed with great economy and saving humor in the later "The Three Bushes." Yet the assertion is more a note of yearning than a point of argument:

> "Love is all
> Unsatisfied
> That cannot take the whole
> Body and soul";
> *And that is what Jane said.*

The stanza implicitly synthesizes the opposing positions expressed in the opening poems: Jane's and the Bishop's, the sailor's and the aesthete's. In addition, it sets the scene for the next poem, "Crazy Jane and Jack the Journeyman," where the effects of love satisfied and unsatisfied are juxtaposed and the very nature of love is seen as unfathomable. But it is the simplicities of "Crazy Jane on the Day of Judgment" that are so ambiguously suggestive: the almost delphic opening proclamation we have quoted, the clever self-characterizing of an intense and articulate woman in the second stanza, the description of a day of black desolation (but without explanatory context) in the third, and the return to questions of love's mystery in the fourth. The tonal shifts, together with the alternating ironic refrains—"*And that is what Jane said*" and " '*That's certainly the case,*' *said he*"—make for a remarkable play of cosmic bemusement, earthy realism, shivering despair and abandonment, and self-ironic deflation.

If one responds to this tonal mixture primarily, rather than attempting intellectual translation, the poem is perfectly attuned to the passionate opening of "Crazy Jane and Jack the Journeyman," which follows it.

> I know, although when looks meet
> I tremble to the bone,
> The more I leave the door unlatched
> The sooner love is gone,
> For love is but a skein unwound
> Between the dark and dawn.

This first stanza could almost be an embittered woman's worldly comment on the inevitable results of giving oneself to a man. Its tone radiates a mystical desolation (despite an attempt at rational control) encompassing both sexual excitement and its transitoriness. Then, in a characteristically Yeatsian abrupt turn, the second stanza sustains this affect at first—repeating the mystical love-skein image and suggesting that the spirit must shed it in sadness—and then takes an entirely new tack of bright exultation:

> A lonely ghost the ghost is
> That to God shall come;
> I—love's skein upon the ground,
> My body in the tomb—
> Shall leap into the light lost
> In my mother's womb.

The leap after death into the immortal "light lost / In my mother's womb" is impossible—in Jane's passionate, pagan "theology"—without one's having thoroughly explored and exhausted mortal love in one's lifetime. Otherwise, the spirit cannot free itself for that leap into transcendent joy:

> But were I left to lie alone
> In an empty bed,
> The skein so bound us ghost to ghost
> When he turned his head
> Passing on the road that night,
> Mine must walk when dead.

We have here the philosophical center of the Crazy Jane sequence, and the best gloss on the poem is Plato's *The Symposium.* Jane is in some senses a modern version of Socrates' far calmer instructress, Diotima, who (in Benjamin Jowett's translation) tells the world's aptest male pupil:

> For he who has been instructed thus far in the things of love, and he who has learned to see the beautiful in due order and succession, when he comes toward the end will suddenly perceive a nature of wondrous beauty . . . not growing and decaying, or waxing and waning . . . not fair in one point of view and foul in another . . . And the true order of going or being led by another to the things of love, is to use the beauties of earth as steps along which he mounts upwards for the sake of that other beauty . . .

For the reader of Yeats, incidentally, the relation between this passage and the "fair" and "foul" lines in "Crazy Jane Talks with the Bishop" is striking. But our immediate concern is Diotima's idea of using the "things of love"—the "beauties of earth"—to mount upwards toward that "wondrous beauty." (We may well recall the shell imagery in "Crazy Jane Reproved," and the birch tree of "Crazy Jane and the Bishop.") Diotima's vision of transforming delight in eternally beautiful forms is close, of course, to Jane's blissful certainty that she will "leap into the light lost / In my mother's womb." This leap may be contrasted with the difficulty of reaching celestial love while still under the spell of the physical evoked in "The Delphic Oracle upon Plotinus," the closing poem of *Words for Music Perhaps*. "Salt blood blocks his eyes"—that is the predicament of Plotinus, and the usual human predicament when love has not been fully experienced and so we are blocked off forever from ultimate transcendent vision. It would probably be forcing the poem to suggest that Plotinus' purely abstract thought will never enable him to live fully enough to be readied for the higher phase—but Yeats *was* capable of just this kind of paradoxical reversal, giving spiritual priority to the supposedly wanton: Jane over the Bishop.

To return to the poem at hand, we are not talking about impatiently sloughing off obstacles to higher things, but rather of experiencing the "things of love" fully so that the higher beauty will be reached in the only possible way: through "due order and succession." If the process is cut short (a violence symbolized in the Bishop's banishment of Jack), then the situation of the first four poems in the sequence must obtain. In the first, Jack's ghost walks. In the second passion is inadequately valued. In the third Jane is still chained to the things of earthly love. And in the fourth she predicts similar results from being "left to lie alone / In an empty bed." No wonder the curses of the first poem are so violent.

The fifth poem, "Crazy Jane on God," has something of the distancing of "Crazy Jane Reproved," but with its deep, visionary purity prefigures the rich but wryly acquiescent tones of Poems XIV–XX. The God of the title and refrain—"*All things remain in God*"—is like the beautifully creative Heaven of "Crazy Jane Reproved," except that here all the violence of historical and personal change is transmuted into song—"My body makes no moan / But sings on." The next poem, "Crazy Jane Talks with the Bishop," holds the same place in this sequence that "Meeting" does in "A Woman Young and Old." Again we have two furious old people confronting each other, bound together in the one case by their former love, in this case by their passionate commitment to their respective ideas of love. This sixth poem dramatizes the conflict alluded to in "Crazy

Jane and the Bishop," but now we have a dialogue at once stinging and exuberant, rather than frenzied raging. And this time Jane has the last word. In the first poem she could only react passionately to being accused of living with her lover like "beast and beast." Here she preempts the Bishop's Christian authority itself. She overrides his arid conception that Love prevails only in a "heavenly mansion, / Not in some foul sty" with her own earthy version, in which "Fair and foul are near of kin, / And fair needs foul." Her own boldly sexual reading of the meaning of Christ's conception and birth ends the dialogue triumphantly:

> "A woman can be proud and stiff
> When on love intent;
> But Love has pitched his mansion in
> The place of excrement;
> For nothing can be sole or whole
> That has not been rent."

We have already noted Diotima's paradoxical comment on fair and foul in *The Symposium.* Beneath the violence of the stanza just quoted, so different in tone from Plato despite its philosophical cast, lurk certain similarities. Part of Diotima's discourse on love touches precisely on the aspect shared by man and beast, the "bodily lowliness" that goes along with the "heart's pride":

> "For love, Socrates, is not, as you imagine, the love of the beautiful only." "What then?" "The love of generation and birth in beauty."

For Crazy Jane, then, being "rent" is necessary for both physical and spiritual (including creative) generation and wholeness. As the refrain has it in "Crazy Jane Grown Old Looks at the Dancers," love is "like the lion's tooth." In this seventh poem, all the more savage intensities of the sequence are aesthetically converted into passionate, murderous dance-movements that powerfully counterbalance hatred and love. The word "love" appears in the refrain only, but the whole implication is that great hatred springs from but one source: equally great love. Even if one disagrees with this interpretation, it is still undeniable that the poem exudes the excitement of artistically controlling a ferocious energy. Here we have the artistic analogue to the creation of the shell's "elaborate whorl," and indeed the human blazes with emotional and sensuous awareness far beyond the creation that "Made the joints of Heaven crack."

Words for Music Perhaps opens with its strongest poems, the "Crazy Jane" group we have just discussed. We shall not go into the others in such detail. Of the next thirteen poems (VIII–XX), "Her Anxiety" (Poem X) and "Three Things" (Poem XV) are probably the most successful. Again, in this 1929 group, we have a movement from youth to old age. "Her Anxiety" focuses the youthful concerns in its two stanzas and refrain:

> Earth in beauty dressed
> Awaits returning spring.
> All true love must die,
> Alter at the best
> Into some lesser thing.
> *Prove that I lie.*
>
> Such body lovers have,
> Such exacting breath,
> That they touch or sigh.
> Every touch they give,
> Love is nearer death.
> *Prove that I lie.*

Yeats catches both the mutual absorption of lovers and its supposed inevitable lessening with the passage of time brilliantly in the second stanza. Yet one might almost wish he had contented himself with his crisply superb opening stanza, whose lyrical start softens the epigrammatic, staccato quality of what follows. At their simplest, the first two lines merely observe that wintry beauty will give way to the even greater beauty of spring. Or perhaps these lines annihilate the succession of seasons of growth and decay between one spring and another. After all, earth dressed in the beauty of one spring can look forward to renewal with absolute certainty, whatever wintry death may intervene. If we favor the first reading, however, "Earth in beauty dressed" provides a curiously erotic image for the bare winter landscape awaiting the spring, analogous to the relationship of human lovers in the next stanza.

In either case, the certainty of spring's return goes counter to the course of love, which—since it resides in mortal, aging bodies—must with equal certainty "alter . . . into some lesser thing." But there is also a resonance (reinforced by the ambiguity we have mentioned) in the first two lines that gives the poem an added poignancy; the spring and rebirth that winter awaits are unavailable to the old. The refrain's demand has been satisfied time and again, incidentally, in "A Man Young and Old," "A Woman Young and Old," the Crazy Jane poems, and here in such poems as "Young Man's Song" (IX), "His Confidence" (XI), and "His Bargain" (XIV). The "proof" has been furnished in many different ways, all powerfully moving because, despite all the dauntless giving of the lie to time's destructiveness, the poignancy of empirical experience remains undiminished and is powerfully expressed in the very language of the counter-assertions. The heart remains "offended" even as it insists on the immortality of a loved woman's true beauty, as we are shown in "Young Man's Song":

> "She will change," I cried,
> "Into a withered crone."
> The heart in my side,
> That so still had lain,
> In noble rage replied
> And beat upon the bone:
>
> "Uplift those eyes and throw
> Those glances unafraid:
> She would as bravely show
> Did all the fabric fade;
> No withered crone I saw
> Before the world was made."
>
> Abashed by that report,
> For the heart cannot lie,
> I knelt in the dirt.
> And all shall bend the knee
> To my offended heart
> Until it pardon me.

Platonic vision (recalled in words taken from the higher-spirited earlier poem, "Before the World Was Made," in "A Woman Young and Old") transcends earthly ruefulness but does not banish it. The point is made as beautifully but more bluntly in the closing lines of "His Confidence" ("Out of a desolate source, / Love leaps upon its course"). "His Bargain" goes further than all of these poems in its defiance of fatality. Or, at any rate, we are shown how seriously we are to take the earlier allusions to a time "before the world was made." The commitment to unchanging love antedates even "Plato's spindle" and eternity itself—and yet, subtly, the poem's brave show is belied by its tone:

> Who talks of Plato's spindle;
> What set it whirling round?
> Eternity may dwindle,
> Time is unwound,
> Dan and Jerry Lout
> Change their loves about.
>
> However they may take it,
> Before the thread began

I made, and may not break it
When the last thread has run,
A bargain with that hair
And all the windings there.

"His Bargain" has, with a combined gentle dignity and romantic intensity, brought the spheres of spiritual and profane love together—an acceptance and affirmation of the wildly symbolic offering of the woman who speaks in the preceding poem, "Her Dream." She had dreamed that she had "shorn [her] locks away" and "laid them on Love's lettered tomb"—a surrender to the inevitable death of love, and a mourning sacrifice of her sense of her own womanly beauty. But the cosmos has other plans:

But something bore them out of sight
In a great tumult of the air,
And after nailed upon the night
Berenice's burning hair.

In both poems love dies but its death is not accepted. Together "Her Dream" and "His Bargain" introduce a reprise, in a less violently passionate key, of the exaltation of love that reached its height earlier on in this sequence with "Crazy Jane Talks with the Bishop" and "Crazy Jane on God." It is the key of mellowed reconciliation.

The new, gentler synthesis can be clearly seen in "Three Things." Here the singing, wave-whitened bone of a woman is like Jane's body singing of past glory in "Crazy Jane on God." The bone celebrates life's sensuous pleasures as if they were the true source of all joy, worldly or otherwise. The three-fold pleasure involves that of her child, men she has gratified, and her own total fulfillment—sexual and spiritual—with her "rightful man." A similar zest for life informs "Lullaby" (XVI), "Those Dancing Days are Gone" (XIX), and the folk refrain of "I am of Ireland" (XX):

"I am of Ireland,
And the Holy Land of Ireland,
And time runs on," cried she.
"Come out of charity,
Come dance with me in Ireland."

The dancing motif is picked up in the first of the closing group of five poems ("The Dancer at Cruachan and Cro-Patrick"), all written in the last half of 1931 and employing a male complement to Crazy Jane, Tom the Lunatic, in all but the twenty-fifth poem. Tom celebrates the sexual principle seen as the vital center of deity and artistic creativity. His last song, "Old Tom Again," is a magnificently paradoxical prophecy of the creative imagination's triumphing over the temporal:

> Things out of perfection sail,
> And all their swelling canvas wear,
> Nor shall the self-begotten fail
> Though fantastic men suppose
> Building-yard and stormy shore,
> Winding-sheet and swaddling-clothes.

Following this incantatory echo of Jane's envisioned leap into the light lost in her mother's womb, comes "The Delphic Oracle upon Plotinus." Just as "A Woman Young and Old," "A Man Young and Old," and the seven Crazy Jane poems end with poems involving a certain amount of aesthetic distancing, so now ends *Words for Music Perhaps*. The Old Tom poems have moved rather far from the personal into epigram and visionary statement; the struggle to realize the ideal through the flesh becomes almost pure symbol in the closing poem:

> Behold that great Plotinus swim,
> Buffeted by such seas;
> Bland Rhadamanthus beckons him,
> But the Golden Race looks dim,
> Salt blood blocks his eyes.
>
> Scattered on the level grass
> Or winding through the grove
> Plato there and Minos pass,
> There stately Pythagoras
> And all the choir of Love.

2. Between *Words for Music Perhaps* and *Last Poems and Two Plays*

Placed near the end of *The Winding Stair and Other Poems*, "Vacillation" (1931–32) is on the border between long poem and sequence. Mostly it seems a meditation centered on a series of images in poems of varied form, its tonalities ranging from musing to self-exhortation to humble gratitude to remorse to powerful conjuration of the past to self-affirmation: stages of thought as Yeats worries at the proper way to conduct a heroically artistic life. Beyond the personal, however, complex moments of realization are flung into this progression that stab into awareness of the mystery of human existence: the compressed vision, almost abstract, of man's whole condition in Poem I; the strange sexual-sacred images that glisten in Poem II; the moment of pure, blazing bliss reported in IV; the refrain "Let all things pass away" counterpoised to fresh images and violent ones of sheer life-experience in VI; the teasing quarrel with a theologian in VIII. Poem III, a clarion call to artists, ends on a note we remember from "Upon a Dying Lady" and foreshadows "Under Ben Bulben":

> Test every work of intellect or faith,
> And everything that your own hands have wrought,
> And call those works extravagance of breath
> That are not suited for such men as come
> Proud, open-eyed and laughing to the tomb.

Other such foreshadowings of *Last Poems* are of interest here—for instance, the remorseful fifth poem anticipates the powerful "The Man and the Echo." Only Poem IV, however, approaches Yeats at his best, and "Vacillation" as a whole seems mainly a preliminary structuring on its way to *Last Poems*.

The twelve "Supernatural Songs" (c. 1934) are more successful, mainly because they include "What Magic Drum?" (VII), "Meru" (XII), and the fairly successful epigrammatic poem "The Four Ages of Man" (IX). The first seven poems, in the visionary mode, introduce a new character, the mystic Ribh. He adds a more religious-philosophical-historical dimension to Crazy Jane's and Tom the Lunatic's preoccupations with love and the nature of God. Unlike the Bishop, who used an "old book" to destroy Jack's and Jane's love-making, Ribh at the opening of the sequence is reading his "holy book" in the light provided by the celestial love-making of Baile and Aillinn. (This is an expanded and reoriented version of the cleaving of souls in the closing stanzas of "A Last Confession," in "A Woman Young and Old.") Ribh stresses the sexual nature of Divine creation even more vigorously than Tom and denounces "every thought of God mankind has had" (V). Notice how the line "Hatred of God may bring the soul to God" will be transmuted in "Under Ben Bulben," in Yeats's exhortation to artists to "bring the soul of man to God. / Make him fill the cradles right.") Then, in "What Magic Drum?" Ribh-Yeats fuses male and female, bestial and human, into a wondering, gently awe-struck vision of the Godhead utterly in contrast to the imagined violent rough beast slouching "towards Bethlehem to be born" in "The Second Coming":

> He holds him from desire, all but stops his breathing lest
> Primordial Motherhood forsake his limbs, the child no longer rest,
> Drinking joy as it were milk upon his breast.

Through light-obliterating garden foliage what magic drum?
Down limb and breast or down that glimmering belly move his mouth
and sinewy tongue.
What from the forest came? What beast has licked its young?

"What Magic Drum?" ushers in a series of poems closely linked to historical cycles (compare "Leda and the Swan" at the opening of the "Dove or Swan" chapter of *A Vision*). In this concern "Supernatural Songs" resembles the civil-war sequences and "Vacillation" more than the love sequences, but with the sexual basis an integral part of historical concern—especially in "Whence Had They Come?" and "Conjunctions." Of the epigrammatic poems in "Supernatural Songs"—"There" (IV), "The Four Ages of Man" (IX), "Conjunctions" (X), and "A Needle's Eye" (XI), the ninth is most successful in its cumulative power, as body, then heart, then mind win, until the final blow falls:

Now his wars on God begin;
At stroke of midnight God shall win.

This poem and too many others in "Supernatural Songs" gain much of their interest from the reader's familiarity with Yeats's theories of history and personality as presented in *A Vision*. The sonnet "Meru," like "What Magic Drum?," is enriched by but not dependent on such knowledge. The basic conception is close to that of the sixth section of "Vacillation," with its refrain "Let all things pass away," and to "The Gyres" and "Lapis Lazuli," but the emphasis in "Meru," as in the civil-war sequences, is more on desolate awareness—the "desolation of reality"—than on the tragic joy such awareness may bring:

Hermits upon Mount Meru or Everest,
Caverned in night under the drifted snow,
Or where that snow and winter's dreadful blast
Beat down upon their naked bodies, know
That day brings round the night, that before dawn
His glory and his monuments are gone.

After "Supernatural Songs," the only sequence Yeats published before *Last Poems and Two Plays* was the unique ballad-and-song cluster centered on "The Three Bushes," in the 1938 volume *New Poems*. Written in 1936, it is a small planetary system: a fast-paced ballad about a lady, a lover, and a chambermaid with small units of subjectively concentrated feeling whirling about it, as it were. The ballad is followed by "songs" of the individual characters: three by the lady, one by the lover (hardly the "laughing, crying, sacred song, / A leching song" his hearers ask him to sing in the ballad), and two by the chambermaid.

The six songs dwell on the central sexual experience, the substitution at night of chambermaid for lady, from the three perspectives. Poetically, this structure is an intriguing device for affective exploration, sorting out as it does the essential lyrical elements from their surface ordering in the tale and allowing them to take over from any priorities of suspense or "interpretation" the ballad itself might suggest. The affective exploration also engages with old preoccupations of the poet having to do with body and soul, both in the meaning of love and in the priorities of the artist. The lover, a maker of songs, needs both kinds of love—spiritual and physical—to create. And the lady's deception of him partially resolves the dilemma of "A Last Confession" (Poem IX in "A Woman Young and Old"), in which the woman gives her soul and loves "in misery" but has "great pleasure with a lad" she loves "bodily." It is only partially resolved for her, obviously, since she must wait for death to have complete union—her success symbolized by three bushes on the three graves, growing inextricably together:

> And now none living can,
> When they have plucked a rose there,
> Know where its roots began.
> *O my dear, O my dear.*

This little sequence is a beautiful reduction of emotional motifs to their essentials before the great final effort in *Last Poems*.

3. *Last Poems and Two Plays*

Thus far we have been discussing Yeats's sequences in their final form. We have not concerned ourselves with earlier printed arrangements or paid any attention to Yeats's various working versions. But it might be useful to remind ourselves that there may be a good deal of shuffling of poems before the final order, or at least the provisionally final order, takes shape. At this stage, new poems or sections may well arrive: here a strengthening of intensity, there a muting contrast, or, very possibly, some further poetic exploration of the implications of two newly juxtaposed poems. One crucial decision may imply a whole flotilla of smaller ones, so that deciding to place a poem exactly *there*, or discard this one entirely, means that something must be dropped or rewritten, or a cluster of poems shifted in position. To hark back for a moment to an earlier chapter, we should note here that—given the special circumstances of Emily Dickinson's life as an essentially unpublished poet—it is more than remarkable that she should have carried the process as far as she did.

Often this process goes on after the first publication. Whitman is only the first in a very long line of sequence shufflers. And to take one small instance in Yeats's work: he made changes in the Crazy Jane grouping between the 1932 Cuala edition of *Words for Music Perhaps* and the 1933 edition of *The Winding Stair and Other Poems,* reversing the order of "Crazy Jane and Jack the Journeyman" and "Crazy Jane on the Day of Judgment" and adding "Crazy Jane Talks with the Bishop" between "Crazy Jane on God" and "Crazy Jane Grown Old Looks at the Dancers." Eventually, however, the original pressure that brought the sequence into being exhausts itself; and even poems resulting from a similar pressure but, say, five years along in the poet's career, will just have to form their own constellation. *This* one is finally filled, although the poet probably seldom feels it is more than a tentative approximation of the best of several possibilities.

When we come to someone's last poems (Sylvia Plath's and Anne Sexton's are cases in point), we are extraordinarily lucky if the poet has had the time or forethought to arrange them for us. True, if he or she had lived just a few days or weeks longer a different order might have emerged; but that contingency, in turn, would demand decisions similar to those we have been discussing. All this is, naturally, by way of preamble to a consideration of *Last Poems and Two Plays* (1939), Yeats's hidden lyric sequence with dramatic complement. It has been very well hidden indeed since 1940, when Macmillan brought out *Last Poems and Plays,* garbling the sequence (originally published in Dublin by the Cuala Press) and reversing the plays' order as well.

The garbling was as follows. First, the nineteen "last" poems were tacked onto the earlier *New Poems* (Cuala Press, 1938). Second, their proper order was disregarded. Three of the opening four poems were shifted to the end *and* rearranged. They include the crucial *opening* poem, "Under Ben Bulben," which was made to *close* both the sequence and the "Lyrical" section of *The Collected Poems*. The original fifth poem, "Three Marching Songs," was dumped, with remarkable insensitivity, further on between "News for the Delphic Oracle" and "Long-Legged Fly." "Hound Voice" and "John Kinsella's Lament for Mrs. Mary Moore" were gratuitously reversed. "The Man and the Echo" leapfrogged over two poems to introduce the new closing group ("Cuchulain Comforted," "The Black Tower," and "Under Ben Bulben"), effectively burying "Politics," Yeats's original choice for the last poem in his last sequence. And third, three poems from *On the Boiler* (Cuala Press, 1939) were inserted: "Why Should Not Old Men Be Mad?," "The Statesman's Holiday," and "Crazy Jane on the Mountain." That the order of the 1939 *Last Poems and Two Plays,* though published six months after Yeats's death, was indeed Yeats's own has been demonstrated conclusively by Curtis Bradford, who located a table of contents in Yeats's handwriting.[1] There is no indication at all that Yeats would have agreed to the 1940 version.

The order of the sequence under discussion, then, is (1) "Under Ben Bulben," (2) "Three Songs to the One Burden," (3) "The Black Tower," (4) "Cuchulain Comforted," (5) "Three Marching Songs," (6) "In Tara's Halls," (7) "The Statues," (8) "News for the Delphic Oracle," (9) "Long-Legged Fly," (10) "A Bronze Head," (11) "A Stick of Incense," (12) "Hound Voice," (13) "John Kinsella's Lament for Mrs. Mary Moore," (14) "High Talk," (15) "The Apparitions," (16) "A Nativity," (17) "The Man and the Echo," (18) "The Circus Animals' Desertion," and (19) "Politics." The lyric section of the book is followed by two plays: *The Death of Cuchulain* and *Purgatory*.

Yeats's sequence, as opposed to Macmillan's nonsequence, moves not toward but away from the prophetic, exhortatory, heroically Irish grappling with death in "Under Ben Bulben" and the poems through "In Tara's Halls." After these it shifts to the heroic, sexual, artistic, classical transcendence of "The Statues," "News for the Delphic Oracle," and "Long-Legged Fly"; to the more personal sexual flaunting and heroic memories of poems 10–14; and to the closing group, in which the poet, under pressure of the terrifying unknown, struggles with some ultimate definitions of his life and art. Eschewing heroics and "high talk" at the last, the poems from "The Apparitions" through "Politics" have their own bravery, dignity, and powerful affirmation of passionate intensity. Here also a rigorous and somewhat appalled self-appraisal balances a touching desire for communion between one human being and another, no matter what the cost to one's pride. This overall progression was of course nullified by the Macmillan editors' apparent desire to give a coolly upbeat heroic ending to Yeats's *oeuvre*. Although we know the last poems he wrote were "Cuchulain Comforted" and "The Black Tower," his aesthetic decisions about where to place them in his sequence are another matter entirely. So was his decision *not* to round off his work with the stoically distanced epitaph closing "Under Ben Bulben"—"*Cast a cold eye / On life, on death! / Horseman, pass by*"—but rather with a passion for life still flaming from his page: "But O that I were young again / And held her in my arms!" (Few life-

1. "Yeats's *Last Poems* Again," in *Yeats Centenary Papers,* ed. Liam Miller (Dublin: Dolmen Press, 1966), pp. 259–88.

delirious poets, who hope as well that their art will not perish with them, really favor snapping it shut with an epitaph.)

We are thrown abruptly into Yeats's last sequence, somewhat in the manner of the opening of "A Woman Young and Old" ("She hears me strike the board and say"). The title of that poem, "Father and Child," establishes the dramatic situation. However sudden and striking its first impact, "Under Ben Bulben" is not as specific, so that we are not sure initially who is being asked by whom to swear to what:

> Swear by what the sages spoke
> Round the Mareotic Lake
> That the Witch of Atlas knew,
> Spoke and set the cocks a-crow.
>
> Swear by those horsemen, by those women
> Complexion and form prove superhuman . . .

The cluster of mystifying allusions in these lines contributes to their assault on our capacity for awe and unreasoning response to a challenge. The audience being asked to take the oath turns out, further on, to be a sophisticated modern one that needs to summon up primitive energies: "You that Mitchel's prayer have heard, / 'Send war in our time, O Lord!' " It contains "poet and sculptor," "Irish poets"—in fact all the "indomitable Irishry" willing to preserve heroic and artistic values fast vanishing from Ireland and the world:

> Irish poets, learn your trade,
> Sing whatever is well made,
> Scorn the sort now growing up
> All out of shape from toe to top,
> Their unremembering hearts and heads
> Base-born products of base beds.
> Sing the peasantry, and then
> Hard-riding country gentlemen,
> The holiness of monks, and after
> Porter-drinkers' randy laughter;
> Sing the lords and ladies gay
> That were beaten into the clay
> Through seven heroic centuries;
> Cast your mind on other days
> That we in coming days may be
> Still the indomitable Irishry.

By singing "whatever is well made," poets will be fulfilling the commands given in the preceding section:

Poet and sculptor, do the work,
Nor let the modish painter shirk
What his great forefathers did,
Bring the soul of man to God.
Make him fill the cradles right.

The key to spiritual and physical renewal lies in the "other days," in the artists' preservation of the heroic Irish legends and myths and celebration of the intensely alive peasants and aristocrats, with proper attention as well to both "holiness" and randiness. Presumably what the artists are to swear to is that they will follow this artistic program with its social and political and spiritual implications: seek to implement the "profane perfection of mankind" that is the "purpose set / Before the secret working mind." The jaunty rhythms and randiness of some of the language in "Under Ben Bulben" are far in tone from "Her Vision in the Wood" (the eighth poem in "A Woman Young and Old"), where the wounded man is her "heart's victim and its torturer," but the artist's role in providing images of perfection is the same:

Michael Angelo left a proof
On the Sistine Chapel roof,
Where but half-awakened Adam
Can disturb globe-trotting Madam
Till her bowels are in heat . . .

The majority of the poems in this sequence are fueled by Yeats's conviction that it is up to the artists following him to continue to provide symbols from the "other days" to counter modern degeneracy. (See also *The Death of Cuchulain* and *Purgatory,* and the very explicit *On the Boiler* for other versions of this preoccupation.) Fortunately the poems are poems, not political rhetoric, and this pragmatic program is rarely as intrusive as it is in "Under Ben Bulben." True, one can find it easily in the lusty swashbuckling of "Three Songs to the One Burden" and "Three Marching Songs." But it is completely subsumed in the two excellent poems sandwiched between the sets of three: "The Black Tower" and "Cuchulain Comforted."

Of course, "The Black Tower" concerns "oath-bound men" waiting for their own "right king" to reappear, but the poem's archetypal evocativeness and complexity involve far more than a staunch holding against modern political and social barbarity. The "burden" of the immediately preceding songs is the heroic refrain *"From mountain to mountain ride the fierce horsemen,"* and the last of the three "Songs" is of the 1916 Easter uprising, cast in heroic terms. (The first song celebrates the lusty sexuality of Crazy Jane and her ilk; the second projects a spiritual oasis in the midst of a people given over to the "devil's trade"—a center of power essentialized in the refrain.) The ancient men of the old black tower follow right on the heels of the modern heroes who had "gone out to die / That Ireland's mind be greater, / Her heart mount up on high." The suggestion is strong that men such as Pearse and Connolly were inspired by the enduring image of the oath-bound men of the black tower, those who wait patiently but probably in vain to hear the "king's great horn" again. The poem is a brilliant portrayal of one aspect of a mentality in communion with the past and the great dead. In that sense it projects a fundamental human predicament: of dedication in the face of intolerable odds. The essential hopelessness of the situation comes through in the heroes' response to the excitable, unheroic old cook's naive optimism ("But he's a lying hound") and in the impotence of the dead in the final refrain. Yet the refrain has an exultantly ominous note of reawakening as well:

There in the tomb the dark grows blacker,
But wind comes up from the shore:
They shake when the winds roar,
Old bones upon the mountain shake.

"Cuchulain Comforted" moves further into a realm of the dead where neither heroic action nor despicable inaction any longer has meaning. A community of cowards teach the individualistic hero to become one of them; all will have become singing birds together. The scene is an unearthly counterpart to "The Black Tower." In both the heroic spirit exists in a kind of Limbo. In the first, its external emblems are strongly asserted, as though enduring in the minds of men of succeeding generations. In the second, it would seem that only art (the "singing" of the birds) preserves faint traces of the memory. (The two poems together bring Pound's "The Return" irresistibly to mind.) In contrast, "Three Marching Songs" militaristically hurtle back to the present and its call for fanatical political action. "In Tara's Halls" presents us with an awkward little parable of the ruler who knows when it is time to abdicate.

Yeats is at his best again in the next three poems: "The Statues," "News for the Delphic Oracle," and "Long-Legged Fly." Passages in *On the Boiler* throw an interesting light on the first of these. Yeats frequently wrote prose drafts of poems but rarely printed them, and so "The Statues" is unusual in its close correspondence to these scattered passages:

> The old Irish poets lay in a formless matrix; the Greek poets kept the richness of those dreams and yet were completely awake. Sleep has no

bottom waking on top. Irish can give our children love of the soil underfoot; but only Greek, co-ordination or intensity.

. . . civilization rose to its high tide mark in Greece, fell, rose again in the Renaissance but not to the same level. But we may, if we choose, not now or soon but at the next turn of the wheel, push ourselves up, being ourselves the tide, beyond that first mark. But no, these things are fated; we may be pushed up.

There are moments when I am certain that art must once again accept those Greek proportions which carry into plastic art the Pythagorean numbers, those faces which are divine because all there is empty and measured. Europe was not born when Greek galleys defeated the Persian hordes at Salamis, but when the Doric studios sent out those broad-backed marble statues against the multiform, vague, expressive Asiatic sea, they gave to the sexual instinct of Europe its goal, its fixed type.[2]

"The Statues" fleshes out this meditation on modern Irish and Classical Greek cross-fertilization superbly, celebrating the artist who carried into "plastic art the Pythagorean numbers" by describing the effect of such statues on the living:

But boys and girls, pale from the imagined love
Of solitary beds, knew what they were,
That passion could bring character enough,
And pressed at midnight in some public place
Live lips upon a plummet-measured face.

Nothing else in the poem quite measures up to the shock here of the coming together of flesh and ideal in the superb line "live lips upon a plummet-measured face." The rest of the poem supplies the historical and political justification for this erotic center. "News for the Delphic Oracle" has a similar erotic center, but this time placed at the end rather than the beginning. The poem's movement is from the humorously "sighing" Pythagoras, Plotinus, Irish heroes, and that familiar "choir of love" from "The Delphic Oracle upon Plotinus" (the closing poem of *Words for Music Perhaps*), to the intolerable yearning of these soulful characters for the delights of brawny, earthy sex:

Foul goat-head, brutal arm appear,
Belly, shoulder, bum,
Flash fishlike; nymphs and satyrs
Copulate in the foam.

2. William Butler Yeats, *On the Boiler* (Dublin: The Cuala Press, 1939), pp. 28, 29, and 37 respectively.

"Long-Legged Fly," the best poem in the sequence, touches the heroic, sexual, and artistic centers we have been discussing, but does so purely presentatively and lyrically. In it the meditating mind so strongly present in "The Statues" is relegated to the first line of each stanza and to the haunting refrain, "*Like a long-legged fly upon the stream / His [Her] mind moves upon silence.*" Similarly, the lengthy description of the Isles of the Blest and their denizens in "News for the Delphic Oracle" is not as sharply effective as the placing here of three historical figures—Caesar, Helen, and Michelangelo—in concrete, intense vignettes that present them at the height of their own forms of inspiration. The "Caesar" of the first stanza, cast as the preserver of civilization, is caught not in the midst of slaughter but meditating his strategy of conquest. Helen, Yeats's prime symbol of the sexual force in action in history and art, is glimpsed at the onset of puberty—"part woman, three parts a child"—practicing a dance straight out of the lusty Irish countryside. Finally, Michelangelo is described in the midst of that creative act which will, in the words of "Under Ben Bulben," "disturb globe-trotting Madam / Till her bowels are in heat":

> That girls at puberty may find
> The first Adam in their thought,
> Shut the door of the Pope's chapel,
> Keep those children out.
> There on that scaffolding reclines
> Michael Angelo.
> With no more sound than the mice make
> His hand moves to and fro.
> *Like a long-legged fly upon the stream*
> *His mind moves upon silence.*

All three figures are in utterly self-absorbed reverie, silently concentrating all their energies toward whatever form of creation is their *métier*—military action, physical beauty, art. Whatever the violent and generative impact of the results of such thought, the act of creation itself belongs to another realm, on which the outside world—soldiers, barking dogs, neighing ponies, the young girl's companions, noisy children—must not be allowed to intrude. *Yeats* may speculate on the results of Caesar's plans, Helen's perfection of her beauty, Michelangelo's painting, and all their analogues in every generation—as he does in the first line of each stanza. But he has depicted the creative figures themselves as yielding totally to the act, not to its results. Hence the singular purity of effect compared with the exhortations in "Under Ben Bulben."

"A Bronze Head," the next poem, introduces personal reminiscence into this sequence for the first time. Yeats is meditating on what is probably a representation of Maud Gonne but could be that of any woman who gave herself to some form of extremism, destroying imaginative possibility and wholeness of being. In the first stanza he contemplates the contrast between the work of art and the terrifying emptiness of the old woman who, like "man" in "Meru," has evidently ravened, raged, and uprooted until she has achieved a similar "desolation of reality"—"*Hysterica passio* of its own emptiness." From there Yeats slides into a consideration of which stage of the woman's life was the "real" one, and of his own prescience in youth of what she would become. In the final stanza he endows her with his hatred of modern degeneracy. The images are variations on ones familiar from the earlier sequences and from "Under Ben Bulben," the six songs, and "The Statues":

> Or else I thought her supernatural;
> As though a sterner eye looked through her eye
> On this foul world in its decline and fall;
> On gangling stocks grown great, great stocks run dry,
> Ancestral pearls all pitched into a sty,
> Heroic reverie mocked by clown and knave,
> And wondered what was left for massacre to save.

Fortunately, this is the end of the relatively unalloyed eugenics theme in the lyrics, although in *Last Poems and Two Plays* one encounters strong statements of it in the old man's prologue to *The Death of Cuchulain* and in *Purgatory* as well. And with "A Stick of Incense," a candidate for one of Yeats's worst poems, unalloyed sniggering crudeness exits as well.

Of the next five poems—"Hound Voice," "John Kinsella's Lament for Mrs. Mary Moore," "High Talk," "The Apparitions," and "A Nativity"—the fourth and fifth most effectively sound the notes that will dominate the end of the sequence. "Hound Voice" is a personalized, not very effective draft, in a sense, of "The Black Tower"; and the "Lament" is a small vaudevillian masterpiece whose bawdy surface comedy barely conceals its far-reaching elegiac strain. "High Talk" starts with a flamboyant bit of boasting—"no modern stalks" upon stilts higher than the poet's—but at the same time there are several disquieting elements. These are, first, failure to match the accomplishment of one's forebears; second, the necessity to start all over again—to "take to chisel and plane"—and finally, the closing passage that recalls, although it does not match, the image of the swan leaping into the desolate heaven in the third section of "Nineteen Hundred and Nineteen." Once again we have the heroic spirit braving death:

Malachi Stilt-Jack am I, whatever I learned has run wild,
From collar to collar, from stilt to stilt, from father to child.
All metaphor, Malachi, stilts and all. A barnacle goose
Far up in the stretches of night; night splits and the dawn breaks loose;
I, through the terrible novelty of light, stalk on, stalk on;
Those great sea-horses bare their teeth and laugh at the dawn.

In contrast, "The Apparitions" mocks such fantasizing, projecting an aging figure desperately assuring himself that he has a "full" heart and the strength to bear the "increasing Night / That opens her mystery and fright." His appalled awareness of oncoming death is carried by the refrain, "*Fifteen apparitions have I seen; / The worst a coat upon a coat-hanger.*" Personal fear is touched on here, as it is to some extent in the closing couplet of "A Nativity":

Why is the woman terror-struck?
Can there be mercy in that look?

"The Man and the Echo," the first of the powerful closing triad, limns the struggle of the "old and ill" poet to put his actions "in one clear view" and stand in "judgment on his soul." The result is terror, not peace. The "echo" has no message except death and mystery; and instead of the "terrible novelty of light" in "High Talk," the final image is of the poet deep in the dark cleft, listening to a death cry:

Up there some hawk or owl has struck,
Dropping out of sky or rock,
A stricken rabbit is crying out,
And its cry distracts my thought.

This darkest moment in the sequence is followed by a strenuous consideration of the sources of Yeats's art—as "The Man and the Echo" is concerned to some extent with the disastrous effect of his art and actions on certain individuals. Throwing aside the symbols and "stilted boys" of his earlier work, the poet returns to the "heart" that had originally provided the impetus for his work until his creations took all his love, "and not those things that they were emblems of." The staggering difference between then and now is conveyed by the contrast between the dream-images that once grew out of a heart "embittered" by need and the squalid reality that is the heart's ultimate, "foul" workshop:

Those masterful images because complete
Grew in pure mind but out of what began?
A mound of refuse or the sweepings of the street,

Old kettles, old bottles, and a broken can,
Old iron, old bones, old rags, that raving slut
Who keeps the till. Now that my ladder's gone,
I must lie down where all the ladders start,
In the foul rag-and-bone shop of the heart.

The pure mind—like that conveyed so powerfully in "Long-Legged Fly"—is only half the creative process. At the end of his life, Yeats savagely sums up the leavings of a lifetime, which are now only the contents of his heart. The closing stanza, especially, points up the infuriating dichotomy between the enduring art images and the all-too-mortal flesh and leads brilliantly into the last poem, "Politics," with its poignant focusing on the dominating desire of the old man—not for political or artistic achievement, but for youth and love:

And maybe what they say is true
Of war and war's alarms,
But O that I were young again
And held her in my arms!

This heart's truth, affirmed in the midst of impending war, involves only the individual: *his* anguish; *his* longing; *his* affirmation of love. Here is the kind of intensity of feeling on which Yeats's art has been based throughout. No *artist* can both "cast a cold eye / On life, on death" and create "masterful images" that outlive the poet and affect the future.

Part Three
American Flowering

Chapter Seven

Edgar Lee Masters, T. S. Eliot, and "Voice"

1. Masters's *Spoon River Anthology*

A lingering influence of Browning in modern poetry, American especially, has been a preoccupation with *voice:* a poem anchored in a single dramatic voice, a group of poems in which several or many voices interact. It may be argued that Browning's voices, in his dramatic monologues but also in his character-swarming *The Ring and the Book,* with its attempt to pierce through the masks of appearance to the ambiguities of reality, should be taken as affective keys rather than as theatrical representation. But it is clear that following in Browning's track led one strong American poetic presence, Edgar Lee Masters, into a sort of swamp. If *A Shropshire Lad* is redundant, the redundancy prize among longer works that might conceivably have been shaped into beautifully articulated sequences must nevertheless go to the American. His *Spoon River Anthology* (1915),[1] almost contemporary with Hardy's *Poems of 1912–13* and Yeats's "Upon a Dying Lady," contains some 244 poems, intended to embody a mob of at least as many voices. But that embodiment is by intention only, since not even Shakespeare could have created so many distinctive characters in a series of relatively short poems. As a thinker about poetry, T. S. Eliot foundered on the convenient theory that voice is the key to lyric structure; but his practice in *The Waste Land* belied the theory, as we shall see, and shook off the hold of this extremely limited conception.

1. New York: Macmillan. Reprinted frequently.

To return to *Spoon River Anthology*, all but two of the poems of Masters's remarkably ambitious work are lyrical monologues (a certain number of them far more narrative in structure than most of the rest) that strive for the compressed poignancy of the Greek Anthology. Except when the speaking characters are very sharply defined—and even then to some extent—an overriding tone of gnomic wryness or ironic resignation tends to blur the voices and make them alternative personae for the puppeteer-ventriloquist who trots them out in order. The speakers are the dead in the cemetery of Spoon River, Illinois, a mythical town on a real river. Or, as the prefatory poem, "The Hill," puts it: "*All, all, are sleeping on the hill.*" This elegiac overture, combining notes of the *ubi sunt* tradition with a direct American colloquialism and with humane but facile pathos and a crude rhetoric, strikes the book's basic tonal chord:

> *Where are Elmer, Herman, Bert, Tom and Charley,*
> *The weak of will, the strong of arm, the clown, the boozer?*
> *All, all, are sleeping on the hill.*

The last of the monologues is "Webster Ford," in which we can see comparable elements, more subtly developed. It begins:

> Do you remember, O Delphic Apollo,
> The sunset hour by the river, when Mickey M'Grew
> Cried, "There's a ghost," and I, "It's Delphic Apollo";
> And the son of the banker derided us, saying, "It's light
> By the flags at the water's edge, you half-witted fools."

Here we have an added visionary dimension such as marks the closing twenty-six or so monologues. The Grecianized refrain "O Delphic Apollo," which persists throughout the poem, is in harmony with the evocative nature-images but clashes with the vernacular language and the local American reference to "the son of the banker." In a poem like this one, Masters achieves something like lyric transcendence. The speaking sensibility in "Webster Ford" is in balanced touch with voices not kindred to his own. Through two resonant images, one of them put into another character's mouth, he suggests an atmosphere in which something miraculous is possible. And he sustains an exalted awe in his address to the elusive, terrifying god. No wonder the youthful Ezra Pound saw Masters as one of the interesting poets of the moment;[2]

2. See D. D. Paige, ed., *The Letters of Ezra Pound 1907–1941* (New York: Harcourt, Brace and Company, 1950), p. 43 and *passim;* and Ezra Pound, "Webster Ford," *Egoist*, II (1 January 1915), 11–12, and "Affirmations: Edgar Lee Masters," *Reedy's Mirror*, XXIV (21 May 1915), 10–12. Pound included ten poems by Masters in his *Catholic Anthology* (London: Elkin Matthews, 1915).

Pound's own art had to do with just such visions and balances. It is possible, too, that Masters's use of vernacular was in part a loving parody of Homeric diction—something Pound would have recognized with enthusiasm.

We may linger briefly over this aspect of Masters, since only Pound seems to have caught its possibilities. He had first been impressed by a single poem, "The Unknown," which in 1913 Masters published in *Reedy's Mirror* using the pseudonym "Webster Ford." (The pseudonym is interesting evidence that Masters probably saw the poem he called "Webster Ford" as a self-portrait.) In "The Unknown," which was also included in *Spoon River Anthology*, the speaker is an unidentified boy who may have reminded Pound of Homer's Elpenor—the "man of no fortune, and with a name to come," as Pound describes him in Canto 1. The imagination haunting this poem bears some rudimentary resemblance to Pound's partly because we *see* the speaker's shade in Hades instead of just hearing him speak:

> Ye aspiring ones, listen to the story of the unknown
> Who lies here with no stone to mark the place.
> As a boy reckless and wanton,
> Wandering with gun in hand through the forest
> Near the mansion of Aaron Hatfield,
> I shot a hawk perched on the top
> Of a dead tree.
> He fell with guttural cry
> At my feet, his wing broken.
> Then I put him in a cage
> Where he lived many days cawing angrily at me
> When I offered him food.
> Daily I search the realms of Hades
> For the soul of the hawk,
> That I may offer him the friendship
> Of one whom life wounded and caged.

One must admit that the *situation* here catches the imagination far more than does either the rhythm or the phrasing. Lines 13–14 ("Daily I search . . .") are an exception, though. If only minimally, they envision the restless, damned state of the speaker. And despite its limitations, the poem does reach toward a renewal of mythical consciousness in the midst of literal-minded American life and, to some extent, in terms of that life. Masters had caught sight of a double opportunity. He could relate his alienated, yet intimately knowledgeable poetic sensibility to the everyday Midwestern world and its crucial issues, and in so doing he could bring the Classical heritage to bear in the most immediate way.

But *Spoon River Anthology* only suggests marvelous possibilities, rather than realizing them. Masters was half-capable of the real thing. In short bursts of frankness and yelling rhetoric, and in brief elegiac and compassionate moments, he was a poet. Also, he was haunted by echoes of past poetry and by turns of idiomatic speech that can stick in the mind like burrs. His is the art of a passionately committed autodidact. His legal practice had given him myriad closeups of violence, treachery, squalor, and sexual predicament, and it sometimes seems that he was virtually addicted to the raw meat of this side of human experience. This tack assorted ill with his populist idealism and with the visionary longings that fill the last part of his book. In his array of emotional starting points, he has affinities not only with Pound but with such predecessors as Hardy and with the American writers Mark Twain, Ambrose Bierce, Stephen Crane, and even, in his combined inspirationalism and disillusionment, Whitman and Dickinson.

But lyrical structure, and there *are* signs of it here, gets buried in the book's bulk. As the voices drift by, characterizing their speakers' lives or telling the manner of their deaths or proclaiming or hinting their sense of existence and its meanings, too much becomes numbingly confused with too much else. In Masters's own mind the basic connections were not lyrical but thematic and fictional. "When the book was put together in its definitive order," he wrote retrospectively in his essay "The Genesis of Spoon River," in the January 1933 issue of *The American Mercury,* "the fools, the drunkards, and the failures came first, the people of one-birth minds got second place, and the heroes and the enlightened spirits came last, a sort of *Divine Comedy.*" And he was proud of having put together "nineteen stories developed by interrelated portraits." A husband, say, speaks from the grave; eventually so do his wife, her lover, one or more of their children, their hired hands, and so on. Murderers exchange versions (over the long haul) with their victims, and corrupted judges and officials with their corrupters. Cynical soldiers speak out of the horror of the Civil War or filth of the Spanish-American War, reminding us in our later moment that innocence had been lost long before the Lost Generation repeated the same experiences. Happily innocent women, victims of rape, whores; reactionaries, radicals, saints—an endless proliferation have their say across these pages. Half the book's appeal is sheer gossip. If you want gossip aplenty—thoughtful gossip, moralizing gossip, incantatory gossip—*Spoon River Anthology* gives you a Bible-ful of it.

The direct historical result was that the book became a *succès de scandale.* Masters was forced to drop his law practice. *Immoral! Radical!* The book is still something of a shocker—everything but great or even sustained poetry. As for its *poetic* effect, Masters pointed a way to many later poets. We see traces, important ones, of *Spoon River Anthology* in works by Horace Gregory, Kenneth Fearing, Robert Lowell, Robert Penn Warren, and others. Yet Masters's own poems that hold are few in number. Even sophisticated critics who wade through the whole bog emerge anesthetized, mumbling praises of the more sentimental poems ("Anne Rutledge," "Lucinda Matlock") in a desperate desire to be "fair."

Two longer pieces follow the monologues and conclude the book. They are "The Spooniad," a mock-epic fragment, and "Epilogue," an allegorical verse-drama echoing Goethe's *Faust* and Hardy's *The Dynasts* but with the virtues of neither. In the former poem Masters seeks to rescue the book from the foggy spirituality into which it finally blunders by providing a hard, sardonic, and unexpected thrust of satire. The action is boldly cruel and gross, the atmosphere humanly demeaning and charged with political intrigue, and the style racily comic in its Homeric and Vergilian treatment of small-town doings. "The Spooniad" gives the book the vigorous center of reference it needs, and would have been quite effective had the whole collection been reduced drastically. The "Epilogue," on the other hand, is a dead loss. It tries to fasten philosophical wings to a work already too wearily overloaded to soar anywhere. The problem of using many voices in the making of a sequence had to be solved in a less mechanical and less quantitative fashion than the one attempted by Masters. And of course Eliot, working on a smaller scale, solved it in *The Waste Land* seven years later—as Pound in his early cantos had also begun to solve it. The solution, as we shall see, required subordinating "voice" to tonal dynamics.

2. Conrad Aiken on *The Waste Land*

Earlier on, we mentioned Aiken's idea of "symphonic form" and certain suggestive ways in which his conception of the long associative poem approaches ours of the modern sequence.[3] His review of *The Waste Land* points up key differences as well:

> . . . the poem must be taken—most invitingly offers itself—as a brilliant and kaleidoscopic confusion; as a series of sharp, discrete, slightly related perceptions and feelings, dramatically and lyrically presented, and violently juxtaposed (for effect of dissonance) so as to give us an impression of an intensely modern, intensely literary consciousness which perceives itself to be not a unit but a chance correlation or conglomerate of mutually discolorative fragments. . . .

3. See page 78.

> . . . *The Waste Land* is a series of separate poems or passages, not perhaps all written at one time or with one aim, to which a spurious but happy sequence has been given. This spurious sequence has a value—it creates the necessary superficial formal unity; but it need not be stressed, as the Notes stress it. Could not one wholly rely for one's unity—as Mr. Eliot *has* largely relied—simply on the dim unity of "personality" which would underlie the retailed contents of a single consciousness? . . .
>
> We reach thus the conclusion that the poem succeeds—as it brilliantly does—by virtue of its incoherence, not of its plan; by virtue of its ambiguities, not of its explanations. Its incoherence is a virtue because its *donnée* is incoherence. Its rich, vivid, crowded use of implication is a virtue, as implication is *always* a virtue—it shimmers, it suggests, it gives the desired strangeness. . . . We "accept" the poem as we would accept a powerful, melancholy tone-poem.[4]

The achievement that Aiken ascribes to Eliot is the achievement he himself strove for, which was not quite Eliot's or Pound's or that of any of the other great sequence-makers. The basic difference lies in the "violently juxtaposed" essential units of a sequence. While they do serve the purposes Aiken mentions—"effect of dissonance," "impression" of a certain kind of consciousness, emergence of a "powerful, melancholy tone-poem"—his notion of an only "spurious sequence" and merely "chance correlation" is as far off the mark as if he had applied the terms to "Tintern Abbey." The tentative and improvisational aspect of art does not mean that the order and ingredients of a great work in its final form are simply accidental, however much serendipity has been involved. We have riddled our quotation from Aiken's review with ellipses not only for the sake of compression but also because the full passage, like most of the review, is devoted to showing why Aiken considers much of *The Waste Land* pretentious in its effort at scholarly and cultural allusiveness. He objects to Eliot's notes, and to certain elements in the poem as well, for trying to superimpose a rational order on what should be—and, basically and magnificently *is,* he asserts—purely ambiguous, incoherent, suggestive. If a little too common-sensibly insistent, Aiken argues the point beautifully and gracefully out of his fine instinct for a poet's poetics. But he does limit poetry to something short of the greatest expectations for it. He does not see the problem of creating a significant context for the keenest reach of realizations as a problem at all—just as a barrier to a necessary reliance on "implication" alone. For him as often for Stevens and for a later poet like John Ashbery, all one needs

4. The review is reprinted in Conrad Aiken, *Collected Criticism* (New York: Oxford University Press, 1968), pp. 176–81.

is a fluid sense of the experiencing self swimming through the streaming colors of its own sensitivity in time and space and occasionally smiling or wincing or seizing momentary glimpses of its identity. With problems in this range the borders shift, and the characterization of a long work as a sequence may be a borderline matter as well.

We must remind ourselves: we are dealing with a genre in process. Each major effort by a poet is an adventure fraught with artistic dangers. The three basic dangers are self-evident. The first two, the risks the greatest sequences run, are that separate poems may well resist mobilization and that a sequence may become overloaded in a number of ways—too much informative context fed into it, too many disparate directions, repetitive elements, distortions of emphasis. The third risk is that a longer structure may unfold narcissistically, as in the recurrent self-mirroring of so much work in the Aiken line. (In his charming essay "The Realm of Resemblances," Wallace Stevens alludes to this danger as intrinsic to metaphorical thought.) A great deal depends on a poet's capacity for earthy and self-challenging engagement with impersonal and destructive forces and on the implied moral premises of his or her existence. These forces and premises are plastic dimensions of language and of art.

Thus, while Aiken was right in pointing out that *The Waste Land*'s power does not derive from its footnotes and its riddling allusions, Eliot did find himself compelled to suggest certain touchstones of religious or cultural tension and awareness. They permeate the poem's imagery and tonalities and are aspects of its moving sensibility. Again, Whitman did not need all the rhetorical blather of *Song of Myself*, yet did need its democratic and ego-regarding assertiveness and what we may call its political emotion along with its language of sensuous excitability and uncontrol. Even Emily Dickinson, with her inward emotional concentration, consistently struggles with the alienating but informing pressures of a world of moral and religious assumption. Tragic or joyous revelation came to her in explosions of insight that have a private, even quirky coloration. Nevertheless, we see a finely free intelligence at work, reacting to those external pressures in every turn of her phrasing. These are three very unalike souls, but all would surely resist the limitations—the relative shallowness—implied in the expression "tone-poem."

We may, however, think of Aiken's conception of "symphonic" structure more richly than he himself usually did. If we see it as involving the kinds of volatile curve of movement and of balancing we have observed in the structure of sequences and also as relying on the radiant centers of affect of the Imagists and of such nineteenth-century French poets as Baudelaire and Rimbaud, we shall have in view the major elements informing the modern sequence. Clearly, these models do not provide a prescription for making sequences. Rather they suggest how much is demanded in the way of improvisation, and how much depends on the poet's whole set of mind and character and artistic genius. The particular design of a vital sequence will always come as a surprise.

3. "Voice" and *The Waste Land*

The reason for the dimension of surprise just mentioned is plain enough. Any lyrically informed work of some length and substance is fundamentally a chaos of intimate notations bound in a tangle of small effects. At the same time, however, it is magnetized into structure by its points of highest intensity and by its ordering in time: the literal succession of its tonalities. So it must be surprising in detail and idiosyncratic in formation while meeting something like an expectation of encompassment or fulfillment. Let us for a moment take a cue from Aiken and observe the first section of *The Waste Land,* "The Burial of the Dead," simply as a succession of tonalities. Like all the great touchstones of literature, the poem is generally treated as everything and anything but the work of art, made up of tiny affective elements combined into more complex units, that it actually is. Observations such as we now offer concerning "The Burial of the Dead" must be the starting point of criticism concerned with the work's essential dynamics.

"The Burial of the Dead" is made up of four major tonal units, separated by the stanza breaks. Discursively, these units have *nothing* in common; emotionally they are disparate but interact. The first unit, lines 1–18, begins with a poignant lyricism: pain, nostalgia, desire, mingled with a sense of a dwindled existence. This tonal complex, at first impersonally expressed, is picked up in the eighth line by what seems a particular voice—Marie's—that embodies the initial poignancy but is more sharply immediate, a dramatic and anecdotal closeup highly suggestive of the post-Decadent, postwar atmosphere of the "lost generation"—the atmosphere of the 1920s. The second unit, lines 19–42, begins as what seems a thundering moral condemnation of lives like Marie's, idle, self-indulgent, punctuated by moments of thrilling excitement. The thunder resolves into terror-tinged, sternly compassionate evangelism: confrontation of a pitiless desert landscape of the soul. Then comes a snatch of romantic Wagnerian song, and then a richly erotic love-duet; still overwhelmed by a lost moment of reciprocity, it swells the earlier kindred musics. The third unit, lines 43–59, introduces jazzy and colloquial rhythms and satirical irony. Yet it has its ominously foreboding side as well, echoing the previous unit's terror and counterpointing its prophetic righteousness and moral panic. The fourth unit, lines 60–76, releases the full pent-up horror sensed in the preceding sections through images and exclamations that are at once exceedingly morbid and manically comic.

Even in this necessarily oversimplified account we have just given, it can be seen that the succession of minute evocations of which "The Burial of the Dead" is composed does not reveal *a* subject matter or dominant speaker but rather a matrix of varied feeling. In a general way one can detect the occasional rise of a purer music of clear, rather melancholy intensity at certain points, and the final unit puts the whole in a context of supernatural desolation—hell in the midst of life, the triumph of death and damnation. Piercing loss, longing, guilt, and irony have culminated in a wild outburst of disastrous and hysterical vision. To stress and confirm the affective progression, certain passages and phrases from songs and poems of the past are introduced.

Now, what is clear and what obscure in this composition? Each of the intimate notations and effects is perfectly clear in itself, and so is the movement of feeling. But the *thought*—as statement, though not as an energy of awareness—is obscure, and so is the central sensibility. "The Burial of the Dead" proffers neither a coherent argument nor a consistent perspective. Yet it does employ intelligence of a high order and does create its own reciprocities. Similarly, it has no "voice" yet deploys voices freely, suggesting a wide range of sensitive consciousness. Without anything like an adequate theory, Eliot nevertheless, in *The Waste Land,* mobilized the resources of language as no poet in English had done before him—without resort to externally imposed narrative, dramatic, or logical structuring.

Instead of a theory to guide it, *The Waste Land* had a state of vast psychological openness at its command, reinforced by an atmosphere of artistic openness in its historical moment and by the sense of a community of rigorous literary experimenters—men and women breaking new ground yet profoundly in touch with living tradition. One could plunge deep into the soul's storehouse of private memories and images with a certain confidence of contact with human preoccupations as conserved in the long dream of our consciousness embedded in history, myth, literature, and genetic heritage. The way to bring all this complex of association into action was through a poetry of lyric immediacy. Such poetry carries with it a possible terror of being swept beyond formal or tonal control; its elusive centers of feeling energize a poem into volatility while concealing themselves. Otherwise the poet's work with language would be superseded by empirical self-analysis; art would be superseded by introspective thought. Under such conditions one has to let the process carry where it may; tentative ordering will be achieved through the almost instinctual balancings of affects and tonal streams and through something like the principle of montage in film-making. Eliot never again took such risks as he did in *The Waste Land,* never again exploited his full energies as there: his wit, his arrogance and pretentiousness, and his extraordinary virtuosity, along with his essential genius as a lyric poet possessing in addition dramatic imagination and a flair for caricature.

The tonalities set up at the start, in "The Burial of the Dead," carry throughout the poem in many swirling contexts. We think particularly of the vulnerable yet powerfully intense notations of life-awareness of the first four lines, the sharply focused recollection of Marie's excitement, the pain of loss and the passion-drenched memory of the "hyacinth girl" passage (lines 35–42) and the vision of terror just a few lines earlier (27–30), and the grotesquely beautiful and violent vision of the city crowds and living dead in the "Unreal City" passage that closes "The Burial of the Dead." These tonal leitmotifs cut through *The Waste Land* in many contexts: high-romantic, squalid, sardonic, but especially—at the poem's heights, beyond its atmospheric and dramatic evocations—in moments of lyrical purity. We find such moments in the three passages (beginning, respectively, "*Datta,*" "*Dayadhvam,*" and "*Damyata*") juxtaposed in lines 401–23, just before the end of "What the Thunder Said." Here one finds long echoes of such earlier lines as "He said, Marie, / Marie, hold on tight. And down we went" and

> —Yet when we came back, late, from the hyacinth garden,
> Your arms full, and your hair wet, I could not
> Speak, and my eyes failed. I was neither
> Living nor dead, and I knew nothing,
> Looking into the heart of light, the silence.
> *Oed' und leer das Meer.*[5]

These lines "connect," without logical continuity, with the passages in "What the Thunder Said" which we have referred to—first, the one that centers on "The awful daring of a moment's surrender" (lines 402–10); and then, the exquisite passage of reverie that condenses the spiritual isolation haunting the poem as a whole into one obsessive, syntactically overlapping set of images:

> *Dayadhvam:* I have heard the key
> Turn in the door once and turn once only
> We think of the key, each in his prison
> Thinking of the key, each confirms a prison

5. Our text of reference is *The Complete Poems and Plays of T. S. Eliot* (London: Faber and Faber, 1969).

Only at nightfall, aetherial rumours
Revive for a moment a broken Coriolanus

When the third passage of pure lyric transcendence is introduced, it seems at first a delighted soaring above the sense of pain and risk and loss of the other moments we have cited. It envisions everything transformed, a gaiety made up of self-discipline, shared experience, love that answers freely to love without disappointment. Its tone is at once elated and—finally, however—even more poignant than anything that has gone before, since the vision turns out to be of something that "would" have happened:

Damyata: The boat responded
Gaily, to the hand expert with sail and oar
The sea was calm, your heart would have responded
Gaily, when invited, beating obedient
To controlling hands

These three piercing yet ambiguous poetic moments, like the earlier ones we have noted, magnetize the chaos of materials in *The Waste Land* into a structure of subtly powerful longing, regret, and effort at spiritual synthesis. The more down-to-earth dramatic and narrative parts of the poem give grain and body to the work: for instance, the whole of "A Game of Chess," with its pictures of bad marriages and brute indifference; and, in "The Fire Sermon," first the sordid scene of mechanical sex and then the passively humble, pathetic speeches by three seduced workingclass girls. But these sensually anchored parts are lyrically reoriented by the free-floating intensities of the purer moments. Each of the earthier parts has its lyrical dimension, obviously, and this is both heightened and transposed by the special moments in the first and in the final sections of the poem. How far Eliot was from the mark, then, in his note to the long scene on the typist and her lover, in which the voice of Tiresias intrudes into the already overcrowded little room, talking English and emanating from anywhere and nowhere and sounding very much like a well-read young American poet mocking his own omniscient air and possibly prurient curiosity—a mannerism conceivably borrowed from Edwin Arlington Robinson's "Eros Turannos." Groping for an explanation that will top his own show of erudition in quoting Ovid's Latin account of the prophet's knowledge of how it feels to be a woman as well as a man, Eliot tells us that "what Tiresias *sees*, in fact, is the substance of the poem."

We shall return to this intriguingly misleading—*mischievous*—note shortly. But here it seems useful to stress the poet's method over his solemnly whimsical half-pedantry: his net for people who want to be in the know about what's *behind* a poem. In essence, his practice was to draw on—to mimic or brilliantly adapt—a staggeringly large number of the non-literary ways in which language is spoken or written or chanted or sung (and to borrow whatever he needed from the literary tradition as well). Despite his tendency to think of a poem's words as belonging to a particular voice, to a great extent he divorced the language of his own poems from dramatic speech, in Shakespeare's or Browning's or Masters's sense. He used language, not for narrative or dramatic purposes of character illustration or suspense and resolution, but first and foremost to build the poem's tonal centers. Granted, the suggestion of a particular character achieved by mimicry of an identifiable voice can contribute mightily to the creation of affects. But a poem's needs are prior to character definitions or indications.

Had he not emphasized "voice" so in his criticism, Eliot might have forestalled some current critical confusions over the language of poetry. The key document is his 1953 essay "The Three Voices of Poetry," which opens:

> The first voice is the voice of the poet talking to himself—or to nobody. The second is the voice of the poet addressing an audience, whether large or small. The third is the voice of the poet when he attempts to create a dramatic character speaking in verse; when he is saying, not what he would say in his own person, but only what he can say within the limits of one imaginary character addressing another imaginary character.

Nowhere in his essay does Eliot suggest that a poem is in essence other than a mode of verbal communication, although he nearly trips over the idea with his conception of the poet talking "to nobody." The problem lies in his view of language as necessarily representing speech: one "speaks," "talks," "addresses," in or out of *propria persona*. But what if one simply uses words as artistic materials to create the poem's dynamics, as one manipulates the materials of other arts? "Voice" becomes simply a device, although, in the lyric and meditative tradition, a potent and predominant one.

If Eliot's dramatic predilections kept him from making such a distinction, it is possible that the prevalence of critical theories based on the conception of "point of view" in prose fiction also interfered. Such a theory looms behind the famous note to line 218 of *The Waste Land* that we have referred to earlier on:

> Tiresias, although a mere spectator and not indeed a "character," is yet the most important personage in the poem, uniting all the rest. Just as

> the one-eyed merchant, seller of currants, melts into the Phoenician Sailor, and the latter is not wholly distinct from Ferdinand, Prince of Naples, so all the women are one woman, and the two sexes meet in Tiresias. What Tiresias *sees*, in fact, is the substance of the poem.

Were one absolutely required to personify the zone of consciousness of *The Waste Land*—as Eliot's immersion in the philosophy of F. H. Bradley might well have inclined him to do—"Tiresias" is certainly the only name within the poem that suggests the requisite scope of awareness. Yet, despite the suggestion that Tiresias is, in some way, experiencing the poem, Eliot nevertheless shrinks from having him *speak* the poem. He drew no theoretical conclusions from his practice, yet *The Waste Land* perfectly embodies the conception of a poem we have been advancing as a dynamic projection, by means of compressed and patterned language, of specific qualities and intensities of emotionally and sensuously charged awareness. The implicit poetics of the poem's making frees it from the necessity for speaker or experiencer or observer. The poem in process creates the semblance of experience and is nowhere a report by someone of events taking place outside the poem. Its interacting tonal centers create a semblance of consciousness—hardly Aiken's "dim unity of personality," since we are speaking of a poetic construct, not a human being—and this tends more towards the infinite than finite. To slap the name "Tiresias" on the whole relationship among the poem's affects distorts its actual role as a contributing element to the poem's sexually and prophetically charged dynamics.

In fact, it is worth reminding ourselves that Tiresias' presence is insisted on in only nine lines in *The Waste Land*, all of them in "The Fire Sermon." Here his voice serves as a refrain element throughout the description of the typist's seduction:

> I Tiresias, though blind, throbbing between two lives,
> Old man with wrinkled female breasts, can see . . .
>
> I Tiresias, old man with wrinkled dugs
> Perceived the scene, and foretold the rest—
> I too awaited the expected guest. . . .
>
> (And I Tiresias have foresuffered all
> Enacted on this same divan or bed;
> I who have sat by Thebes below the wall
> And walked among the lowest of the dead.)

These lines, if we receive them as unqualifiedly serious, tinge and humanize the tawdry scene with deep resonance and sympathy, counterpointing the narrative of loveless modern coupling with deep memories of another realm of experience. As a result, it is almost as if the numbed and inarticulate typist had been able to find her own voice. (One wonders how strongly this and the poem's other scenes of sexual degradation contributed to Williams's vision in *Paterson*, which so compassionately seeks a means of expression for similarly disadvantaged women.)

The point is that the poem uses Tiresias, not the other way around. For the moment only, it assumes what Eliot calls a "quasi-dramatic" mode or an "assumed voice." The immediate effect is of a subjective, lyrical lament erupting three times in the course of the relatively brisk and objective narrative. This section, with its refrain-like interruptions, is fairly sustained; but of course elsewhere the poem's tonal shifts are extremely rapid. Indeed, it is just such shifts that made *The Waste Land*—and Pound's *Cantos* and Joyce's *Ulysses*—so explosively original. The relatively consistent personae or central sensibilities of the poems of Hardy, Yeats, Whitman, and even Dickinson no longer appropriated the whole aesthetic limelight.

The Waste Land has suffered (or enjoyed) more annotation—much of it illuminating in its way—than any other poem of this century, but we may be sure that Pound's seven jealousies ("Complimenti, you bitch. I am wracked by the seven jealousies," he wrote Eliot during the course of *The Waste Land* correspondence) were not aroused by the show of erudition. That other young man had his own information-arsenal. Both were sons of Browning, but Eliot had gone straight to the heart of the use of knowledge, the use of voices, the use of images and sudden transitions in poetry: the montage technique that rides a chaos of inner awareness without reducing its energies. The editing of *The Waste Land*, with Pound's important cooperation since he was heading in the same direction, involved decisions much like cutting-room decisions for a film. We have noted something of the affective juxtapositions of Part I, "The Burial of the Dead": their glorious independence of surface continuity and their consequent free interaction at the psychic level that alone counts. Not to labor the matter, we may look at the more complex Part III, "The Fire Sermon," which—the method having been established in the relatively simpler opening section—has some eight separate units of affect, a continual change of tone and focus held under control through the driving sexual preoccupations that disturb it and through certain counter-notes of healthy calm amid the turbulence, fear, and revulsion. (The same complex of elements reveals itself in much of Pound—indeed, the basic elements in it are so close to the common life that great artistry is involved in their aesthetic objectification and uniquely challenging character in the best work of both poets.)

A quick look, then, at the hurtle of affects in "The Fire Sermon" as they energize themselves into structure. The first movement (lines 173–

202) introduces an extended verbal music of loss, spiritual dispossession, horror, randy song, interspersed notes of rasping satire, and a counterpoising joy at youthful innocence. The passage begins with an impersonal image of late autumn near the Thames, a simple personification that recalls comparable images in the *Inferno* and sets up resonances of private desolation and unfulfilled longing ("the last fingers of leaf / Clutch and sink into the wet bank"). The passage has many other echoes, deeply evocative yet ironic in the context as well, as it sinks into revulsion and hopelessness, then goes happily bawdy—an exuberance perfectly compatible with sharply contrasting tones because the literary intelligence here has not set up a *viewpoint*—and then picks up a beautiful, wondering line from Verlaine's "Parsifal." Clearly there is neither a single dominating voice nor a single dominating idea here. When, in lines 203–6, birdsong noises are introduced, together with an allusion to Tereus' rape of Philomena, we respond to the compression of all that has gone before into a few notes of sweet, painful simplicity evoking the whole range of innocence, violation, and the continuing vitality of mythical consciousness. A short and then a long movement (207–14, 215–56) follow, presenting two conditions of grossness and violation in contexts of tragic import and some buffoonery to contain extreme feeling and prevent its turning into sentimental or merely solemn expression. Then comes the charmed passage (257–66) evoking the good, unpretentious life of simple workingmen in London, buoyed by music and other "splendour" in the world around them, including the presence and colors of the church of Magnus Martyr; and after this two views (267–91)—really, combined songs and impressionistic passages—of the Thames. The first Thames-stanza exquisitely evokes the modern Thames-scene; the second, the river as it was in Elizabethan days, with the vast glamor of the presence of royalty (albeit behaving naughtily). Then come the speeches of the women whose innocent trust has been violated—speeches that are really songs of humility at the same time and are the tonal culmination of the section in their touching sweetness that echoes the brief birdsong passage earlier on. These plaintive song-confessions (292–306) are echoed from the male side, as it were, in the closing five lines of the section, which make a chant of confession and passionate projection (though not actually of contrition) of some phrases of St. Augustine.

We have tried, in the long preceding paragraph, to convey the rapid unfolding of affective complexes in "The Fire Sermon" and at the same time to suggest that the surface discontinuity in fact removes obstacles to the play of tonal interrelationships. The shifts of style and verse-form are similarly conducive to affective harmony. While the long opening lines promise a certain gravity, the tone is leavened very quickly, partly by the display of sheer virtuosity that admits comic and ironic effects amid the tragic ones. The echoes from the whole poetic tradition prepare us, too, for the almost purely lyrical and incantatory tones that take over in the latter part of the section. The half-formality of the drawn-out seduction scene introduces an eighteenth-century plainness and mockery that is both enhanced and dissolved by the more lyrical passages. We should stress only one more aspect of *The Waste Land* here: Eliot's discovery, or breakthrough, cleared a way powerfully and brilliantly—and that is all we need to know. The sequence had come into its own, although no one—not even Eliot, or Pound with his "complimenti"—saw the full import of what had happened.

4. Intractable Elements: After *The Waste Land*

The great curve of Eliot's later development lies between *The Waste Land* and *Four Quartets*, with "The Hollow Men" (1925) and "Ash Wednesday" (1930) lighting up the movement in between. "The Hollow Men" is a marvelous, sustained echo—*que ce cor a longue haleine*—of *The Waste Land* that yet introduces an altered emphasis in Eliot's methods. "Ash Wednesday" has a certain continuity with this altered emphasis, while looking ahead to the more meditative and "positive" interests of *Four Quartets*. The character of Eliot's unfolding work compels us to face, too, the struggle with intractable elements endemic to the poetic sequence. In terms of the pressures that are at work in lyrical structure, these elements have to do with the fact that the sequence is intrinsically a response to more than the sensibility can readily handle: a destructive psychological state, for instance; or the palpable anxiety induced by social changes that seem to render past beliefs and ideals of personality obsolete; or, contrariwise, the passion to repossess the lost past—a people's old ways of speech and folkways and acute sense of place. At its most extreme, the sequence copes with a sense of utter chaos. In terms of form, the genre engages in an open struggle to encompass these pressures by dynamic improvisation towards aesthetic equilibrium. Because powerful empirical forces of the mind and history and raw human predicament are involved—forces that are in reality uncontrollable—the creative effort represents an engagement with intractable problems of formal improvisation of the sort that are both the glory and the burden of the greatest art.

The Waste Land was superbly triumphant in this engagement, but it could after all be written only once. Moreover, after *The Waste Land* Eliot drew back from the path he himself had opened. His tonal range narrowed—there is nothing in the later poems quite like the seduction scene or the pub monologue, for instance. And the sense of fragmentation, often connected with a dramatic bit abruptly introduced and dropped, lessened. These changes sprang largely from his eschewal of numerous diverse quasi-dramatic voices; in the later work he rarely, however sketchily, attached affective shifts to obvious types or personalities. Voice had been his chief *laissez-passer* into the realm of pure lyrical dynamics. But now, lacking the dramatic compression that highly charged closeups or conversational moments had encouraged, his tonal centers tended to expand over a substantial number of lines, with fewer sudden transitions. (Pound, on the other hand, was to develop, brilliantly, the art of the highly compressed tonal center. Ten lines of a canto can strike as many different notes.)

The Waste Land did share with the two sequences that followed it the struggle with intractable elements we have mentioned. Both "The Hollow Men" and "Ash Wednesday" reflected that struggle in their slow, painful process of assembly, just as *The Waste Land* had done. Some years after the fact, in his 1959 *Paris Review* interview, Eliot discussed the evolution of "Ash Wednesday" as typical of all his sequences:

> Yes, like "The Hollow Men," it originated out of separate poems. As I recall, one or two early drafts or parts of "Ash Wednesday" appeared in Commerce and elsewhere. Then gradually I came to see it as a sequence. That's one way in which my mind does seem to have worked throughout the years poetically—doing things separately and then seeing the possibility of fusing them together, altering them, and making a kind of whole of them.

The intricate history of "The Hollow Men" well illustrates this process. It involved *rearranging* poems quite as much as "fusing" and "altering" them to make "a kind of whole." The result was a lyrical coda to *The Waste Land*, converting that work, as it were, into a purer, more coherent, more selective stream of awareness. "The Hollow Men" derived from several shorter sequences whose elements were either passages originally meant for *The Waste Land* or else obvious afterbeats. It isolates one essential if complex tonal strain: an active terror of moral and emotional desiccation, and displacement, that amounts to a death-struggle with anomie. Various passages in *The Waste Land* anticipate "The Hollow Men" in projecting the same state, often through comparably surrealistic symbols of disgust and decay ("rats' alley," "a heap of broken images, where the sun beats"). But no one voice there corresponds to that of the "hollow men." The closest analogue, though hardly a voice or a chant in the same sense, would be the crowd flowing over London Bridge in "The Burial of the Dead" (lines 60–63):

> Unreal City,
> Under the brown fog of a winter dawn,
> A crowd flowed over London Bridge, so many,
> I had not thought death had undone so many.

Though created in the wake of a greater work, "The Hollow Men" is a remarkable sequence of a special sort. Within its narrower range it embodies, in purified form, the kinds of wrestling with intractable pressures we have particularly specified: with psychological and cultural crisis and with the need to repossess and redirect past realities. All are present in *The Waste Land* and, directly or by implication, in "The Hollow Men." In the latter work the dominant psychic tone is of "paralysed force, gesture without motion." The "place" is negative—"this broken jaw of our lost kingdoms" (our irretrievable past certainties). "Making a whole" of "The Hollow Men" was a matter of relating these elements in an emotionally orchestrated structure qualitatively unlike *The Waste Land.*

One vital difference is that, despite the macabre animated-cartoon figures disguised as scarecrows in its first two sections, the new poem offered no dramatic characters of any sort and no clearly individual human voices. The "hollow men" are an expressionistic chorus line. An "I" appears momentarily in the first line of Part II ("Eyes I dare not meet in dreams") and again at the start of the second stanza ("Let me be no nearer / In death's dream kingdom"). Both "we" and "I" in this sequence strongly suggest the sensibility behind the poem, a sensibility eager to communicate a psychological state. But within the poem itself they serve only to present a set of songs that range from morbidly comic and self-deprecating to pitifully timid and disoriented to mystically visionary, horror-filled, and sardonically childish. Perhaps the difference from *The Waste Land* should be put in more relative terms, however. We can certainly say that the gradual working-through of "The Hollow Men" represents a triumph of lyrical over dramatic *emphasis* in the deployment of tonalities.

The text is tellingly linked with *The Waste Land.* Both sequences, for example, are connected with "Doris's Dream Songs," a small sequence of three poems that Eliot published in 1924 but later redistributed.[6] Here he juxtaposed a rejected passage from the original version of *The Waste Land,* somewhat revised, and the poem that eventually, with changes, became Part III of "The Hollow Men." These poems—the nightmare vision of barbaric savagery that begins "The wind sprang up

6. For the progression of publication discussed in this paragraph, see Donald Gallup, *T. S. Eliot: A Bibliography* (London: Faber and Faber, 1969), Items A8, C158, C158a, C160, and C162.

at four o'clock" and the passively poignant "This is the dead land"—formed a deeply troubled song-cycle with the opening poem, "Eyes that last I saw in tears." This poem, in turn, was attached the next year to another such tiny cycle that included the poems that were to become Poems II and IV of "The Hollow Men." Called "Three Poems," this grouping was superseded the same year by a group that substituted Poem I for "Eyes that last I saw in tears" but retained II and IV. The new grouping had the title "The Hollow Men"—a title of course retained when, in the volume *Poems 1909–1925* (1925), the entire sequence of five poems appeared in their present order. Parts I (originally published in 1924 as "Poème") and V provide the most energetically human and raspingly bitter images in the sequence and therefore, also, a structural focus of keenly sardonic vibrancy at its beginning and end. It is noteworthy too that the latest section written—Part V—adds notes of self-depreciative buffoonery (to counteract over-solemnity) such as we find also at the close of "Gerontion" and of *The Waste Land.*

It should be clear—to simplify the issue—that "The Hollow Men" is a lyrical distillation (though not an exact equivalent) of *The Waste Land.* It attempts to convey the state of desolation and damnation evoked by the longer work without the complex elaboration and proliferation of centers of attention of its predecessor. With stirring force, it renders the essence of the modern-Dantean tonality Eliot has again and again sought to catch in his poems. Part IV, for instance, projects a kind of delicate sympathy suggesting the dim pathos of schizophrenic withdrawal through its subtle version of the dark banks of Dante's Acheron:

> In this last of meeting places
> We grope together
> And avoid speech
> Gathered on this beach of the tumid river
>
> Sightless . . .

The almost pathological dimension of the affect here provides a humanizing turn, a delicate modulation that skirts the danger of merely echoing Dante's allegorical imagination. A similar context shadows the remarkable self-mocking chant at the very beginning of "The Hollow Men":

> We are the hollow men
> We are the stuffed men
> Leaning together
> Headpiece filled with straw. Alas!
> Our dried voices, when

We whisper together
Are quiet and meaningless
As wind in dry grass
Or rats' feet over broken glass
In our dry cellar

These lines, and much else in "The Hollow Men," clearly pick up from imagery in *The Waste Land:* its effects of parched dryness, for instance, and its vision of "bones cast in a little low dry garret, / Rattled by the rat's foot only, year to year." The key differences, we have already suggested, are the elimination of the multitude of scenes and dramatic voices and the purer stress on lyrical evocation of one particular state: a consistently arid, passive, desolate, inert, circumscribed hopelessness. This overall consistency subsumes the various effects, from fractured liturgical phrases and nursery-rhyme parody to analytical observations, confessions of fear, and the wistful questioning and confession of Part III. In a passage like the following, the utterance is like that of a Greek chorus mimetically dancing and singing the innermost reveries of human fear and need:

Is it like this
In death's other kingdom
Waking alone
At the hour when we are
Trembling with tenderness
Lips that would kiss
Form prayers to broken stone.

The language is all suggestion—an alerting to the volatile and vulnerable state of sensibility underlying the poem. At the same time, however, Eliot is—in this poem as in "Ash Wednesday"—also on his way to the dominant method (as he felt) of *Four Quartets:* "simplification of language and . . . speaking in a way which is more like conversing with your reader." Eliot is drawing a distinction (in the *Paris Review* interview) between "Ash Wednesday" and *The Waste Land* on the one hand and *Four Quartets* on the other, crediting his playwriting experience with the shift of a style that enables him to discuss even "difficult" subject matter in a "simpler way." From this viewpoint, "The Hollow Men" would fall somewhere between, with its lack of vividly presented scenes or even the suggestion of another figure to whom the poet may address himself. At least the Lady of "Ash Wednesday" is *there,* in all her archetypal splendor. But "The Hollow Men"—with its empty men and eyes, its sunlight, its fading voices, its possible multifoliate rose, its Shadow, its liturgical "Thine"—is non-peopled in any concretely dramatic sense. Yet it evokes human loss, absurdly meaningless human existence, Chekhovian dolor—all without moving into anything like discourse as Eliot's later work sometimes does.

It is odd that the making of plays should have accompanied a lessening dramatic emphasis in Eliot's poems. In pointing to the development of a more conversationally easy diction, Eliot overlooked the danger of becoming overly discursive at the expense of poetic power. "Ash Wednesday" is still in touch with the centrally lyrical impulse that shapes "The Hollow Men." At the same time, it gives the general impression of an inward turning and pondering that will not be controlled by that impulse to the same degree. We feel the immense importance to the poet of communicating his insights into his own condition (and their religious implications) to a willing reader. The poetry has begun to become that self-conscious "intolerable wrestle / With words and meanings" described in "East Coker." This is not to say that Eliot does not succeed superbly, and far better than his windy successors in the thinking-in-verse mode, but he was on his way to the problems of *Four Quartets.* Just as in its fashion "The Hollow Men" is a lyrical purification of *The Waste Land,* so "Ash Wednesday" lyrically foreshadows the willed visionary transcendence and didactic aspect of *Four Quartets.*

To consider the possibly negative side of this foreshadowing first: Certain types of verbal and rhythmic self-indulgence show up in "Ash Wednesday" that recur—and more frequently—in *Four Quartets* as well. We do not, it is true, see the deliberate slack discursiveness that appears regularly in the latter work, in lines like these from "The Dry Salvages":

> It seems, as one becomes older,
> That the past has another pattern, and ceases to be a mere sequence—
> Or even development: the latter a partial fallacy
> Encouraged by superficial notions of evolution . . .

But at certain points, as in this passage from Part V of "Ash Wednesday," another sort of formal lapse imposes itself on us:

> Where shall the word be found, where will the word
> Resound? Not here, there is not enough silence
> Not on the sea or on the islands, not
> On the mainland, in the desert or the rain land,
> For those who walk in darkness
> Both in the day time and in the night time
> The right time and the right place are not here

No place of grace for those who avoid the face
No time to rejoice for those who walk among noise and deny the voice

The forced character of these lines, which gain a factitious tension by way of their repetitions and almost comically-doggerel internal rhymes, could hardly have escaped an ear like Eliot's. Clearly, he sought to create an atmosphere of loss and confusion—like Hopkins's in "The Wreck of the *Deutschland,*" perhaps, but without the active energy of an outcry like "where, where was a, where was a place?" The internal rhymes, if heard sympathetically, fall but hollowly and embody a negative of that state in which "the word" does indeed "resound" and promise joy and grace. This is true as well of the equally mannered passage opening the same movement:

If the lost word is lost, if the spent word is spent
If the unheard, unspoken
Word is unspoken, unheard;
Still is the unspoken word, the Word unheard,
The Word without a word, the Word within
The world and for the world;
And the light shone in darkness and
Against the Word the unstilled world still whirled
About the centre of the silent Word.

O my people, what have I done unto thee.

One is tempted to answer: "Thou hast driven us to the edge of vertigo with thy self-parodying portentousness—that's what thou hast done." But again, the issue is not quite that simple. The sickening redundancies and self-echoings can be felt as conveying, through insistence and despite an odd self-mockery, an ultimate affirmation of the desolate spirit. We are inclined to think that here, as in the comparable passages whose over-stylization, redundancy, or discursiveness grate on us in *Four Quartets,* the method calls attention to the intractability of the material without solving the problem artistically. The lyrical encompassment of that material (the problem of faith for a sophisticated intelligence that has not been vouchsafed revelation, and the parallel psychological problem of morale within a dwindling life-perspective) would be the solution.

As we have already suggested—and it is a question we shall pursue again further on, with other poets as well—this challenge of the intractable is recurrent in the sequence. Meanwhile, we must recall the important positive accomplishment of "Ash Wednesday" (although simple confrontation of the intractable at the expense of surface perfection is an act of artistic courage in its own right). Eliot in his early forties was still basically the lyric poet of "The Hollow Men," even in the way that "Ash Wednesday" continues the struggle with anomie and finds visionary glimmerings of glory, however faint or distant, to cling to. But now the proportions have changed—with the notes of radiance stronger and of greater duration. Indeed, even the fifth movement ends with a beautiful music of humble prayerfulness that swells, in the sixth and concluding part, into rich near-affirmation.

The sequence begins with a poem of reconciliation to the loss of vital powers—a song in the spirit of devout humility that is followed, in Part II, by an exuberantly witty and imaginative allegorical song expressing the happy self-abnegation of a heap of dried bones. This curiously delightful tour de force is interrupted by a prayer—cradled tenderly within the buffoonery—to the Virgin; the prayer, gravely lyrical, contains the paradoxical essence of the passages from Part V we have quoted, but in a mood of quietistic piety rather than vertiginous insistence. Then Part III leaps into three intensely personal, interlinked, dream-visions of psychological states—the first two of horror and panicky escape from vileness and grossness, the third a fullblown yet ambiguous vision of sexual sweetness and temptation. Here we are very much in the realm of Eliot's sensually alive, powerfully directive early imagery. "Ash Wednesday" gains its essential conviction and strength from this shortest movement of the poem:

> At the first turning of the second stair
> I turned and saw below
> The same shape twisted on the banister
> Under the vapour in the fetid air
> Struggling with the devil of the stairs who wears
> The deceitful face of hope and of despair.
>
> At the second turning of the second stair
> I left them twisting, turning below;
> There were no more faces, and the stair was dark,
> Damp, jaggèd, like an old man's mouth drivelling, beyond repair,
> Or the toothed gullet of an agèd shark.
>
> At the first turning of the third stair
> Was a slotted window bellied like the fig's fruit
> And beyond the hawthorn blossom and a pasture scene
> The broadbacked figure drest in blue and green
> Enchanted the maytime with an antique flute.
> Blown hair is sweet, brown hair over the mouth blown,
> Lilac and brown hair;

Distraction, music of the flute, stops and steps of the mind over the
third stair,
Fading, fading; strength beyond hope and despair
Climbing the third stair.

Lord, I am not worthy
Lord, I am not worthy

but speak the word only.

In this movement moral and intellectual awareness has the quality of experienced reality, an affect that radiates backward and forward throughout the sequence. Then the reciprocally climactic fourth movement rises into the enchantment of ecstatic, time-transcendent vision. It centers on a Beatrice-like figure in a garden, who goes "in white and blue, in Mary's colour" and whose paradisal perfection holds despite "the years that walk between"—the years one would have thought had demonstrated the impermanence of all fresh hopes:

In blue of larkspur, blue of Mary's colour,
Sovegna vos

Here are the years that walk between, bearing
Away the fiddles and the flutes, restoring
One who moves in the time between sleep and waking, wearing

White light folded, sheathed about her, folded.

The echoes here of the tantalizing imagery, with all its sensuousness, of the third stanza ("At the first turning of the third stair") in the previous movement serve as a modulating element in bringing the sequence into a new key—of exquisite spiritual exaltation. (The balancing of affects in the two passages is, incidentally, an especially poignant foreshadowing of the rose-garden "memory" in "Burnt Norton," with its ambivalent evocations of innocent childhood desires and intensities and, very possibly, of adult sexual ones as well.) The vision of Part IV introduces a glow of joyous, redemptive possibility into the earlier tonalities—shrunken self-regard, painfully won humility, buffoonery, and hauntingly terrified, longing-filled memory—that otherwise dominate the sequence. Parts V and VI, afterbeats of this vision, confirm the permanence of this glow despite their renewed language of humility, fear, and nostalgia.

In all Eliot's sequences, the anguished seriousness of the sensibility, and also its genius for a counter-vision of spiritual delight, shine through. Their concern is so obviously with matters of great psychological and human import that his religious or moral tendentiousness is usually secondary. But we are still left with questions about the degree of success of sequences after *The Waste Land*—in particular, of *Four Quartets* as a whole. It seems generally true of the modern sequence that the powerfully sustained effect of a *Waste Land* is the exception. This "should" not be so. Intrinsically—since it is made up of short and independent, yet interacting, units—the sequence solves the problem of writing a successful long poem by meeting the objections so cogently raised by Poe. For one thing, the form militates against that loss of sustained intensity which Poe deplored as inherent in long continuous works. For another, its surface fragmentation prevents our reading poetry as primarily narrative, discursive, or self-expressive. Ideally, the changes of tonal key, perspective, formal surface, and context from unit to unit commit a sequence to lyrical rather than extraneous structure. Nevertheless, a sequence can be tedious and lose its bearings. One too often finds prosy transitions, stretches of low-charged discourse, passages slack in energy, and wearisomely frenetic patches force-fed to sustain a factitious liveliness. In all these instances, the problems of the intractable have defeated the poet's art to some extent.

We do not deny that earlier forms of the long poem faced similar problems and solved them brilliantly in their historical moments. We would just insist that, in the modern circumstance, we have come to need immediacy of presentation above all, and are soon impatient with exposition, explanation, or moralizing that drains away a work's presentative energy and mobility of alertness and feeling. Nevertheless, poets have wished to provide illuminating contextual material, albeit presentatively, and they have fed such material into their structures. The process can be compared to flashbacks or newsreel insertions in the course of films. Examples are the documentary items, such as letters or speeches, inserted by Pound and Williams, and the autobiographical prose in Lowell's *Life Studies*. Even the notes to *The Waste Land*, half tongue-in-cheek or not, would be fatal if considered integral to the text. In short, a work can be overloaded even with presentative materials intended to provide a suggestive frame rather than direct exposition. Gloomy thoughts such as these arise when one ponders the task of the sequence-maker. Like all true poet-warriors, Eliot in *Four Quartets* bore arms magnificently though not always, finally, to victory.

5. The Meditative Sequence: *Four Quartets*

Every sequence of note has its own unique character. In *Four Quartets* it is the strong meditative bent, together with directly personal self-scrutiny by the poet (a sort of recurring analysis of what his lifework has amounted to), that is the distinctively new direction. When combined with the strong quasi-mystical Christian ethos that had always been present in Eliot's writing, giving its coloration to all the various tones of sadness, elated vision, anomie, and even tough realism, the new orientation released discursive and rhetorical tendencies hitherto kept on a very short leash. This development was decisive, and yet there are very interesting parallels with the earlier work. A brief comparison of the opening sections of two poems will illustrate.

April is the cruellest month, breeding
Lilacs out of the dead land, mixing
Memory and desire, stirring
Dull roots with spring rain.
Winter kept us warm, covering
Earth in forgetful snow, feeding
A little life with dried tubers.
Summer surprised us, coming over the Starnbergersee
With a shower of rain; we stopped in the colonnade,
And went on in sunlight, into the Hofgarten,
And drank coffee, and talked for an hour.
Bin gar keine Russin, stamm' aus Litauen, echt deutsch.
And when we were children, staying at the arch-duke's,
My cousin's, he took me out on a sled,
And I was frightened. He said, Marie,
Marie, hold on tight. And down we went.
In the mountains, there you feel free.
I read, much of the night, and go south in the winter.

(*The Waste Land*, 1–18)

Midwinter spring is its own season
Sempiternal though sodden towards sundown,
Suspended in time, between pole and tropic.
When the short day is brightest, with frost and fire,
The brief sun flames the ice, on pond and ditches,
In windless cold that is the heart's heat,
Reflecting in a watery mirror
A glare that is blindness in the early afternoon.
And glow more intense than blaze of branch, or brazier,
Stirs the dumb spirit: no wind, but pentecostal fire
In the dark time of the year. Between melting and freezing
The soul's sap quivers. There is no earth smell
Or smell of living thing. This is the spring time
But not in time's covenant. Now the hedgerow
Is blanched for an hour with transitory blossom
Of snow, a bloom more sudden

> Than that of summer, neither budding nor fading,
> Not in the scheme of generation.
> Where is the summer, the unimaginable
> Zero summer?
>
> If you came this way . . .
>
> ("Little Gidding," 1–21)

The poems are twenty years apart. The same tonality that opened the 1922 work is sounded again at the start of "Little Gidding." "April is the cruellest month" and "Midwinter spring is its own season" are similar: contemplative assertions of a paradox. Nevertheless, both lines, and the rest of the opening passages, epitomize the shift in poetic style and in the quality of experience being portrayed. The psychological center has a new positioning in "Little Gidding," whose lines are measured, ruminative, controlled, and staidly playful (as in the delightful "Sempiternal though sodden towards sundown"). Gone is *The Waste Land*'s poignancy of unwilling arousal to new life, with all the pressing images of difficult awakening; and gone also are the abrupt transitions, such as the leap to reminiscence in "Summer surprised us." Instead, "Little Gidding" advances at a leisurely pace, and its words insist on their appropriateness to one particular meditating mind. We have moved from an open, volatile ambience to a sensitively philosophical one. *Four Quartets* is *thoughtful,* in stretches, as Eliot's earlier sequences, especially *The Waste Land,* are not. And this mode of thoughtful communication damps down the more strictly aesthetic controlling impulses to a degree that may be considered unfortunate. Eliot, of course, felt just the opposite. We have already cited his most relevant observation, but should quote it here at somewhat greater length:

> . . . writing plays made a difference to the writing of the *Four Quartets.* I think that it led to greater simplification of language and to speaking in a way which is more like conversing with your reader. I see the later *Quartets* as being much simpler and easier to understand than *The Waste Land* and "Ash Wednesday." Sometimes the thing I am trying to say, the subject matter, may be difficult, but it seems to me that I am saying it in a simpler way.

The opening lines of "Little Gidding" are a perfect instance and, incidentally, a clear reminder of the fact that it tends to be the beginnings of the *Quartets* in which Eliot's simpler, more conversational style comes off most superbly. In general, the dominant mode has changed radically from the earlier highly charged float of reverie, with its welling up of images laden with significance. In the "Little Gidding" passage we have a strenuous meditation that initially involves connecting a sudden perception of a cold, sunny day as "midwinter spring" with intimations of eternity and immortality. "Little Gidding" is one of the great examples of what Wallace Stevens called the poem of the mind in the act of finding what will suffice. Every image is rigorously explored, and its right to be in the poem established. As with the other *Quartets,* this poem is carefully anchored in place and time, and its meditation on meaning takes off from very concrete situations.

Thus, whereas *The Waste Land* opens with the contrast between April's agitation and winter's somnolence, and with symbolically uneasy awakenings, "Little Gidding" places us literally in the midwinter landscape, responding to the paradoxical effect of light on ice and snow. The correspondences and differences are closely intertwined. The dull roots, stirred by spring rain, of *The Waste Land* become the dull spirit stirred by intense glow; the little life fed with dried tubers becomes the soul's sap that quivers between melting and freezing; cruel April becomes a spring beyond the seasons—not of time's covenant. All these reoriented images are carefully prepared for by the grandly sweeping generalizations of the first three lines, from which all the other implications of the passage grow in a way different from the progression of *The Waste Land.*

Intimacy of tone, not casual in any sense but allowing us to overhear the musing mind at work, marks the start of "Little Gidding" and, in fact, of each of the *Quartets.* In "Burnt Norton" we overhear a self-hypnotic speculation, at the start, on time and timelessness; it is at once drily philosophical and incantatorily—even prayerfully—rocking, full of repetitions, rhythmic balancings, and phrasal echoings imbued with a certain awe. In "East Coker," the opening passage, with its initial refrain "In my beginning is my end," is very like that of "Burnt Norton" except that it is less purely abstract. The language becomes increasingly concrete, pressing, and vivid in each of the succeeding openings, as the first few lines of each will show:

> Time present and time past
> Are both perhaps present in time future
> And time future contained in time past.
>
> ("Burnt Norton")

> In my beginning is my end. In succession
> Houses rise and fall, crumble, are extended,
> Are removed, destroyed, restored, or in their place
> Is an open field, or a factory, or a by-pass.
>
> ("East Coker")

I do not know much about gods; but I think that the river
Is a strong brown god—sullen, untamed and intractable,
Patient to some degree, at first recognised as a frontier;
Useful, untrustworthy, as a conveyor of commerce;
Then only a problem confronting the builder of bridges.

("The Dry Salvages")

Midwinter spring is its own season
Sempiternal though sodden towards sundown,
Suspended in time, between pole and tropic.

("Little Gidding")

Eternity, divinity, are evoked in all four passages, but the pitch of intensity, the urgency to connect, rises from beginning to beginning. The endings of the poems, too, project increasingly affirmative tones after the balance of delighted challenge against desolation at the close of "Burnt Norton":

Sudden in a shaft of sunlight
Even while the dust moves
There rises the hidden laughter
Of children in the foliage
Quick now, here, now, always—
Ridiculous the waste sad time
Stretching before and after.

("Burnt Norton")

Old men ought to be explorers
Here or there does not matter
We must be still and still moving
Into another intensity
For a further union, a deeper communion
Through the dark cold and the empty desolation,
The wave cry, the wind cry, the vast waters
Of the petrel and the porpoise. In my end is my beginning.

("East Coker")

. . . And right action is freedom
From past and future also.
For most of us, this is the aim
Never here to be realised;
Who are only undefeated
Because we have gone on trying;

We, content at the last
If our temporal reversion nourish
(Not too far from the yew-tree)
The life of significant soil.

("The Dry Salvages")

Quick now, here, now, always—
A condition of complete simplicity
(Costing not less than everything)
And all shall be well and
All manner of thing shall be well
When the tongues of flame are in-folded
Into the crowned knot of fire
And the fire and the rose are one.

("Little Gidding")

It is an interesting, if possibly diversionary, thought that these four endings, combined as separate stanzas of a single poem, would work together rather well—that is to say, about as well as the four *Quartets* work together, with the pressure toward affirmation finally bursting through at the end to override the depressive side of the work by sheer force of will and mobilization of rhetoric. The four beginnings could not work in the same way—all but the last prepare for a finer music and more moving effects to come, while the opening lines of "Little Gidding" are already creating that music and impact. They are not epitomes of their individual poems as are the endings, which condense the whole gravamen of what has preceded them so beautifully. The successive endings, too, crystallize the moral and religious impetus of the *Quartets*—their extremes of mood and morale in the first; their self-challenge, next, in the face of advancing age and "the dark cold and the empty desolation"—the pull towards depression and anomie that persists from Eliot's earliest poems onward; then, their ripened faith, and the humility freed of bitterness that can take comfort simply in helping to "nourish" the "life of significant soil"; and, finally, the visionary blaze of reassurance at the close of "Little Gidding."

But if the successive endings epitomize, with elegant intensity, the evolving moral and religious tonal sets of the sequence, a subtle psychological correlative unfolds as well in Part I of each *Quartet*. We may think of this aspect as a double stab of memory or shared subjectivity—the one laying bare acutely private feelings, the other equally personal in origin but tending toward the archetypal or historical. The process is paralleled in Part I of *The Waste Land*. First, the "hyacinth girl" passage comes through as a piercingly personal slash of dialogue (or half-dialogue: a girl's remembered words and her lover's evocation—a kind of long-delayed reply to her—of their moment of impassioned entrancement):

> "You gave me hyacinths first a year ago;
> "They called me the hyacinth girl."
> —Yet when we came back, late, from the hyacinth garden,
> Your arms full, and your hair wet, I could not
> Speak, and my eyes failed, I was neither
> Living nor dead, and I knew nothing,
> Looking into the heart of light, the silence.

The second memory comes in the closing stanza of Part I: the archetypal scene on King William Street in London, with the mad outcry to "Stetson," that is a vision of death-in-life and the endlessly recurring horror of war and the pity of man's condition in general. This scene too is basically a personal memory but quickly turns wildly symbolic. That is, the literal memory is of a filthily murky London morning, with vast crowds of people "flowing" over the streets. They seem to the observer to be moving without will: living automatons that summon up Dante's living dead in Hell's antechamber; and it seems that he too is one of them, and that he shouts his macabre questions at a reflex image of himself in the crowd; and that the voices of other poets who have had comparable visions of the hideous depths of human damnation converge with his own. This is memory in the impersonal sense, palimpsest upon palimpsest of overlaid experience, imagination, and shared vision.

In "Burnt Norton" the two thrusts of intimacy are, apparently, more closely related, in accordance with the carefully knit meditative structure of the *Quartets*. By the same token, their integration with the poem's whole meditative process is self-evident; their relation to the rest of the text seems "natural" by comparison with the montage and collage technique of *The Waste Land*. After the opening meditation on time's mysteries, which points so emphatically to the continuing reality of unrealized possibilities, the first pure notes of open subjectivity—as laden with sweet ruefulness as "April is the cruellest month" and "mixing / Memory with desire"—surge into view. They suffuse the meditative abstractions that have gone before with a powerful nostalgia; and then a speaker—"I"—enters the language to redirect it, making leisurely haste to remind the reader that *every* mind contains similarly redolent memories of missed experience:

> Footfalls echo in the memory
> Down the passage which we did not take

> Towards the door we never opened
> Into the rose-garden. My words echo
> Thus, in your mind.
> But to what purpose
> Disturbing the dust on a bowl of rose-leaves
> I do not know.
> Other echoes
> Inhabit the garden. . . .

"Footfalls" that "echo in the memory," a "rose-garden" never seen—these pressed flowers of regret are far from the dramatic transport that began in the "hyacinth garden." Yet, gentler though its affect is, the rose-garden passage registers its pang of unforgotten disappointment sharply. The impact is especially intense because of the sudden shift from the hypnotic rocking in the cradle of the abstract just before this passage. Eliot has adapted his dynamics to the meditative context with true virtuosity. The pang of unforgotten disappointment presents itself in a single unbroken sentence. The active urgency or purposefulness, falling rhythms, internal rhyming, and parallel constructions in this sentence hold the details of unconsummated experience—footfalls, unentered passageway and door and garden: notations of anticipated but missed fulfillment—in eternally poised (but hardly Keatsian) stasis. Then the first strong caesura ("Into the rose-garden. My words echo") intrudes on the spell, only to extend it cunningly through a transition that enlists our general assent: "My words echo / Thus, in your mind."—The rhetorical maneuver reminds us that we all have our perhaps repressed rose-gardens: lost hopes that yet persist within the existential present by virtue of having once been felt to be really possible. The rhythmic break also startles the poem out of reverie and into a kind of objectivity and philosophical perspective connnected to the opening meditation on time: "to what purpose / Disturbing the dust on a bowl of rose-leaves / I do not know." Finally, though, the private dimension, the quietly emphatic sense of irrevocable loss, remains dominant. (One might even imagine that "your mind" is not the general reader's so much as a particular person's for whose desired love the "rose-garden" is a metaphor.)

The poignant moment, however, is interrupted by three swift transitional notations: an ambiguous assertion ("Other echoes / Inhabit the garden"), an almost pedagogical question ("Shall we follow?"), and then, again almost pedagogically, an authoritative directive from the realm of folklore and archetype ("Quick, said the bird, find them, find them, / Round the corner"). Thus we are eased into what will become a reversal of the remembered deprivation. We are going to find a memory of the rose-garden after all. This will be the second memory-passage of this movement of "Burnt Norton," a full-blown imagined mystical memory of a moment of transcendent timeless reality, apparently "experienced" by sheer force of the transforming will. Eliot does not pretend fairy-tale magic here, or rely on a facile suggestion of literal supernatural vision. Instead, the full transition, which includes a reprise of the notations we have just observed, but one fully acknowledging the workings of self-deception, provides a new context of psychological sophistication:

> Other echoes
> Inhabit the garden. Shall we follow?
> Quick, said the bird, find them, find them,
> Round the corner. Through the first gate,
> Into our first world, shall we follow
> The deception of the thrush? Into our first world.
> There they were . . .

And so we have arrived, by mild associative forcing, and have entered "into the rose-garden" on a rushing current of oxymora and negatives: "the deception of the thrush" first, and then the sight of "invisible" presences, the hearing of "unheard music," and the emergence out of emptiness and "dry concrete, brown edged" of a state of epiphany:

> And the pool was filled with water out of sunlight,
> And the lotos rose, quietly, quietly,
> The surface glittered out of heart of light . . .

A great deal might be said about the passage that ends Part I of "Burnt Norton" and whose climactic lines we have just quoted. We might note the phrase "heart of light," for instance, which appeared also in the first memory-passage of *The Waste Land,* with equally complex resonances: romantic ecstasy in a context of desolation in both settings, despite many differences. (Eliot was much taken with Conrad's *Heart of Darkness.* He uses the expression "heart of light" with the same combined seriousness and irony as are suggested by Conrad's title.) But the main thought we would advance now is that the pattern we have been following, the double stream of shared memories or moments of intimate awareness, provides an independent energy in *Four Quartets.* It mingles with the meditative and rhetorical streams—is indeed deployed so as to reinforce them. Yet it provides an entire, other complex of tensions, based on a pressure to reconcile an overwhelming regret, or personal sense of the isolating power of moments of felt reality, with some larger, self-engulfing perspective. The attempt to resolve this complex of tensions is one of the great structural issues of the *Quartets.*

Thus, in "East Coker," a curiously alienating yet magnetized closeup of a hot afternoon in a country village is succeeded by the invocation of a wedding dance, centuries back, of the village folk—a race-memory that sinks the aimless moment of existential concentration into anonymity. Then, in Part I of "The Dry Salvages," a fiercer tone enters. It begins innocently, with childhood memories of living near the Mississippi, a "strong brown god":

> His rhythm was present in the nursery bedroom,
> In the rank ailanthus of the April dooryard,
> In the smell of grapes on the autumn table,
> And the evening circle in the winter gaslight.

The sensuous awareness of the river-god's nearness—embodied in a child's very physical sense of immediate realities—is another token of the powerful sense of place of the speaker. In its turn, this absolute repossession (not necessarily blissful) of a childhood world gives way to immersion in contemplation of the sea's endless destructiveness and the deceptiveness of time. These impersonal energies render the most particularly individual memories irrelevant, subordinating them to obliteration and death. So the child's closeness to the river, in part a cherished family memory but in part an indissoluble memory of gross luxuriance (the "rank ailanthus"—with perhaps a reference as well to this tree's unpleasant smell)—yields to a depressed acceptance to be echoed further on, in Part II:

> Time the destroyer is time the preserver,
> Like the river with its cargo of dead negroes, cows and chicken coops,
> The bitter apple and the bite in the apple.

Finally, the marvelous discovery of "midwinter spring" at the start of "Little Gidding," with its unique excitement of apperception, is connected with—and subordinated to—the locating of a historical site, utterly unprepossessing on its surface, where a religious community is said once to have experienced revelation. Theirs was an incommunicable triumph, "beyond the language of the living" to portray, and therefore nullifying the significance of the poet's art so richly displayed in the opening lines. This, the most exalted juxtaposition (the miracle of the vision of eternity created by a happily self-deceiving eye out of nature's *trompes l'œil,* and the "genuine" miracle experienced by true believers) of the sequence, is enormously appealing in its orchestration of tonalities of self-transcendence. Still, it is most interesting that the most powerfully sustained passage of directly personal import in the entire sequence comes in the section immediately following. Most of Part II of "Little Gidding" is given to an *Inferno*-like encounter between the poet and a "compound familiar ghost" that is both his own mirror-image and a composite of all great poets from Dante on who have conceived and executed greatly. The message of the "dead master" is of the inevitable failure of the poetic enterprise and humane ideals. Apart from aging into disillusionment and humiliation, "From wrong to wrong the exasperated spirit / Proceeds, unless restored by that refining fire" of purgatorial self-immolation "Where you must move in measure, like a dancer." One does not, in *this* life, actually reach the assurances envisaged elsewhere in the sequence. That is the unresolved state in which the most brilliant section of the entire sequence leaves us, and the state whose complex awareness each of the first movements has brought into the foreground.

We shall not go into further details about the structuring elements of the *Quartets*. Their pressure towards a triumphant rhetorical resolution, supported by a certain amount of half-winning, half-tedious discourse—attempted control by a single voice—has carried the work further toward complete triumph than the blight of didacticism might have been expected to allow. The weaknesses do not kill the movement of the whole process, although they muffle it from time to time. The strengths are undeniable, and there is a special strength, in addition, that comes from facing into an artistic task as intractable as the one Eliot set for himself. Perhaps the problems he confronted are in any case inevitable when a sequence stretches beyond its optimum length (a variable in each instance). At all events, he took the usual risk in *Four Quartets*. Whatever the loss thus incurred, the effort was crucial in creating a poetic emphasis that encouraged such unacknowledged followers as Williams, Stevens, and Olson. *Four Quartets* plunges deeply into the major enterprises of making a modern meditative sequence, and also of relating personal to profoundly historical and regional memories.

Chapter Eight

Ezra Pound I: The Early Sequences

1. Phanopoeia and the Image

Ezra Pound's sequences are likely to remain news for some time in ways that those of his contemporaries won't, and in ways he did not think of when he defined literature as "news that *stays* news." The "controversial" character of his poetry reflects that of the man. To start with the irreducible public issue, he—in and out of his poems—is not easily forgiven his Fascist and anti-Semitic bias. And unless C. H. Douglas's theories of social credit usher in a new, more equitable millennium, or Confucian thought takes over as the mainstay of Western civilization, stretches of the *Cantos* will never be able to shed their impression of tediously pedantic, sheer eccentric crankiness (although we should remember that such crankiness is a genuine enough basis for an affect—as are bloodthirstiness and any sort of vileness—quite as much as are sweetness and light). In addition to such questions involving the roles of rhetoric, belief, and personality in poetry, questions often grounded in cruelly destructive human concerns, Pound's work stirs up others that are perhaps less passionate but of equally central interest for the nature of poetry. Modern poetry in particular has been redirected by his approach to translation (as in *Homage to Sextus Propertius*), to the relation of poet and persona (as in *Hugh Selwyn Mauberley*), to the creative use of documents (the *Cantos*), and to open structure and the possibility of creating a modern epic (the *Cantos* again).

In length and range closer to Joyce's *Ulysses* than to any other work discussed in this book, the *Cantos* are the true testing-ground for a theory of the sequence. Here we have a sequence-succession, with each sequence lyrically structured—like *Paterson* but on a far larger scale. Pound, in fact, carried the implications of Imagism farther than any other poet. Unless we read the *Cantos* as successions of tonal centers or affects—"images" writ large and complex—they are a chaos. It is this point, much neglected, that we wish to emphasize above all.

The method of the *Cantos* surfaced in miniature, if only by implication, in 1912, when one of the century's best known poems, "The Return," appeared in the June issue of *English Review* and also in Pound's volume *Ripostes*. It was set to music in 1913 by Walter Morse Rummel and received wider attention the next year in the *Des Imagistes* anthology, published both in hardcover and as the February issue of *Glebe*. It appeared again in 1917, in Harriet Monroe's and Alice Corbin Henderson's famous anthology *The New Poetry*. It has always been greatly admired, and Yeats's accolade in *A Packet for Ezra Pound* (1929) is still powerfully convincing: "it gives me better words than my own." We offer it here in the version in *Collected Early Poems of Ezra Pound*,[1] which reproduces the highly satisfactory *Ripostes* spacing:

See, they return; ah, see the tentative
        Movements, and the slow feet,
        The trouble in the pace and the uncertain
        Wavering!

See, they return, one, and by one,
With fear, as half-awakened;
As if the snow should hesitate
And murmur in the wind,
                        and half turn back;
These were the "Wing'd-with-Awe,"
                        Inviolable.

Gods of the wingèd shoe!
With them the silver hounds,
                        sniffing the trace of air!

Haie! Haie!
        These were the swift to harry;
        These the keen-scented;
        These were the souls of blood.

        Slow on the leash,
                                pallid the leash-men!

1. Ed. Michael J. King (New York: New Directions, 1976), p. 198. For all other quotations, however (except for the *Cantos*), our primary text of reference is *Personae: The Collected Shorter Poems of Ezra Pound* (New York: New Directions, 1949).

Just the look of the poem on the page was provocative enough for E. E. Cummings to credit it with having deeply influenced his own experiments with form.[2] Its rhythm continues to provoke fruitful discussion.[3] Its concision and concreteness will keep it in any anthology of Imagist work, along with such shorter poems as Williams's sixteen-word "The Red Wheelbarrow," H.D.'s tiny "Oread" and "The Pool," and Pound's own two-line "In a Station of the Metro":

> The apparition of these faces in the crowd;
> Petals on a wet, black bough.

Nevertheless the exquisite technique of "The Return" continues to be misunderstood, as evidenced by the complaint of one critic, William H. Pritchard, that he has difficulty "in coming up with much of interest to say" about it:

> Put in the language we have found useful in speaking about Frost or Yeats, there is no specifiable "voice" in "The Return" and nothing to be gained from investigating the "tones" in which things are said. It is an undramatic poem, revealing of no "speaker" with emotions and attitudes toward something, which emotions and attitudes "develop" over the course of the lines.[4]

With engaging clarity these "no's" reveal the irrelevance of a critical approach that asks lyric poetry to conform to drama. In a play, indeed, all the words must be spoken by a particular character, and—in any given performance—in a particular tone, and they may very well be helping the plot along by alerting us to some appropriate development in the speaker's "emotions and attitudes." But a lyric poem (or, for that matter, a narrative one) is not dramatic speech, although it may suggest such speech so powerfully that it is relatively difficult to recognize all the other techniques the poet is using to create the poem's moment by moment tonalities. Still, even for readers who have not broken the habit of concentrating only on "speakers" and "their" emotions (rather than on the emotional charge of the *language*) in lyric poems, there is no special problem presented by "The Return." From their critical standpoint, after all, the "speaker" in the poem is beholding the return to fullblooded

2. See Richard S. Kennedy, *Dreams in the Mirror: A Biography of E. E. Cummings* (New York: Liveright, 1980), p. 106.
3. See, for example, Stephen J. Adams's "Pound's Quantities and 'Absolute Rhythm,' " in *Essays in Literature,* 4 (1977), 95–109, and S. M. Gall's "Pound and the Modern Melic Tradition: Towards a Demystification of 'Absolute Rhythm,' " in *Paideuma,* 8 (1979), 35–47.
4. *Lives of the Modern Poets* (New York: Oxford Univ. Press, 1980), p. 151.

presence of the faded gods and heroes of Grecian antiquity, and the analogy with the shades summoned up by Odysseus and quickened by the blood of ritually sacrificed beasts is perfectly clear. The shifts to the most vivid intensity and away from it again, and from a tone of compassionate wonder to ecstatic outcry and then to elegiac sadness, are absolutely charged with "emotions and attitudes" that " 'develop' "—all handled with a virtuosity totally subordinated to the movement. Pound had not failed to write a poem with dramatic elements, then. Rather, he had purified his sense of structure beyond dependency on them.

What "The Return" and other Imagist poems accomplished, precisely, was to shift our attention away from a search for what some dramatic character, identified or not, might be *saying* to something more nakedly present: what the poem before us is creating. It is undoubtedly easiest to recognize the visual aspects of such a nakedly present creation—the picture or "image" the words conjure up. Yet obviously Pound's sense of the image as an "intellectual and emotional complex in an instant of time" suggests more than the visual element stressed in his definition of "Phanopoeia": "a casting of images upon the visual imagination." The "intellectual and emotional complex in an instant of time" is well on the way to our conception of the tonal center or affect. A visual component may predominate in some types of poetry, just as the poet's mimicry of the myriad ways language is used in ordinary life predominates in other types—for instance, in dramatic monologues. These are two important possible components among many. As for "The Return," it is neither wholly speech nor wholly a succession of images cast upon "the visual imagination." But if we see it as a progression of tonal centers, we shall indeed find something "of interest" to say about it—and we shall point ourselves toward the evolution of the sequence in our century.

First, there is the arrangement of the poem on the page. From the vantage point of the 1980s, it seems extraordinary that space was rarely used to reinforce meaning or to provide rhythmic clues before "The Return." At least for poetry in English, indentation was used either to emphasize a rhyme-scheme or, in pattern poems, to force a piece into a shape regardless of its organic nature. Conceivably Pound took as his model Mallarmé's brilliant, bizarre, and original use of space and typography in "*UN COUP DE DÉS*," published in 1897.

The strategy of "The Return" is equally striking. In stanzas one and four the parallel spacing illuminates the pathetic disparity between the plight of the wavering figures in the present and the power they once had. The indented lines in stanzas two and three look like broken continuations of the one-word last line of the first stanza—"Wavering!"—and the effect is particularly marked in the second stanza, which ends with the one-word line "Inviolable." Pound has drawn a contrast subtly and powerfully and simply by the position of the words on the page. The indented, broken last stanza has a rhythmic weight matching the slow pace of the hounds and their masters. The eye connects this ending with the poem's longest line, the third, and we realize the further linking of these effects by the delayed off-rhyme of "uncertain" and "leashmen," which with finality hammers home the pitiable state of once mighty figures.

The poem is one of Pound's great rhythmic successes. This is "free" verse in which stress, quantity, and pause not only reinforce meaning superbly but suggest an underlying musical measure of surprising regularity that contributes significantly to the poem's impression of controlled artistry. Whether or not one recognizes the musical measure,[5] the wavering rhythm of the first stanza, whose pauses are indicated by punctuation and line breaks, is immediately evident. So is the powerfully sustained rhythm of the third and fourth stanzas. In each case the rhythm reinforces the imagery of movement, becoming an important contributing element to the tonal centers so carefully delineated by the stanza breaks.

If we follow the poem stanza by stanza, something emerges that is rather different from Mr. Pritchard's "guide-book summary." ("It can be said to be about how far away we are historically and spiritually from the Greek gods, from any organic myth.") For one thing, the first stanza is shot through with personal intensity of the sort whose supposed absence Mr. Pritchard deplores. Indeed, the first five words—had "The Return" not predated Hopkins's first edition by six years—might remind any reader of the "ah my dear" of "The Windhover" or the "ah! bright wings" of "God's Grandeur." In all three poems the interjections convey poignant wonder, along with the varied emotional tonalities specific to each poem. Most centrally, these are sensuous delight, self-rebuke, and devout pity in "The Windhover," an almost maternal tenderness and dismay shot through with thankfulness in "God's Grandeur," and compassion and nostalgia in "The Return." In Pound's poem the returning figures have a feebleness and timidity as of old age. They gain strength in line 1 of the second stanza; and, although the rest of the stanza projects their hesitancy and fear, the poem itself advances confidently toward seeing them as the once " 'Wing'd-with-Awe,' / Inviolable." The second stanza's affect, then, is more complex than that of the first. The most important new element is its strong suggestion of a meditating but concretely visionary mind, making comparisons, evoking the emotional state of these figures, and finally drawing conclusions as to who they

5. "The Return" is in 4/4 time. See Gall, "Demystification," pp. 40–41.

are—or rather were. Without the projection of a concentrated, inquiring mind, the sudden eruption of the powerful past into consciousness would seem arbitrary rather than inevitable:

> Gods of the wingèd shoe!
> With them the silver hounds,
> sniffing the trace of air!

This recapturing of the original nature of the wavering figures of the first and second stanzas is presented as being unqualified. They are absolutely here, tangible to the aroused imagination and—with the "Haie! Haie!" of the fourth stanza—ready to leap forth on the hunt. But then the poem pulls back, as we noted earlier, into the elegiac mode ("These were . . .")—the vitality has been only momentarily recovered, and by the fifth stanza we have been gracefully brought back to a broken, impotent present: "Slow on the leash, / Pallid the leashmen!" But now this perception is tinged with amazement—or so the exclamation-point suggests—that the heroes' very nature can have changed so utterly.

"The Return" is just that: a *return*, making tangible to imagination a lost way of perceiving the world. The poem has given life to the dead, in a modern analogue of Book XI of the *Odyssey*. (The phrase "souls of blood," given the Classical heroic context, carries penumbral associations with Odysseus' offering of sacrificial blood to quicken the shades of Tiresias and other illustrious dead.) It is far better than Pound's earlier "Sestina: Altaforte," which resurrects the berserk, war-obsessed Bertran de Born: "Eccovi! Judge ye! Have I dug him up again?" Yet that poem was a precursor of this one in summoning up a past world in the full heat of its own reality. "The Return" embodies Pound's genius for such resurrection more superbly than almost any other of his shorter pieces. It offers the germ of the whole course of his poetic endeavor, stated most simply at the close of Canto 81:

> To have gathered from the air a live tradition
> or from a fine old eye the unconquered flame
> This is not vanity.
> Here error is all in the not done,
> All in the diffidence that faltered . . .[6]

6. Ezra Pound, *The Cantos* (New York: New Directions, 1972), p. 522.

2. *Homage to Sextus Propertius*

In March of 1917 Pound wrote Joyce, whose *Dubliners* and *A Portrait of the Artist as a Young Man* he admired immensely, "I have begun an endless poem, of no known category. Phanopoeia or something or other, all about everything. 'Poetry' may print the first three cantos this spring. I wonder what you will make of it. Probably too sprawling and unmusical to find favour in your ears. Will try to get some melody into it further on." By July he was in a state of acute, self-conscious doubt over the quality of these first cantos, which appeared in the June, July, and August issues of *Poetry:* "I hope to God you won't try to read my beastly long poem in 'Poetry.' I have revised the whole thing, and it is at least better than it was, and will appear in my American edition, which you will receive, if it, you, and I survive till late Autumn."[7]

These three cantos were very early drafts indeed, and different enough from the final versions so that they have been dubbed "ur-cantos." Halfway through the first we find a draft of the passage that will eventually open Canto 3, "I sat on the Dogana's steps" (the ur-canto's Browningesque version: "Your palace step? / My stone seat was the Dogana's vulgarest curb"). The second ur-canto provided some of the language for the Ignez da Castro and Cid passages in Canto 3 and occasional phrases in later cantos. And the third ur-canto closed with the translation from the *Odyssey* that became the opening passage of the *Cantos* in their final form.

The matrix out of which the *Cantos* came naturally involved Pound's urge towards epic length—and he was soon to have the first chapters of *Ulysses* in hand as a prose analogue that perhaps not only suggested the reorientation of the book around the opening fragment from the *Odyssey* but also a way of breaking out of Browning's spell. A powerful second motive was his urge to find whatever was living in past literature as well as present reality. As he said of his *Homage to Sextus Propertius*, completed in 1917: "there was never any question of translation, let alone literal translation. My job was to bring a dead man to life, to present a living figure."[8] The twelve-part *Homage* was his most successful and ambitious work so far, and by far the most important sequence in the 1919 *Quia Pauper Amavi*. This contained, along with the Propertius and the three ur-cantos, the five-part "Langue d'Oc" and the eight-part "Mœurs Contemporaines." Pound had shifted his attention away from

7. *Pound/Joyce: The Letters of Ezra Pound to James Joyce*, ed. Forrest Read (New York: New Directions, 1967), pp. 102, 122. The "American edition" is the 1917 *Lustra*.

8. Letter to A. R. Orage, April 1919. *Selected Letters 1907–1941 of Ezra Pound* (New York: New Directions, 1971), ed. D. D. Paige, pp. 148–49.

individual lyrics and towards lyric sequences. The Provençal translations and imitations are in heavily archaic language. The "Mœurs Contemporaines," although showing considerable wit, are essentially trivial and perhaps best regarded as exercises in preparation for the *Mauberley* portraits. The ur-cantos are dragged down by Browning's stylistic devices, especially the monologue technique. But the *Homage to Sextus Propertius* lives with the quick immediacy of the "silver hounds, / sniffing the trace of air" in "The Return." It is, blessedly, verse at once "free" and under control, improvisatory yet channeled by sustained reference to Propertius' tonal contexts and formal elegiacs:

> If she with ivory fingers drive a tune through the lyre,
> We look at the process.
> How easy the moving fingers; if hair is mussed on her forehead,
> If she goes in a gleam of Cos, in a slither of dyed stuff,
> There is a volume in the matter; if her eyelids sink into sleep,
> There are new jobs for the author;
> And if she plays with me with her shirt off,
> We shall construct many Iliads.
>
> (from V,2)[9]

This passage will suggest the special brilliance of the *Homage*—its melodic virtuosity that enables it to be serious and playful, driven and detached, together. The reference to dactylic hexameters is present, a pressure on the shape of the poem, but with no mechanical deadliness. Rather, the line-lengths are varied according to the dictates of English speech-rhythms and the verse-pattern is strongly affected by the same sense of musical measure we observed in "The Return." Moment by moment the *Homage* is as satisfying as any poetry (except the highest peaks of the *Cantos*) that Pound ever wrote; and we may say that, in 1917, with the collaboration of a poet born almost twenty centuries earlier, he created the first of the famous short sequences of the post-World War I decade. It was published a year before *Mauberley*, three years before *The Waste Land*, and before either of Yeats's civil-war sequences (1921 and 1923).

The "collaboration" with Propertius created a liberating interplay of ancient and modern sensibility, more dynamically open than any single "point of view" would account for. Propertius' own series of poems ("elegies"), mingling its erotic, satirical, and lyrically delighted tonal streams

9. For a text printed on facing pages with the relevant Latin passages, see J. P. Sullivan, *Ezra Pound & Sextus Propertius: A Study in Creative Translation* (London: Faber and Faber, 1965), pp. 115–71. Sullivan has "corrected" Pound's text, not always with his consent. Our quotations follow *Personae*, pp. 205–30.

with a number of others in a manner independent of dramatic consistency, has a comparable openness. An important dimension, for example, is the passionate, often bitter or mocking literary stance the poems strike, "elevating the genre of elegy to the lofty position traditionally enjoyed by epic" (a form that "resolutely resisted the expression of personal attitudes") and "appropriating epic values" for its own ends.[10] The stance has the air of proud political resistance and an equally proud aestheticist bohemianism. All this may be seen in the audacious passage we have already quoted, with its contentious exaltation of sensuousness and feeling. It would be going too far to say that Pound exactly projects Propertius' tones, but his adaptation does isolate (by cutting, compressing, and rearranging) various major tonal streams quite independent of a "unifying voice," streams we can follow in the Latin as well. Three passages from *Homage* I—the first an exquisitely joyous invocation of Greek masters of pure lyricism that opens the sequence brilliantly, the second a disdainful assault on official "heroic" poetry celebrating generals and politicians, and the third an intense avowal of fidelity to the demands of genius—will indicate the complex affective elements at work.

(1)

Shades of Callimachus, Coan ghosts of Philetas
It is in your grove I would walk,
I who come first from the clear font
Bringing the Grecian orgies into Italy,
                                        and the dance into Italy.
Who hath taught you so subtle a measure,
                        in what hall have you heard it;
What foot beat out your time-bar,
                  what water has mellowed your whistles?

(2)

Out-weariers of Apollo will, as we know, continue their Martian generalities,
    We have kept our erasers in order.
A new-fangled chariot follows the flower-hung horses;
A young Muse with young loves clustered about her
                        ascends with me into the aether, . . .
And there is no high-road to the Muses.

(3)

Flame burns, rain sinks into the cracks
And they all go to rack ruin beneath the thud of the years.

10. Joy C. King, "Propertius' Programmatic Poetry and the Unity of the *Monobiblos*." *Classical Journal*, 71 (1975–76), 108–24.

Stands genius a deathless adornment,
a name not to be worn out with the years.

The first of these passages—the opening stanza of the sequence—is an elegantly syncopated melodic adaptation from Propertius' elegiacs. It evokes their subtle quantitative shifts within the controlled imbalance between the two parts of each distich—

Callimachi Manes et Coi sacra Philetae,
in vestrum, quaeso, me sinite ire nemus.
primus ego ingredior puro de fonte sacerdos
Itala per Graios orgia ferre choros.
dicite, quo pariter carmen tenuastis in antro?
quove pede ingressi? quamve bibistis aquam?

(IV.1)[11]

We do not propose to make a comparative study of Propertius' and Pound's prosody and diction, but the most cursory parallel reading of the Latin and the English here will show the highspirited projection of the very feel and movement of the original in the modern poet's lines and half-lines. Even Propertius' delight in addressing his beloved poetic forerunners is conveyed, inevitably with the added dimension of Pound's exuberance at recovering the Propertian immediacy and music. This rich initial affect finds a deeper echo in the third passage, which ends *Homage* I on a note at once steeped in fatalism, reverence for genius, and affirmation of the immortal value of the true lyric poet's art. The second passage (actually, the second stanza, which begins an expansive satirical assault constituting the bulk of the poem) represents the bridge of wit and sardonic hectoring between the purely lyrical affects at the start and at the end of the poem. The turgid time-servers who "expound the distentions of Empire" are contrasted with the true poet content to write "something to read in normal circumstances," something that "devirginated young ladies" can enjoy "when they have got over the strangeness"—something, in short, that will have its day, whenever it may come, simply because of its quality and human appeal. The wryly self-depreciative language (as in the first poem of *Mauberley*) is of course an expression of self-confidence as well as an opening toward the love-poetry that will appear later in the sequence. But all these tonal motives are contained within the larger curve we have noted, starting with the opening lines of joy in the lyric process and ending with the tragically tinged

11. See Sullivan, p. 114, for Latin text used here, and pp. 111–13 for explanation of the text Sullivan uses and for collation of the sections in the Latin and in Pound's sequence.

challenge to mutability and death in the ringing phrase "genius a deathless adornment." This curve, indeed, epitomizes the whole sequence, which ends in poem XII on a note combining the tonalities just described and, precisely, "appropriating epic values" to lyrical ends:

> Varro sang Jason's expedition,
> Varro, of his great passion Leucadia,
> There is song in the parchment; Catullus the highly indecorous,
> Of Lesbia, known above Helen;
> And in the dyed pages of Calvus,
> Calvus mourning Quintilia,
> And but now Gallus had sung of Lycoris.
> Fair, fairest Lycoris—
> The waters of Styx poured over the wound:
> And now Propertius of Cynthia, taking his stand among these.

Because Pound selected passages, rearranged them, and did some improvising on his own, he created a new sequence. It is a translation in the sense that a child is a translation of its mother, umbilically related to the Propertius through its specific materials and through its experiment in giving epic dimension to a series of affective centers. It is a buoyantly varied work. The general movement is charged with dynamic contrasts and reversals of mood and atmosphere, and the very ground of feeling and preoccupation changes again and again while the largest curve of development is sustained. We start, in the first two poems, with a celebration of true lyric poetry. The celebration is soon revealed as a double response—to the poet's natural kinship with the purest streams of his art, and against the pressure to write "epics" to order. Gradually the sense of artistic crisis—an enforced need to make decisions about one's future direction—gives a harder edge to the verse. Increased emphasis is laid on love poetry as central in the lyric tradition, but there is also a sophisticated, urban overlay with overtones (to our ears) of Donne-like and Cavalier and "modern" complexity.

As if in response to these opening poems, the next two—without reference, however, to the intellectual issues and personal decisions that have been introduced—take us with a certain charming absurdity into the world of lovers' concerns and predicaments. In III the lover, a genuine but timorous romantic, has been summoned by his mistress to come to her at night and bravely resolves to do so in spite of the dangers that might beset him on the way. In IV he hears with disbelieving satisfaction a report of how she has missed him in his absence and how jealous she is of possible rivals. The mock-quandary of the first and the skeptical hopefulness of the second share a certain exaggeratedly worldly realism.

Then the fifth poem, from which we took our first quotation, brings together all the tonalities with which the sequence has so far engaged—a mixture of the logically disparate streams of high inspiration, colloquial musicality, withering polemic, bawdy imagination, and nearly tragic seriousness that have entered the work along the way. Thus, the concluding stanza of section two, followed by the very brief final section, of *Homage* V:

> And my ventricles do not palpitate to Caesarial *ore rotundos*,
> Nor to the tune of the Phrygian fathers.
> Sailor, of winds; a plowman, concerning his oxen;
> Soldier, the enumeration of wounds; the sheepfeeder, of ewes;
> We, in our narrow bed, turning aside from battles:
> Each man where he can, wearing out the day in his manner.
>
> 3
>
> It is noble to die of love, and honorable to remain uncuckolded
>   for a season.
> And she speaks ill of light women,
>   and will not praise Homer
> Because Helen's conduct is "unsuitable."

As opposed to "Caesarial" rhetoric, we have the restraint of Classical lyric represented in the catalogue of appropriate concerns culminating in the line "We in our narrow bed, turning aside from battles." The contrast makes for irony, but also introduces the power of understatement to suggest profound and gravely courageous insights. The line yokes lovers' pleasure with epic significance by making the pleasure an antiheroic *choice;* and the word "narrow" even suggests that the choice is a kind of self-discipline, an acceptance of one's destiny—as the quiet line that follows does also. Both lines, with their remarkably echoing sounds making them perfectly reciprocal, are close in tone to epitaph despite their sensual grounding and the implied sexual jest in "wearing out the day in his manner." But the poem is not permitted to end sententiously, however engaging its grave wisdom. After the section break, Propertius' beloved Cynthia is scathingly shown expressing her idiotic moral views on the *Iliad.* Beauty's ignorant ear is not allowed to go scot-free; truly, one can surrender everything to passion except one's brains—another new note, introduced into the poem at the last possible moment.

Homage VI and VII then shift to the most intense inwardness of a sustained kind that we find in the sequence—the first following a pure stream of heavily elegiac tonality from the first line ("When, when, and whenever death closes our eyelids"), the second a parallel stream of sweetly erotic joy ("Me happy, night, night full of brightness"). These pivotal poems of the sequence root the whole work in a non-rational, counter-moralistic realm of awareness and feeling, in which what counts is keen engagement with the volatile subjective life. Our models are gods and heroes, not as conquerors in battle but as blazing embodiments of the passionate nature of men and women: " 'Paris took Helen naked coming from the bed of Menelaus, / Endymion's naked body, bright bait for Diana.' " After VI and VII a certain distancing occurs, in two poems of mostly mocking concern over Cynthia's supposedly mortal illness and then in the final three poems, which play bemusedly over the despairs, delights, and misunderstandings of the love relationship. In the final poem, XII, the mistress has betrayed her lover with his best friend. His reaction is to clown rather than suffer—and yet, as we have seen, the sequence closes with its proclamation of devotion to the art of the love-elegy and its sacred theme: "And now Propertius of Cynthia, taking his stand among these."

We have only touched on the virtuosity and significance of this pioneer modern sequence. Pound's "collaboration" with an ancient author enabled an interesting freedom of structural experiment. It bears a certain (however limited) resemblance to Joyce's more sophisticatedly removed partnership with Homer in *Ulysses,* chapters of which began appearing in *The Little Review,* at Pound's instigation, in March 1918; and to Pound's own idiosyncratic repossession of the *Odyssey* in the early cantos, stirring into life at about the same time. The dynamics of the *Homage* are far less original and daring than those of *Ulysses* and the *Cantos,* in good part at least because Pound stays far closer to his original model, but the resonances are extraordinary nevertheless. The work is self-evidently the prime forerunner of *Mauberley,* in which we find a comparable rhetorical context, implicit advocacy of lyric as the clue to modern "epic" structure, and motif (neither jocular nor satirical, however) of erotic failure.

3. *Hugh Selwyn Mauberley*

Mauberley, in its turn, was the first of the original sequences of the 1920s that changed modern poetry. As we have noted, Yeats's double civil-war sequence and Eliot's *The Waste Land* were not far behind. All have their powerful affinities, although Eliot's work is more than half again as long as either of the other two. (It takes about twenty-five minutes to read aloud, as compared with about fifteen for *Mauberley* and sixteen for the double sequence.) Pound's and Yeats's sequences are interestingly similar in general structure, each consisting of two separate groupings of poems written in a number of different verse forms. In each, too, one of the groupings has a more inward cast than the other and was written later—a subjective recasting of the more public preoccupations revealed in the first grouping. These works are early models of the multiple sequence, made of smaller units that are sequences in themselves. Works like *Paterson* and the *Cantos* were still to come.

In addition, Pound's and Yeats's preoccupations were sometimes very close. Although *Mauberley* cannot match Yeats's power and scope, and the vantage points differ radically, two of its poems are among the best short war poems ever written: "These fought in any case" and "There died a myriad." Other splendid passages abound, despite an allusiveness and somewhat curdled ironic literariness that has its attractions (appointing the reader to membership in an avant-garde elect, on condition of acceptance of a certain tradition as " 'the sublime' " and the one true basis of modern experiment) but is almost parochial in the character of its engagement. Given these limitations, *Mauberley* brilliantly makes use of literary forebears—in "Envoi," for instance—and achieves a striking, emotionally convincing balance between its aggressively embattled first movement and introspective, self-doubting, self-reducing second movement. It is, in the character of its energy and perspectives and "epic" privacy, exactly what Pound described it as being: an Anglo-American, twentieth-century counterpart to the *Homage*. There is a curious sense in which *Mauberley* is more a vital presence in our poetry than a great exemplar of it. For one thing, it demonstrates that the formal development of any sequence is a matter of equilibrium rather than a straight line of development. And for another, it reaches ardently toward contact with the significant past, yet plunges into the world of the present moment and of the self-contradictory subjective life.

These are points of great similarity to Yeats's work. But one important emphasis in *Mauberley* is dissimilar. Its modulation toward a desexualized vision (contrast the horrified preoccupation with female sexuality in *The Waste Land*) sets it apart from Yeats, although both poets vibrate to pre-Raphaelite music in the way they aestheticize womanly beauty. Thus, Yeats's vision of unicorn-borne ladies, their "hearts . . . full of their own sweetness, bodies of their loveliness," parallels and may even echo Pound's vision of his lady in "Envoi":

Tell her that sheds
Such treasure in the air,
Recking naught else but that her graces give
Life to the moment,
I would bid them live
As roses might, in magic amber laid,

Red overwrought with orange and all made
One substance and one colour
Braving time.

In these poems emphasis falls on the artist's vision of beauty—in Yeats's "Meditations in Time of Civil War" on "abstract joy" and the "half-read wisdom of daemonic images"; in Pound's "Envoi," on the poet's desire to immortalize his lady's graces rather than on sharing the "life" she gives to the moment. There is a further parallel between the poet's condition in the fifth part of "Meditations" ("caught / In the cold snows of a dream") and the poet's "anaesthesis"—leading to "maudlin confession"—in the second part of *Mauberley.* But the civil-war sequence does not concern itself with the private problem of displacement of the sexual life by the life of art (which, in turn, is seen as a possible inability to make great art as well). The extreme self-consciousness of poem II in the second part of *Mauberley,* and its unusual tone of personal regret and bitterness, belie the surface preciosity of language. The closing stanzas of this poem have an intensity that mark it as the emotional epicenter of the sequence:

He had passed, inconscient, full gaze,
The wide-banded irides
And botticellian sprays implied
In their diastasis;

Which anaesthesis, noted a year late,
And weighed, revealed his great affect,
(Orchid), mandate
Of Eros, a retrospect.

. . .

Mouths biting empty air,
The still stone dogs,
Caught in metamorphosis, were
Left him as epilogues.

The depth of feeling here has gone more or less unnoticed. Despite what we have called *Mauberley's* desexualized vision, the muted suggestion of sexual malaise prepares the way for kindred but stronger notes in *The Waste Land.* Pound's momentarily mystifying diction is one reason for critical inattention to this bitterly confessional passage. Another is that, as he was later to do in the *Pisan Cantos,* he has kept the expression of direct, intimate personal feeling to a minimum partly by employing "he" instead of "I"—especially useful for hiding the obvious from readers who look for "story" or autobiography or ideas instead of the affective life of a poem's phrasing). The most important reason, however, must be that major attention has gone to the first part of *Mauberley*, the over-extended "Contacts" section (*Hugh Selwyn Mauberley: Contacts and Life*, in the actual order of the sequence), with its anecdotes and character sketches and gossipy vivacity. Sixty years ago, undoubtedly, the bulk of the poems in this section were deliciously shocking to would-be cognoscenti. At least one of them still is: "Mr. Nixon," presumably about Arnold Bennett. And some of them were especially touching in their concern for a still-remembered generation. The seventh poem, for instance, evokes honored figures gently and ruefully, as does the moving sixth poem, "Yeux Glauques"—beautiful despite its allusive embroidery. But with the loss of topical bite, and the lapsing of material charged with a generation's personal memories into data of literary history, the first part of *Mauberley* withers beside Yeats's civil-war sequences and Eliot's *The Waste Land*. The second, more inward "Life" movement—*Mauberley (1920)*—has worn better although nowhere does it match the two war poems of the first part and the "Envoi." Nevertheless, as we have suggested, the whole work remains a model of the sequence—its problems and pressures and involvements. In the remainder of our discussion, we shall stress two aspects: the disturbed sense of sexual dislocation that runs through the work in its muted fashion, and the matter of the *supposed* persona or personae of the two parts.

First, though, a quick note or two on the general structure.[12] Formally, the opening section ("Contacts") consists of three clusters of poems framed by "E. P. Ode pour l'élection de son sepulchre" at the start and "Envoi" at the end. The first cluster, II–V, presents a scornful indictment of a rampantly commercial society that quashes art, passion, and vision alike and has betrayed its proclaimed ideals in the recent World War. The sequence reaches its first peak in poems IV and V—the war-poems, angry and elegiac, that we have already mentioned. At a far less intense pitch, the middle cluster of titled but unnumbered poems provides four vignettes of varied tonality; the first two are the pre-Raphaelite and Nineties evocations we have noted, and the next two the sardonic character sketches "Mr. Nixon" and "Brennbaum" that point up the current state of arts and letters in London. The third cluster, poems X–XII, continues this examination of the age more wryly, focusing on the fate of the stylist ("unpaid, uncelebrated"), on the limited possibilities for genuine, passionate involvement between cultivated men and women ("No instinct has survived in her"), and (like Yeats) on the aristocracy's abdication of its responsibility to support the arts. The first

12. For a fuller discussion of the structure, see M. L. Rosenthal, *A Primer of Ezra Pound* (New York: Macmillan, 1960), especially pp. 29–41.

movement then ends with "Envoi," the splendid love song modeled mainly on Waller that affirms the lyricist's art in the spirit of the ending of the *Homage*, and also bravely reverses the ironically lacerated pretended defeatism of the initial "Ode."

Part II mirrors all this subjectively. In its five poems, of which only the third (" 'The Age Demanded' ") and the fifth ("Medallion") have titles, the language builds a complex image of the "obscure reveries / Of the inward gaze" that the second poem in Part I showed the modern world flatly rejecting. This image suggests not only detachment from the world's "welter" (Part I, poem X) but the shrinking from all vital human connection that amounts to "anaesthesis" and "*apathein*" (Part II, second and third poems). If in the first part of *Mauberley* the artist's failure is placed squarely at the door of a tawdry, bourgeois age, the second part gives us a corrective counterbalance with complexity aplenty. Picking up the undercurrent of sexual dislocation in the earlier poems, it now suggests the need for radical self-reorientation—away from the hyperaestheticism that seduces the poet from seeking an art of power. The suggestion is subtle and negatively couched, yet all along the way the poems imply that only life-engagement can make the artist's work something more than minor. (Their reality-oriented aestheticism recalls Joyce's in *A Portrait of the Artist as a Young Man.*)

To return to the opening "Ode" of the sequence: Its force lies in a mock-heroic treatment of a poet's struggles within, and retreat from, a thoroughly unsympathetic world of literati. At the same time, the essentially inward and non-violent nature of his struggle "to resuscitate the dead art of poetry" is captured in relatively passive, self-depreciative, or even domestic imagery: "wringing lilies from the acorn," playing the role of a "trout for factitious bait," being "held" by "chopped seas," fishing "by obstinate isles," and observing "the elegance of Circe's hair / Rather than the mottoes on sun-dials." Still—and this is a strong indication of the seriousness underlying the mocking surface—the equation of the poet's trials with Odysseus' divinely imposed ones suggests an ideal of aggressive heroism, reciprocal with the dynamically magnetic femaleness associated with the names of Circe, Penelope, and the Sirens. Their specific sexual allure is summoned up in the third poem, which deplores the suppression of erotic life—"phallic and ambrosial"—along with artistic and spiritual vitality in an age of "macerations." Already, however, a dichotomy has been set up between the response to female beauty in aesthetic terms ("the elegance of Circe's hair") and the more strenuous Odyssean response— even "Penelope" in the first poem is not an ardent mate but a literary style! On the other hand the Dionysian "phallic and ambrosial" possibilities inherent in an earlier age, when "faun's flesh" went right along with "saint's vision," breathe the total engagement with existence of Homeric art.

After the intervening war poems, "Yeux Glauques" attempts to locate a thin surviving strand of sexual vitality in late nineteenth-century art, ambiguously and furtively linked with the use of female models and the theme of the prostitute (as in Rossetti's poem "Jenny"). But even that survival was damned by the Buchanans and aestheticized almost out of sight by an etiolated art. Sex itself is a convenience having nothing to do with passion; in the seventh poem we are told that "Dowson found harlots cheaper than hotels," in the tenth the stylist's mistress is "placid and uneducated," and in the delightfully malicious eleventh poem all sexual energy is seen as having been bred out of the upper classes:

> No, "Milésian" is an exaggeration.
> No instinct has survived in her
> Older than those her grandmother
> Told her would fit her station.

Such negative aspects of the Nineties persisted into our century, at least through the Great War. (The persistence does not of course hold for our world six decades later—one important reason *Mauberley* does not altogether connect with modern readers unused to adjusting time-binoculars. Yet its underlying struggle is not against sexual inhibition but against the self-betrayal of hyper-aestheticism.) In the twelfth poem the sardonically named Lady Valentine might be capable of being roused to an equally misnamed "passion," but only by someone of proper social status and wealth—or, as a last resort "in the case of revolution," by a radical poet held in reserve for emergency use. And even "Envoi" is a highly proper, hardly amorous love song within the convention of a poet's chaste celebration of an idealized mistress.

All this is not a matter of the poem's avoiding allusion to sex, or the use of "virile" language. "Don't kick against the pricks," the young poet is advised in an obvious double-entendre. And the opening poem of the second section has even more luminously direct punning, more luminous because the self-depreciative sexual emphasis so undercuts the posturing of the opening "Ode" it echoes and parodies:

> "His true Penelope
> Was Flaubert,"
> And his tool
> The engraver's.

This self-vision, not exactly as a castrate but rather as one whose sex is, as it were, misdirected, is in a sense confirmed in the long prose epigraph to the same poem. The epigraph stresses the extreme delicacy rather than the more elemental drive of love. So, of course, does the poem itself, with its dreamy yet stinging imagery of a drifting, ambrosia-intoxicated failure to connect with earthy human passion. We do not wish to oversimplify. The unresolved complexities of a genuine though tangential psychic self-examination do not reduce to a pair of contrasting attitudes; rather, they remain complexities centered around rival emphases that, however, sometimes merge in the general flow of emotional exploration. Thus, the urge to aestheticize female beauty rather than respond passionately to a woman herself is depicted *splendidly* despite the implicit self-criticism. It enters the next poem as well, in the porcelain imagery that continues the roll call of the various metaphors in which artists have fixed female beauty; and "Medallion," which closes the entire sequence, gives great stress, and with exquisite precision, to this same porcelain imagery. The intervening fourth poem, just as exquisitely, limns the demise of a drifting hedonist of the imagination (another split self-view); but it is "Medallion" that provides the precise objective correlative for the distancing and aestheticizing urge at the center of the sequence's enamored struggle against the perfections of lesser genius:

> Luini in porcelain!
> The grand piano
> Utters a profane
> Protest with her clear soprano.
>
> The sleek head emerges
> From the gold-yellow frock
> As Anadyomene in the opening
> Pages of Reinach.
>
> Honey-red, closing the face-oval,
> A basket-work of braids which seem as if they were
> Spun in King Minos' hall
> From metal, or intractable amber;
>
> The face-oval beneath the glaze,
> Bright in its suave bounding-line, as,
> Beneath half-watt rays,
> The eyes turn topaz.

To resume: in "Ode" the observation of "Circe's hair" is a commendable occupation, as is the conversion, in "Envoi," of the living lady's graces to roses "in magic amber laid, / Red overwrought with orange."

In "Mauberley (1920)" the second poem presents the "fundamental passion" as artistic, not sexual: the urge "to convey the relation / Of eye-lid and cheek-bone / By verbal manifestation" or "To present the series / Of curious heads in medallion." In the third poem there is no essential difference in emotional impact between a "glow of porcelain" and a woman, since she is perceived in terms of color "Tempered as if / It were through a perfect glaze." The poem has prepared us more than adequately for the conversion, in "Medallion," of the living singer (perhaps performing "that song of Lawes"?) to a porcelain image. Pound's use of Salomon Reinach's *Apollo* may well extend to endorsing some of his opinions on specific painters, in which case the allusion to Luini's portraiture porcelainized represents a double withdrawal from robustly passionate concerns. In this context, the "grand" piano would indeed be profanely overwhelming.

Through its imagery and echoes of phrasing, "Medallion" encapsulates the asexual tonal twist sounded in many of the preceding poems. "Face-oval" and "suave bounding-line," for instance, suggest a purely aesthetic frame. Among the many echoes, "clear" recalls the "thin clear gaze" of "Yeux Glauques," and the "sleek head" contrasts with the "faun's head" of the same poem. "Gold-yellow," "honey-red," "amber," and "topaz" hark back especially to "Envoi" and the third poem of the second part. The portraiture naturally reminds us of "Yeux Glauques" and of "Brennbaum." The "basket-work of braids," besides adding a new craft, summons up Circe's elegant hair in "Ode," and the reference to King Minos echoes "Minoan undulation" two poems earlier. Also, the reference to metal takes us back to previous medallions and the engraver's tool, and the phrase "half-watt rays" in the last stanza is a particularly telling way of emphasizing the disparity between this almost twilight existence and brilliant Dionysiac fulfillment. Note too the *glaze-rays* rhyme, which echoes *gaze-glaze* in the third poem of the second part and *gaze-days* in "Yeux Glauques," formally reinforcing the poems' shared preoccupation with diminished potency.

In a sense the movement of *Mauberley* was prefigured in that of "The Return." The recovery of a powerful vision, akin to the literal sense of supernatural presence that was part of life in earlier societies, in the midst of a visionless society and against all the odds, is the painful, heroic aim projected in both works. *Mauberley* as a whole creates an image of existence stunted both by the age and by artistic tendencies that have lost contact with life and with vital tradition. The personal dimension, especially in II, ii, underwrites the sense of deep betrayal implicit in this image. The great effort at corrective reorientation comes in the *Cantos*.

Chapter Nine:

Ezra Pound II: The *Cantos*

1. A Representative Passage

Yeats's civil-war sequences and *The Waste Land*—and to a lesser extent *Hugh Selwyn Mauberley*—share the honors for achievement among shorter sequences; the *Cantos* and *Paterson* among the longer. Looking ahead for a moment, we might point out that, whereas Williams followed Pound's lead in writing a sequence of sequences, *Paterson* has far less literary and historical scope than the *Cantos*, maintains a more relaxed and conversational tone, and is also very much shorter. The *Cantos* as a whole runs some 800 pages in the complete 1972 edition, and one wishes Pound had sacrificed some of the pages he evidently thought necessary for didactic purposes. Yet it is always tricky to decide just which pages—much the same problem as deciding which portion of a Whitman catalogue could be left out. In any case, we can say with reasonable certainty that *The Pisan Cantos LXXIV–LXXXIV* (1948) is the outstanding group in the volume and that the opening group, *A Draft of XXX Cantos,* almost matches it in quality. The opening group comprises a double sequence, the first part of which appeared in 1925 as *A Draft of XVI. Cantos* and the second part in 1930 as the additional fourteen cantos included in *A Draft of XXX Cantos* (expanded from the 1928 volume, *A Draft of The Cantos 17–27*).

The other groupings are *Eleven New Cantos XXXI–XLI* (1934), *The Fifth Decad of Cantos XLII–LI* (1937), *Cantos LII–LXXI* (1940), *Section: Rock-Drill De Los Cantares LXXXV–XCV* (1955), and *Thrones de los Cantares XCVI–CIX* (1959). *Drafts and Fragments of Cantos CX–*

CXVII (1968) makes no pretense to unity. The tiny Canto 120 that ends the *Cantos*, originally part of Canto 115, acts as a poignant envoi and retraction and forces a momentary sense of closure on the volume. Here, for the last time, tones first emphasized in the *Pisan Cantos*—humble aspiration, remorse, fear of a self-imposed, loveless isolation—are sounded:

> I have tried to write Paradise
>
> Do not move
>   Let the wind speak
>   that is paradise.
>
> Let the Gods forgive what I
>   have made
> Let those I love try to forgive
>   what I have made.
>
> (803)[1]

This has something of the flavor of Chaucer's envoi and retraction. But Pound's need for forgiveness has a different source. Chaucer's was based on the supposed waste of his genius on allegedly frivolous and lecherous works: those poems in the *Canterbury Tales* that "sownen into synne," *Troylus and Criseyde*, and most of his writing except his translation of Boethius and "othere bookes of legendes of seintes, and omelies, and moralitee, and devocioun." Pound's prayer stems from appalled recognition of the ruthlessness in many of his cantos. For while the *Cantos* shows admirable impatience and disgust with everything that militates against an earthly paradise of a beautiful, ordered society centered on life-enhancing values, the book is marred by a virulent, vindictive hatred. When that hatred surfaces, it has room neither for pity for the weak nor for loving understanding of the essential complexity of any human being's character and actions. Fortunately the ruthlessness is only one stream among many; moreover, Pound sometimes handles it beautifully and, in fact, humanly. An instance is Artemis' powerful "compleynt" against Pity at the start of Canto 30, in which the archaic spelling of "complaint" and the personification of Pity immediately evoke Chaucer's very tender-minded "Compleynt unto Pity." Again, in Canto 80, Pound's sudden apprehension of ruthlessness ("I have been hard as youth sixty years") is a whiplash of self-reproach. It is the climax of a passage of key intensity in the *Pisan Cantos* that epitomizes his mature

1. Ezra Pound, *The Cantos* (New York: New Directions, 1972), p. 803.

method. It will serve our purposes well, we think, to focus on this passage rather intensively before further comment on the *Cantos* as a whole:

> care and craft in forming leagues and alliances
>   that avail nothing against the decree
> the folly of attacking that island
>   and of the force ὑπὲρ μόρον
>
> with a mind like that he is one of us
>   Favonus, vento benigno
>   Je suis au bout de mes forces/
>
> That from the gates of death,
>   that from the gates of death: Whitman or Lovelace
>   found on the jo-house seat at that
> in a cheap edition! [and thanks to Professor Speare]
> hast'ou swum in a sea of air strip
>   through an aeon of nothingness,
> when the raft broke and the waters went over me,
>
> Immaculata, Introibo
>   for those who drink of the bitterness
> Perpetua, Agatha, Anastasia
>   saeculorum
>
> repos donnez à cils
>   senza termine funge   Immaculata Regina
>   Les larmes que j'ai creées m'inondent
> Tard, très tard je t'ai connue, la Tristesse,
> I have been hard as youth sixty years
>
>   if calm be after tempest
> that the ants seem to wobble
>   as the morning sun catches their shadows
>   (Nadasky, Duett, McAllister . . .
>
> (512–13)

Indubitably such writing offers more surface difficulty than the other sequences we have been exploring. Nevertheless, the dynamics of individual passages such as the one just quarried from the twenty-four page Canto 80 can be grasped fairly quickly once the allusions and foreign expressions have been deciphered—a task rendered relatively painless by the substantial annotation the *Cantos* has received.[2] To demonstrate

2. See the *Annotated Index to the Cantos of Ezra Pound* by John H. Edwards, William W. Vasse, *et al.* (Berkeley and Los Angeles: University of Calif. Press, 1955), for Cantos 1–84; and *A Companion to the Cantos of Ezra Pound,* Vol. I, by Carroll F. Terrell (University of California Press, 1980). Volume I covers Cantos 1–71.

the limits to which Pound has managed to take lyrical structure without losing intensity, we have chosen this extremely compressed passage in which Pound's mixture of Greek, Latin, French, Old French, and Italian is even more polylingual than usual. As we mentioned in discussing *The Waste Land,* he manages still faster shifts of tone than Eliot, so that virtually every line can, when needed, convey a distinct tonality. (One does not need to "decipher" to see the shiftings signaled along the way in the quoted passage: the slightly detached resignation of "avail nothing" and "the folly" in the first strophe; the sense of intimate recognition in the second; the fused notes of danger, pleasure, absurdity, and desolation in the third; the calming incantation superimposed on "the bitterness" in the fourth—and the heightening of this affect in the fifth; and finally, in the sixth, the precariously regained feeling of self-control associated with minute yet infinitely reassuring observations.) The context provided by the *Cantos* as a whole prevents such rapid shifts from irreparably fragmenting the poetry and helps clarify and refine the resonances of each line or phrase.

In the very first line of our quotation, "care and craft in forming leagues and alliances," this contextual support is at work. The initial ambiguity of phrasing disappears when we look back to the preceding (unquoted) strophe, which includes a reference to the *Odyssey.* The poet, a prisoner in the Disciplinary Training Center near Pisa, has recalled the man's name who "first declaimed me the Odyssey" and is trying to remember just where this happened. In the *Pisan Cantos* the active pursuit of memories and minute observation of nature are holds against madness—resources for maintaining equilibrium under the pressures of disaster, imprisonment, and debilitating self-doubt. The opening line, then, evokes the wily Odysseus, but does so in the context of this desperate present need and recourse to recollection and to acute *noticing* ("the ants seem to wobble / as the morning sun catches their shadows"). The attention here is not on details of the *Odyssey* but on the predicament of displacement, vulnerability, and defeat, and on a "care and craft" like those of Odysseus but in the service of keeping one's sanity. In addition, the passage projects three circumstances in which an individual, even a Homeric hero, is helpless. No matter how careful one's preparation or brave one's execution, one cannot overcome fatality ("the decree"), hubris ("the folly of attacking that island"—presumably England), and unpredictable eventuality ("the force *ὑπὲρ μόρον*"—beyond what is destined). The strophe, in short, projects a struggle against impossible odds by one who is neither crude nor thoughtless. Further, the struggle has epic dimensions; we are in the realm of "leagues and alliances," of decrees, of combat, and of Homeric destiny—"the force *ὑπὲρ μόρον.*" This phrase occurs at various places in the *Odyssey;* but, given the tone of this passage and the reference to near-drowning ("and the waters went over me"), it most immediately evokes Odysseus' struggle in the breakers on Skheria: "There unlucky Odysseus would certainly have perished, beyond what was ordained by the fates, had not grey-eyed Athene given him sure counsel." (There is a wry comic turn in the fact that Pound is similarly helped by the poetry anthology he luckily finds "on the jo-house seat" in the DTC, and even—as we see elsewhere—by some fellow inmates.)

The next strophe adds a more personal dimension to heroic triple defeat, and the associative flow hereafter is exquisitely articulated—complex in detail but sure and direct in affect. The first line implicitly links the defeated sensibility to the DTC inhabitants—later to be described, Homerically, as "men of no fortune and with a name to come" (pp. 513, 514). The second line, a Latin sentence fragment, is at once wistful, desperate, nostalgic, and prayerful. "Favonus" is apparently a euphonic Poundian version of Favonius, the mild west wind. Pound may have chose "Favonus" particularly for the connotation of gentleness, which is intensified by the epithet "vento benigno," "with kindly breeze." More frequently Pound uses Zephyrus or Zephyr for the west wind, which is intimately associated with Aphrodite, both in the Sixth Homeric Hymn (which contributes phrases, in Latin translation, to the end of Canto 1) and in the *Cantos*. The most explicit linking of west wind and Aphrodite—or, to retain the Latin flavor, Venus—comes in Canto 74:

> each one in his god's name
> as by Terracina rose from the sea Zephyr behind her
> and from her manner of walking
> as had Anchises
>
> (435)

All these associations bring to bear tonal contexts that have emerged elsewhere in the *Pisan Cantos* and earlier. Here they are associations of gentleness, beauty, and happy desire in the midst of present suffering. Thus, the allusion to Anchises in Canto 74 inextricably links the goddess of love to the west wind, and the connection is reinforced some twenty or so lines further on by the line "a great goddess, Aeneas knew her forthwith." The line is an allusion to line 405 of the *Aeneid,* in which Aeneas recognizes his mother, Venus, by her manner of walking: "*et vēra incessū patuit dea.*" But as with the evocation of the *Odyssey* in the first line of the passage we have been considering in Canto 80 ("care and craft . . ."), so the expression "Favonus, vento benigno" five lines later directs us intimately to the DTC experience. That is, the allusion serves the poem's affective complex—not the reverse—with a more immediate and homely connotation of the "benign." A form of the word has made an appearance in Canto 74, just before the evocation of the goddess, when a fellow prisoner is half-humorously celebrated as a "jacent benignity" (p. 434). This is a Mr. Edwards "of the Baluba mask," a man of "superb green and brown" who has made the poet a packing-case table in secret ("doan you tell no one / I made you that table"). The act of benignity by one of "those who have not observed / regulations" leads the poet to perceive, with Paul, that "the greatest is charity." So all benignities, of whatever rank and color, are in essence being summoned—and responding—when the distraught prisoner confesses (we are back in Canto 80) that he is at the end of his tether: "Je suis au bout de mes forces/." And of course the details of Mr. Edwards's appearance and speech, like the observation of the ants' movement, at once sustain the poet in his travail and serve to humanize him personally.

In the third strophe the DTC milieu is explicitly reintroduced, for the first time since, seven pages earlier, we saw a "young nigger at rest in his wheelbarrow / in the shade back of the jo-house," and the closeup momentarily interrupted a long stream of reminiscence. That interruption fixed a moment in the empirical present in relation to the mind's timeless juxtapositions and prepared us as well for the later "jo-house" scene we have mentioned: the rather comic mutual deliverance from "the gates of death" effected by Pound's retrieval of *The Pocket Book of Verse.* In Pound's control, the poems of Whitman and Lovelace can be put to better use than in the privy. Across the centuries the "live tradition" (the longer human memory of poetry that connects with but encompasses personal memory) can be demonstrated once more as sudden illuminations from those poems enter later cantos: "at my gates no Althea" in Canto 81 (p. 519) and the splendid passage beginning "O troubled reflection / O throat, O throbbing heart" that introduces the climactic lines of Canto 82 (p. 526).

Again, we are in the timeless world of poetic tradition—and, at the same time, very concretely in the here and now of the DTC. In the next few lines the language heightens, matching the gravity of "gates of death" in a superb evocation of rescue from the depths:

> hast'ou swum in a sea of air strip
> through an aeon of nothingness,
> when the raft broke and the waters went over me . . .

Pound manages his poem's communion with the poetry of the past perfectly, impregnating his language with the auras of Jonson's "Celebration of Charis," the *Odyssey,* and Psalms, but without losing personal immediacy. Thus "hast'ou" is used here as reverent address, suggesting the coming of some goddess figuratively swimming through the wire mesh reinforcing the cage in which Pound was incarcerated—as Ino/Leukothea succored Odysseus in Book Five. "Hast'ou" was linked to Jonson's poem at the end of Canto 74: "Hast'ou seen the rose in the steel dust / (or swansdown ever?)." It bears with it the great lyric tradition of celebration of female beauty: "O so white, O so soft, O so sweet is she!" The archaic form suggests a religious context as well—pagan as well as Christian, given the echo of the remarkable Canto 47 which is shot through with implications of fertility rites:

> Hast thou found a nest softer than cunnus
> Or hast thou found better rest
> Hast'ou a deeper planting, doth thy death year
> Bring swifter shoot?
> Hast thou entered more deeply the mountain?
>
> (238)

The "aeon of nothingness" suggests the extremity of the speaker's isolation in time and space. There is an added resonance as well, from *Odyssey* 5, in which Hermes, speeding at Zeus' decree to end Odysseus' exile on the island of Ogygia, must cross "that tract of desolation . . . the bitter sea." And of course the Odyssean comparison is brightly clear in the images of raft and overwhelming sea. Twice in Book 5 the waters go over Odysseus. The first time, the nereid Leukothea saves him after Poseidon has destroyed his raft; the second time, as we have mentioned, it is Athene who holds him from the gates of death.

The swirl of associations in these few lines, each a context for the others and part of the affective complex within which the daily agony and self-orientations of the DTC experience occur, makes it impossible to limit their frame of reference to one dominant literary or intellectual tradition. Thus, the fourth strophe brings yet another context, this time Christian, into the passage, linking Pound's predicament obliquely but inescapably with the Crucifixion (as happened also in Canto 74 at the start of the *Pisan Cantos*) and his vision with that of Christian tradition. We have been somewhat prepared for this shift by the phrase "and the waters went over me," which in imagery and cadence suggests the King James Bible, especially the Psalms, quite as much as the *Odyssey*. But now the Odyssean analogue is dropped for a while and we enter the realm of Christian liturgy: "Immaculata, Introibo"—Immaculate, I shall enter. (The words "*Introibo ad altare dei*"—I shall enter unto the altar of God—begin the Preparation of the Roman Mass.) The feminine form "Immaculata" is appropriate for the female saints Perpetua, Agatha, and Anastasia whose names—meaning "eternal," "good," and "reborn"—help project the sense of a sphere of holiness that they inhabit and the poet would somehow share. Only through sacrifice and pain, apparently, can one enter this eternal realm ("saeculorum")—an obvious affective link between the poet and those pure, femininely unaggressive early martyrs whom we would otherwise think of as his exact opposites. They too have had to "drink of the bitterness" as Christ once did. The association with Christ is adumbrated earlier in the canto (p. 500), in a grotesque passage in Italian that pictures a child nailed to a church floor in the form of a cross. Whether this is sheer fantasy or a descriptive note concerning a detail of church sculpture or painting or a surrealistic projection of the poet's sense of his own victimization—an innocent somehow punished for his immersion in art and imagination—is difficult to judge, especially as this section of the canto is bathed in moon and death imagery. Pagan and Christian imagery mingle in an atmosphere of doom and weeping, and in the same context the word "immacolata" (Italian for "immaculate") is used for the "moon nymph."

A further reach of connotation for the word "immaculate" occurs in the fifth strophe of the passage we have been considering. There the epithet "Immaculata Regina" draws the holiest of intercessors, Mary Queen of Heaven, into the poem's orbit. And we should mention another, earlier use of "Immaculata," in a Confucian image that appears in Canto 74: "Light tensile immaculata / the sun's cord unspotted" (p. 429). These lines echo the ideogram *ming*, at the top of the same page, which signifies radiation or light or clear distinctions. The *Pisan Cantos* mobilizes various symbolic systems and their characteristic imagery and emotionally charged key-terms ("immaculate") against the disintegrating momentum of personal disorientation and loss of power. It superimposes one affective complex upon the other in this effort. Thus, the phrase "senza termine funge" seems to refer to the timeless existence and energy of heaven but is also used in Pound's translation of Confucius' *Unwobbling Pivot*. While in the DTC, he was writing both the *Pisan Cantos* and an English version of his Italian translation of Confucius. The relevant sentences in the 1944 Italian text read: "*La purezza funge (nel tempo e nel spazio) senza termine. Senza termine funge, luce tensile.*" The Pisan English version of this is: "The *unmixed* functions (in time and space) without bourne. This unmixed is the tensile light, the Immaculata. There is no end to its action." "Immaculata" has been added, as "Immacolata" will be to the 1955 Italian edition.[3]

"These fragments I have shored against my ruins"—what is remarkable here is not our extended commentary but the way that so many

3. See Massimo Bacigalupo, *The Forméd Trace: The Later Poetry of Ezra Pound* (New York: Columbia University Press, 1980), pp. 110–11.

elements are compressed, in all their vibrancy of interrelated feeling and awareness, in the thirty-line passage we have chosen: one out of many we might have fixed on to illustrate the way everything counts lyrically in the canto. Meanwhile we are compelled by the unfolding of the strophes to be more and more responsive to the stream (sometimes almost underground) of *personal* emotion and self-mirroring that gives a special continuity to these lines. It is not only the hopeless tone of the first strophe, or the outcry of anguish in the second ("Je suis au bout de mes forces/"—with the slant-line at the end actually *picturing* the state of being totally thwarted), or the anecdotal jocularity combined with tragic overtones of the third, or the bitter language of the fourth and fifth, or the pathos of the opening lines of the sixth, or all of these working together that make for this cumulative affective movement of the personal. By this point in the *Cantos* the very elements of technique and method are saturated with personal tonality: the volatile yet clear intelligence at work in the affective juxtapositions, the diving into phrases in many languages to evoke whole sets of tonality and association, the rhythmic line placements that choreograph the shiftings, the sudden stark stabs of forthrightness in English ("and the waters went over me," "those who drink of the bitterness," "I have been hard as youth sixty years"), and the sheer multiplicity and subtlety of association and sound-echoing. Pound uses his multilingual resources with tremendous effect, and the outbursts of personal directness in English gain a unique power in relation to them.

Two instances of this last aspect: First, the opening prayer in the fourth strophe not only draws on Villon ("*repos eternel donne à cils,*" which becomes "repos donnez à cils"—"grant them rest") but echoes the whole Western elegiac tradition and every requiem mass ("*dona eis requiem*"). It is a luminous node unto itself, but, in response to the preceding strophe, also refers to all "those who drink of the bitterness." Our second instance is the way that the straightforward confession closing the second strophe ("je suis au bout de mes forces") is followed up. Matching the tone of this line, the remorse of the next lines in French prepares us for the harsh English line, the result of having had to "drink of the bitterness," that follows:

> Les larmes que j'ai creées m'inondent
> Tard, très tard je t'ai connue, la Tristesse,
> I have been hard as youth sixty years

Part of the process of making up for all the tears forced from others ("The tears I have created are drowning me / Late, very late I've come to know you, Remorse") is embodied in the sixth strophe, of which we have quoted the opening lines only. With a benignity reciprocal in its way with that of Mr. Edwards, the builder of his packing-box table, the poet ensures "a name to come" for all the "men of no fortune" who are prisoners with him in the DTC. The elementary (if Homeric) kindness of this gesture is akin to Pound's new, self-healing emphasis on minutiae of nature when he focuses on the denizens of the "green world" about him: wasps, birds, lizards, midges, and here the ants that "seem to wobble" but, significantly, are depicted as shakily making their way in the "calm" after "tempest"—tiny Confuciana, as it were, existing in relation to the "unwobbling" ultimate balance of things. This image in the opening three lines brilliantly links tiny, vulnerable social creatures with the emotional human tempest of being overwhelmed by tears—those of others and one's own—and with the tempests that beset Odysseus as the carrier *par excellence* of man's epic burden.—And then comes an affectionate and grateful catalogue of trainees ("Homeric" again) after which the poem continues on its varied, allusive, and yet passionate and immediate way.

We have discussed these thirty lines as a deep and sure sounding of the way the *Cantos* works, since manifestly such a treatment of the whole body of cantos would require volumes. The method is the mature culmination of the process embodied, very simply but also very indicatively, in an early poem like "The Return."

2. The *Pisan Cantos* in Retrospect

At the beginning of this chapter we proposed the importance of reading the *Cantos* tonal center by tonal center, and it should be obvious from the preceding discussion how complex and quickly shifting the individual affects may be. The demands made on a reader's memory, intellectual alertness and curiosity, and imaginative responsiveness can be considerable. Yet many passages, and even a number of whole cantos, can be approached with a more relaxed attentiveness. Canto 1, with its partial translation of *Odyssey* XI; Canto 13, the early Confucian poem beginning "Kung walked"; and Canto 47, more challenging than the other two but with a pure curve of lyrical movement—these are among the cantos that demonstrate Pound's essential ear and the constructive powers that guided him in the making of the several volumes of cantos. Each of these three poems has its special quality and formal movement, readily available to a literate reader even when, as in Canto 47, the overall conception may be somewhat elusive. Other cantos work more by accretion, detail piled upon detail in passionate catalogues that sometimes become objective correlatives for the hatred or disgust or mixture of admiration and impatience with which Pound infuses his more didactic poems (on such matters as usury, political corruption, good government, and ethical systems). In general, it is necessary to take each volume of cantos seriously, as a separate sequence with its own aesthetic that yields to the kind of reading we have been suggesting, and to realize that the *Cantos* as a whole have many interconnections without themselves making a self-contained sequence. Although Pound's original conception in a sense allowed for adjustments (and he became a master at improvising connections and reaching out for strands of relatedness), life and history carried him far beyond what "care and craft" could provide for. "The decree" arranged things differently and made for a destiny beyond what one might have expected originally: ὑπὲρ μόρον indeed.

We shall not here go into the basic positionings of the separate groups of cantos, and of the basic axes of reference in their structure. A number of discussions are available on this subject,[4] and we have purposely refrained from stressing any schematically intellectual approach when the truly pressing—and revealing—issue lies with the *Cantos* as poetic art. It seems preferable, now, to turn for a few pages to questions of evaluation of the *Cantos* as poems and as groups of sequences.

First of all, then—to begin with a self-evident point—the individual cantos are highly variable, and so are the clusters that make up the volumes. But if someone should ask whether any one work, by its intrinsic formation, compels recognition as the classic American poem of mid-century and after, the answer must almost inevitably be: "the *Pisan Cantos.*"[5] Perhaps it should be "the *Pisan Cantos, hélas!*"—one is after all not required to accept the leading ideas of a great work: Milton's theology, say. One needs, rather, to recognize the kind of energies it generates. And indeed we shall want to comment, a little further on, on the way the intractable elements in Pound's art—the sturdy, proclaiming insistence, consciously perverse, on the idealism and soundness of his Fascist commitment, and the copious welter of materials assimilated into his sequence—have contributed centrally to the work's triumph. His experience with the earlier groups of cantos had prepared him for

4. See, among other works, Hugh Kenner, *The Poetry of Ezra Pound* (Norfolk, Conn.: New Directions, 1951) and *The Pound Era* (Berkeley and Los Angeles: University of California Press, 1971); Donald Davie, *Ezra Pound: Poet as Sculptor* (New York: Oxford University Press, 1964); Eugene Paul Nassar, *The Cantos of Ezra Pound* (Baltimore: The Johns Hopkins University Press, 1975); Daniel D. Pearlman, *The Barb of Time* (New York: Oxford University Press, 1969); and M. L. Rosenthal, *Sailing into the Unknown: Yeats, Pound, and Eliot* (New York: Oxford University Press, 1978), which also has an extended discussion of Canto 47, on pp. 12–25.

5. Williams's *Paterson* is of course a close second, though lacking the compression and driving central pressure of the *Pisan Cantos*.

this greater success by teaching him to handle large masses of associative material, partly by deploying them as interacting streams of tonality and partly by a varied, uneven rhythm of opening out and then closing in. He had learned how to let the flow of association take over almost beyond control and how to channel its accumulated power into moments of extraordinary compression and focus like the one that ends Canto 74 after an incredibly swelling flood of memory and bits of related information:

> Serenely in the crystal jet
>   as the bright ball that the fountain tosses
> (Verlaine) as diamond clearness
>   How soft the wind under Taishan
>   where the sea is remembered
>   out of hell, the pit
>   out of the dust and glare evil
>   Zephyrus / Apeliota
> This liquid is certainly a
>   property of the mind
> nec accidens est   but an element
>   in the mind's make-up
> est agens and functions   dust to a fountain pan otherwise
>   Hast 'ou seen the rose in the steel dust
>   (or swansdown ever?)
> so light is the urging, so ordered the dark petals of iron
> we who have passed over Lethe.
>
> (449)

This passage, so strategically located at the end of the very long opening canto of the *Pisan* group, should remind us forcibly that we are not dealing with Pound's ideas generally or with some dehydrated construct but with a work of high poetic authority. The lines bring into pure, concentrated view the entire set of Pound's sequence. One cannot praise them enough. But even if one found fault with them, one could hardly ignore their music, their surge of controlled power, and their shifting depths. Canto 74 as a whole is an enormous gathering of elements of present experience and predicament, self-therapeutic observation and memory, and emotional staving-off of self-abasement by defensive counter-affirmations. It wells up, as it were, out of the depths of personal and historic disintegration and humiliation: a summoning up of every context available to the commanding sensibility that launches the sequence into its complexly sustained orbit. The concluding passage, just quoted, is a rich symbolic correlative of the canto's process. All the elements of the poem make up the "bright ball" that the "fountain" (the "liquid" that is "certainly a property of the mind"—the fluid ambience of one's awareness and feeling) "tosses" so "serenely in the crystal jet." Aesthetic conversion of all the external and internal turbulence creates the "serenity" of art's passionate distancing; this is both the suggested aim of the poem and its actual result if successful. The imagination at work imposes the vision of the sacred Oriental mountain Taishan on the hills near Pisa, while "How soft the wind" appreciates a momentary sensuous delight, physically real or evoked by a thought, and at the same time prepares the way for the wide mythical context of a long-ago world in which winds were named as supernatural entities: "Zephyrus / Apeliota." Then comes the fierce Odyssean and Dantean language of high adventure and torment of lines 5–7 (just before the naming of the winds), followed by the sweetly speculative "serene" placement of the poem's whole process in lines 9–12. The exquisite echo of Jonson's "Celebration of Charis" in the next two lines both confirms the lyrical intensity within which the passage moves and emphasizes the mind's deep formation of organic patterns through sheer sensitivity to the innate reciprocity of reality and insight. The closing lines reveal the intrinsic ordering and fatality of this process and—at the very last moment—suggest that a tragically and chthonically informed spirit is returning from Erebus to the world of the living.

Looking back over the passage again, we may note a great deal more about its artistry. Each line, for instance—beginning with the very first words: "Serenely in the crystal jet"—emphasizes two musically balanced phrasings. Again, notice how the passage picks up Verlaine's lovely image of the mind's delight ("bright ball that the fountain tosses") and deepens it with tones of nostalgic memory, loss, pain, and death-knowledge, and of intense meditation holding disintegration at bay. The movement is not intellectual. It is a carrying forward, an accumulation borne by echoing and gathering sound-patterns despite the variations of line-length and meter. We move rapidly from one emotional tonality to another: from "Serenely in the crystal jet" to "How soft the wind under Taishan" to "where the sea is remembered" to "Hast 'ou seen the rose in the steel dust" to "we who have passed over Lethe." That is, we move from the sweetly restless calm of dreaming contemplation to the darkened sense of a tragic voice that has returned from the lands of the dead. At the end, souls returning from those lands, solemnly innocent, are impelled by a necessity of whose nature they are unaware. The aesthetic process itself has become a mode of fatality.

The *Pisan Cantos* has not, obviously, changed since first appearing in 1948. But now, perhaps, we can see better how brilliantly Pound used the writing of it to control the turbulence within himself and in the events all around him following Mussolini's defeat and his own arrest. The changes in perspective lie in the fading away of a special set of political circumstances and terms of debate. Pound's offensive politics, and his supposed insanity (suspect because it saved him from being charged with treason), no longer stand between us and the poem in the old way. Pound is dead, we have been through Korea and Vietnam, Churchill and Roosevelt and Stalin are in the dust with Ezra, and the recent and current international brews of OPEC monopoly, hostages, refugee misrulers, and military and terrorist politics could be poured right into the *Cantos* without more than a momentary quiver in the structure of reciprocities already there. The dreaming mind, the suffering yet courageous and humorous self, and the chaotic welter of circumstance within the Pisan sequence provide a certain mirroring reciprocity with our daily circumstances; and this fact has a good deal to do with the emergence of the work as our volatile classic, the type of what has become of the epic tradition in our day. The literal Ezra Pound, who stood in front of his poem blocking the view when it first appeared, has been transformed into the archetypal figure, wise fool and victim-hero, who lives in scattered moments throughout its pages. We meet him at the very start brooding nobly over the history of Europe ("The enormous tragedy of the dream in the peasant's bent shoulders"), and at the very end speaking with stoical pathos:

> If the hoar frost grip thy tent
> Thou wilt give thanks when night is spent.
>
> (540)

The combination of nobility, staunchness, and pathos, whatever other feelings move through the sequence as well, defines a governing tonal state, projected most movingly, perhaps, in a few lines of Canto 76:

> πολλά παθεῖν,
> nothing matters but the quality
> of the affection—
> in the end—that has carved the trace in the mind
> dove sta memoria
>
> (457)

The man in the poem does not define its entire range of sensibility in a direct way. Voices of other people, including other poets, are heard; and the play of affects creates many centers within which the illusion of a moving sensibility roots itself temporarily. The comic tone at the beginning of Canto 77, for instance, introduces the hearty perspectives of a certain tradition of satirical American dialect humor:

> And this day Abner lifted a shovel. . . . .
> instead of watchin' it to see if it would
> take action
>
> (464)

When, further on in Cantos 80 and 83, Pound playfully teases and mimics Yeats, the gaiety of spirit he thus introduces also brings with it attention to the brutalities art must cope with ("Your gunmen thread on moi drreams")—a connection similar to that suggested by the delight he takes in the speech of uneducated fellow prisoners in the DTC, speech that sometimes is merely randily exuberant but at other times touches very serious chords and merges with tonalities presented in more exalted or sophisticated language. The lyrical structuring of the whole work is intricately reciprocal. Pound seems our only modern master who, like Milton, has been able to hold a complex of sound-resonances and a range of emotional levels and degrees of intensity in mind and under control over a very extended composition. While the *Pisan Cantos* has epic scale in the sweep of its consciousness and memory and in its engagement with concerns drastically affecting the destiny of millions, its poetic power derives from the lyrical conversion of these elements: the passionate realization of murderous and impersonal force working through human vehicles, and the interaction of this realization with the other elements of the composition.

The receding of the static of the late 1940s, then, together with growing familiarity with the *Cantos*, enables most of us to see the significant artistry of the Pisan sequence more cleanly than before. We are able to sort out the independent life of its phrasing without so much distraction now, and to recognize its essential engagement with three major internal pressures working on it profoundly. The first of these is the effort to sustain morale in the face of defeat, privation, and breakdown. This effort, the key to the work's innermost structure, employs every possible means. It summons up memories that confirm the self's continuity with a rich past. It superimposes imagined scenes on the bleak literal scene of present humiliation. It fixes on tiny observed details as a means of holding the sensibility steady. It gains distance through wry humor. It falls back on a familiar rhetoric and invective on the one hand, and withdraws into quietistic acceptance on the other:

> and when bad government prevailed, like an arrow,
> fog rose from the marshland
> bringing claustrophobia of the mist

beyond the stockade there is chaos and nothingness
Ade du Piccadilly
Ade du Lesterplatz
Their works like cobwebs when the spider is gone
encrust them with sun-shot crystals
(501)

Thus one moment of self-sustaining effort in Canto 80. In addition to this primary pressure in the *Pisan Cantos,* the sequence receives a second in its recurrent visions of a humane and beautiful ordering of life. And the third pressure comes from the sense of shock—the impact of the downfall of a European world whose wreckage still burns with its furies of torment, its Miltonic conflicts, its betrayals and confusions. These pressures, and the streams of tonality arising from them, define the work's organic body.

What about the enormous, grotesque torque of the Fascist rhetoric? Its passion contributes violently to the poem's life: sometimes as viciously slashing as a street mob amok; sometimes conveyed as high tragic eloquence—or as an aspect of entranced idealistic vision: "To build the city of Dioce." Probably you cannot do the modern epic without political passion that *counts,* as Milton's counted in *Paradise Lost.* Cromwell's Latin secretary would have found it hard to sort out the political from the religious pressures in his work, and the Irish after the sack of Drogheda and the Act of Settlement must have found it even harder to do so. Pound's position is similar: great commitment and imagination, zeal for very pure moral and social conceptions—all horribly bound to injustice and bloody violence, and beyond them to the impersonal cycles of what he called "the process" and Milton called "the ways of God."

Time has clarified the issue because the process, or God's ways, long ago forced us to look beyond the terminology and the specific historical perspectives of World War II and the events leading up to it. Nor do we need to make up our minds about what to do with Pound himself any longer. Meanwhile, the sequence of eleven poems has grown on us. The pearl secreted by the poet in the DTC near Pisa emerges as an archetypal work that bespeaks the active spirit smashed back by history and holding its own against paranoia and forgetting. The spirit that carries through must face the consequences and yet, as it were, sustain the whole of whatever a living tradition is. Its memories, aroused knowledge, and ways of experiencing all converge in the suggested states of sensibility related to one another by the work. Lost worlds of the past—Pound's own earlier life, the innumerable evocations of ancient and unfamiliar cultural touchstones and even of an earlier American folk culture—accumulate around what appears to be the mind of the poem.

So we have the construction of a complex chamber of sensibility, within which resonate the tonalities of a desperately suffering yet exquisitely alert and creative self. The scale is at once vast and intimate. The movement is not toward a change but toward full exploration and confirmation of a condition staked out very early on. There is no heroic action, but heroic awareness and heroic survival instead. There is no restoration of power by an Achilles or an Odysseus; just the ironic adoption of Odysseus' pseudonym—OU TIS, no one—while the imagined companionship of divinities and of vivid presences dwelling in the mind is invoked to warrant the facts of being alive and having lived. The political perversity itself, even, is a holding action of this sort, contributing to a major feeling of the sequence: the feeling of having been wrenched out of orbit by destiny and then abandoned—left weary and indignant and distraught and nameless.

Allow us to reiterate two points on the structure of the *Pisan Cantos* as a modern classic that we think need to be stressed. First, the barrier presented by the political perversity (which Pound deliberately thrusts at us at the earliest possible moment in Canto 74) is a model of the kind of intractable elements with which the most serious modern art must contend. It creates a disorientation of perspective that Pound himself saw he must meet, and he therefore chose to bring it rapidly into the foreground. He introduces a momentary chaos of association around it, and then displaces the political emphasis by summoning up all his lyrical genius and all the energy of his yearning to repossess lost experience and to dominate his predicament by sheer imaginative force. Meanwhile, he suggests his own desperate psychological case as well, through enormously touching, very brief notations. That is to say, the sequence outdoes itself in its effort to engulf its own grossly overweening outburst concerning Mussolini the "twice crucified." To encompass the gross and the monstrous, or the openly guilty, has since these cantos first appeared become a virtual program, a compulsion, for later poetry at a certain depth.[6] What is at stake is the confidence that we are human despite every obstacle and disgrace. This profound compulsion has its parallels in European literature, of course. In its less valuable manifestations it infests both poetry and criticism like a parasitic invasion, bent on wiping out its hosts. The artistic enterprise keeps being called into question. Good. This must inevitably happen, when existing expectations of method and aim seem too limiting. But difficulty is a challenge contrived for themselves by artists who have a sense of direction. "Make it new!" does not really mean "Kill it!"

Our second point for reiteration is a related one. Faced with the disorientation and powerlessness of his predicament, Pound needed all his

6. *Life Studies*, *Kaddish*, and *Crow*—for three obvious examples among many.

artistic resources to give the multiple aspects of his poem full play and yet provide a self-contained form, prismatically unified. The powerful rush toward endless proliferation of memory and association could easily have proved resistant to poetic control. But through his alternations of letting-go and drawing-in, a rhythm making use of both his energetic copiousness and his talent for astringent, precise, isolated evocation, he mastered a problem no one else could have handled. As he put it—and would not Odysseus have agreed?—

> Here error is all in the not done,
> all in the diffidence that faltered . . .

Adam, we must grant, would have disagreed—but only after it was too late to avoid being an epic hero.

3. Problems of the *Cantos*

First, then, we have suggested that the *Pisan Cantos* is the only other sequence, except for *The Waste Land* and the much longer *Paterson,* that can match Yeats's civil-war sequences as an expression peculiar to our century. In all these works the undertow of personal despair and disorientation, although different in specific character in each instance, is a compelling and crucial ingredient, at once private in its source and yet intrinsically bound up with cultural and political malaise. *A Draft of XVI Cantos* lacks the sense of acutely experienced personal suffering of this kind, although dark and tragic notes are present. In Canto 4 (1920), for instance, we find the foreshadowing of the *Pisan* poems we have mentioned. One might have been able to foresee, just possibly, not of course the shape of World War II or the Holocaust or the likelihood of an intelligence as lively and fine as Pound's being seduced by Fascism, but the probability that our modern American classic would turn out, somehow one day, to be something like the *Pisan Cantos.* Thus, at the start of Canto 4, we see side by side the imagery of a ruined civilization in the wake of a great war (World War I's ravages are at least subliminally evoked in the image of Troy as "a heap of smouldering boundary stones"), celebratory outcries picked up from Pindar and Catullus, a vision of Dionysian revelry that is not coarse but paradisal, and even lines having to do with old men's reveries and memories:

> Palace in smoky light,
> Troy but a heap of smouldering boundary stones,
> ANAXIFORMINGES! Aurunculeia!

Hear me. Cadmus of Golden Prows!
The silver mirrors catch the bright stones and flare,
Dawn, to our waking, drifts in the green cool light;
Dew-haze blurs, in the grass, pale ankles moving.
Beat, beat, whirr, thud, in the soft turf
        under the apple trees,
Choros nympharum, goat-foot, with the pale foot alternate;
Crescent of blue-shot waters, green-gold in the shallows,
A black cock crows in the sea-foam;

And by the curved, carved foot of the couch,
        claw-foot and lion head, an old man seated
Speaking in the low drone . . . :
            Ityn!

(13)

In lines 5–12 of this passage, we see subtly revealed Pound's special gift of poetic second sight, a gift for seeing and hearing another world, whether mythical or supernatural or purely imagined, superimposed upon a literal scene—the "green cool light" of dawn, when the "dew-haze blurs"—without obliterating it. Here the effect is of a kind of ecstasy of multiple awareness, while in the *Pisan Cantos* it is a valiant and pathetic effect as well. Yet in both contexts the gift is a prime means of coping with the overwhelming assault of that same multiple awareness: the source of so much delight, yet an obsessive pressure, too, that requires channeling and direction if it is not to overpower and destroy the sense of artistic control.

A curious dimension of this gift of Pound's is the fear that it is, in itself, dangerous. It is probably wrong to put this observation too bluntly, however; for its psychological component may be more a kind of fear of giving oneself too intensely to something like mystical vision—another kind of loss of control. The fear, if we are right in suggesting its presence, seems to be one basis of Pound's fascination with the figure of Actaeon, punished by Artemis/Diana despite the innocence of his error when he comes upon the naked goddess bathing. The fascination is one of deep identification, as we can see in the passage in Canto 4 in which the poet too stumbles into her domain, the forbidden realm of female mysteries. The visionary situation here is very different from that in the *Pisan Cantos* when Aphrodite's face appears to the poet in his tent, a reassurance *after* he has begun to be punished for the social, political, and aesthetic vision that has brought him, finally, into a military prison camp. In the earlier canto the poet has shared Actaeon's experience, partly because his head is full of that poor hunter's metamorphosis, as described by Ovid, and of other such metamorphoses as well, and partly out of empathy with the Provençal poet Peire Vidal, said to have been set upon by hunting dogs as Actaeon had been:

> The valley is thick with leaves, with leaves, the trees,
> The sunlight glitters, glitters a-top,
> Like a fish-scale roof,
>   Like the church roof in Poictiers
> If it were gold.
>   Beneath it, beneath it
> Not a ray, not a slivver, not a spare disc of sunlight
> Flaking the black, soft water;
> Bathing the body of nymphs, of nymphs, and Diana,
> Nymphs, white-gathered about her, and the air, air,
> Shaking, air alight with the goddess,
>   fanning their hair in the dark,
> Lifting, lifting and waffing:
> Ivory dipping in silver,
>   Shadow'd, o'ershadow'd
> Ivory dipping in silver,
> Not a splotch, not a lost shatter of sunlight.
> Then Actæon: Vidal,
> Vidal. It is old Vidal speaking,
>   stumbling along in the wood,
> Not a patch, not a lost shimmer of sunlight,
>   the pale hair of the goddess.
>
> The dogs leap on Actæon,
>   "Hither, hither, Actæon,"
> Spotted stag of the wood;
> Gold, gold, a sheaf of hair,
>   Thick, like a wheat swath,
> Blaze, blaze in the sun,
>   The dogs leap on Actæon.
> Stumbling, stumbling along in the wood,
> Muttering, muttering Ovid:
>   "Pergusa . . . pool . . . pool . . . Gargaphia,
> "Pool . . . pool of Salmacis."
>   The empty armour shakes as the cygnet moves.
>
> Thus the light rains, thus pours, *e lo soleills plovil*
> The liquid and rushing crystal
>   beneath the knees of the gods.
>
> (14–15)

Without pressing a hovering association into an emphatic interpretation, we suggest that the envisioning sensibility at this point in the poem has merged with the Actaeon figure. Primarily, it is an act of imagination that is involved, but it has been enhanced and even enabled by immersion in the *Metamorphoses*. Gargaphia, for instance, is the pool where Actaeon saw Diana bathing, and Pergusa the lake associated with Proserpina's ravishment by Pluto (called Death—*Mors*—by Ovid), and Salmacis the fountain marking the place where the son of Venus and Mercury, Hermaphroditus, was united forever with the nymph Salmacis as a half male, half female creature. All these victims (Actaeon, Proserpina, Hermaphroditus) were innocents betrayed by a force beyond normal destiny: *ὑπὲρ μόρον*, in the Homeric phrase echoed so powerfully in Canto 80 from the *Odyssey* to fix Pound's own situation.

Even the exhausting struggle against madness and breakdown, which is the great emotionally muscular effort of so much of the *Pisan Cantos*, is foreshadowed concretely in the passage under discussion, as elsewhere in the early cantos. The momentary introduction of Peire Vidal, troubadour and mad lover, into the passage is an obvious link between Actaeon and the modern poet of erotically visionary sensibility. As his *Vida*, compiled probably a century or more after his death at the beginning of the thirteenth century, puts it,

> And he loved besides, Loba de Pennautier . . . Peire Vidal called himself Lop because of her, and carried the badge of wolf. In the mountains of Cabaret, shepherds hunted him with dogs, greyhounds and great mastiffs, as if the man were a wolf. In fact, he wore a wolf-skin, giving that scent . . . And the shepherds hunted him down with the dogs and beat him so badly that he was taken for dead . . .[7]

So Vidal, as far as Canto 4 is concerned, is "stumbling" along into disaster together with Actaeon and with the modern poet, while Ovid's words map out the dread direction they are taking.

> The dogs leap on Actaeon.
> Stumbling, stumbling along in the wood,
> Muttering, muttering Ovid . . .

There is a curious dream-parallel to this passage, with its composite doomed Actaeon-figure pieced together out of many psychic and literary or mythical sources, in Pound's poem "The Coming of War: Actaeon," which appeared in the March, 1915 issue of *Poetry*. The title itself evokes the turbulence and seriousness associated with Actaeon's name, and the poem presents his funeral cortège in stately, foreboding, distanced, and awestruck language. "An image of Lethe," the poem begins—and one

7. Paul Blackburn, transl., *Peire Vidal* (New York/Amherst, Mass.: Mulch Press, 1972), p. 19.

thinks ahead to the closing line of Canto 74: "we who have passed over Lethe." Indeed, the identification of deep poetic vision with the dream-state in "The Coming of War: Actaeon"—in so grave a key—suggests, if we allow the personal element in the emotional unfolding of this extraordinary poem, a tragic self-projection by Pound that anticipates, in its bitter elegiac fatalism, a major mood of the *Pisan Cantos.*

We turn now to another complex emotional stream of symbolic imagery in the *Cantos*—a further illustration of problems created by Pound's associative prolificacy as well as of its magnetism (although thus far we have stressed only the latter). This is his remarkable use of water-imagery, which also of course creates an ambience of metamorphosis as well as endless possibilities for visual effects. Most important, perhaps, is its endlessly sexual resonance as Pound often employs it. The opening passage of Canto 4, already quoted, will illustrate:

> Crescent of blue-shot waters, green-gold in the shallows,
> A black cock crows in the sea-foam . . .

So too the fountains and pools of the Actaeon-passage we have been discussing are each associated with female beauty, either menacing or menaced, and with sexual (or anti-sexual) violence. At this late, post-Freudian date, the female symbolism of water imagery needs no elaboration, but these instances clearly establish Pound's passionate engagement with it. Even earlier in the *Draft of XVI Cantos,* we see the engagement established quickly and beautifully and explicitly in Canto 2, in a context of Dionysian creativity that is also a pure instance of Pound's need, or compulsion, to superimpose mythical or magical scenes on literal ones:

> I have seen what I have seen:
> Medon's face like the face of a dory,
> Arms shrunk into fins. And you, Pentheus,
> Had as well listen to Tiresias, and to Cadmus,
> or your luck will go out of you.
> Fish-scales over groin muscles,
> lynx-purr amid sea . . .
> And of a later year,
> pale in the wine-red algae,
> If you will lean over the rock,
> the coral face under wave-tinge,
> Rose-paleness under water-shift,
> Ileuthyeria, fair Dafne of sea-bords,
> The swimmer's arms turned to branches,

Who will say in what year,
    fleeing what band of tritons,
The smooth brows, seen, and half seen,
    now ivory stillness.

And So-Shu churned in the sea, So-shu also,
    using the long moon for a churn-stick . . .
Lithe turning of water,
    sinews of Poseidon,
Black azure and hyaline,
    glass wave over Tyro . . .
(9–10)

The beautiful strophe at the end of the first Pisan canto, holding its diverse elements in balance, recalls this contextual imagery of sexually charged watery metamorphosis. In its opening lines the sense of the imagining mind's maintaining exquisite equilibrium among all the pressures and preoccupations of existence is uppermost:

Serenely in the crystal jet
    as the bright ball that the fountain tosses

Seven lines further along, the meditation grows more abstract without losing touch with the metaphor of living water:

This liquid is certainly a
    property of the mind
nec accidens est   but an element
    in the mind's make-up . . .

In between these focusing moments, notes of hell-terror and death-consciousness enter, associated with both the water-imagery and the metamorphic process imposed by the dreaming mind. The phrasing here is strikingly reminiscent of that in "The Coming of War: Actaeon" and is a distant echo of the "hell-cantos" (14 and 15) and the war-canto (16) that conclude the *Draft of XVI Cantos:*

How soft the wind under Taishan
    where the sea is remembered
out of hell, the pit
out of the dust and glare evil

Toward the end of the passage, the sexual notation is introduced, softly yet decisively, in the phrase "or swansdown ever," from Jonson's highly erotic "Celebration of Charis." It is, as in the earlier cantos, a notation fraught as well with the drive of irreversible fatality and with the knowledge of death:

> Hast 'ou seen the rose in the steel dust
> (or swansdown ever?)
> so light is the urging, so ordered the dark petals of iron
> we who have passed over Lethe.
>
> (449)

In pursuing these few connections (among many others) between Pound's early cantos and other poems and the rest of the cantos, especially the Pisan group, we have been noting the special sort of organic cohesion he creates. It inheres in the method rather than in any scheme, and certainly not in any master-plan followed meticulously through. The separate volumes of cantos are individual sequences but hardly coalesce in a seamless curve stretching over a half-century, each building into the next without benefit of revision or reorientation.[8] The many connections Pound sets up among the separate volumes are improvised by constant echoings and recollections of earlier effects, often in altered context, as we have just been trying to show. This sort of deep continuity occurs with any writer, so that in a very large sense the corpus of anyone's books does constitute a sequence. But the more normal and useful concept is that a sequence does not expand indefinitely; rather, it presents itself in a single curve or restricted system of affective movement or tension, however proliferative its internal dynamics may appear to be in the sheer energy of what goes on. One problem presented by the *Cantos* is the contradiction between the fact that it does not, cannot, present itself as such a curve or system and the fact that it does bring the whole of what has gone before into the surround of each succeeding sequence. Despite the essential independence of each separate sequence, that gathering surround rarely permits the individual sequence to stand out clearly. The exceptions—after of course, the opening group—are the 1940 volume combining the Chinese cantos (52–61) and the Adams cantos (62–71), because of its thematic emphases; and the Pisan group, because of its intrinsic psychological and moral struggle that makes it, supremely, the culmination of the heroic artistic effort underlying the *Cantos*.

Now, the kind of proliferation of tonal connections we have just been following—taking us along unexpected paths leading, say, from the revealed gift of superimposing a beautifully detailed vision on a literal scene to a fear of tragic discovery or disaster to the multiple associations with

8. See Rosenthal, *Sailing into the Unknown*, pp. 79–115

Actaeon and Vidal to the liquid medium of metamorphosis and generative sexuality, with a good many other touchstones of association along the way—is active throughout the *Cantos*. It is a fascinating proliferation and, as we have noted, establishes a process of interwoven evocation whose character forever summons up the whole body of awareness throughout the work. Unfortunately it does invite us—nay, *requires* us—to bring our own knowledge into the text as well, and here the accursed philistine critics do have a vulgar point, alas, to make. They will never be *right*, just because the *Cantos, nevertheless*, has so very much to offer. But they do have a point. Readers of Pound, for instance, will be aware from his criticism how much he admired Vidal's poems.[9] "We find," he says in the essay on Arnaut Daniel, "—we find in Provence beautiful poems, as by Vidal when he sings 'Ab l'alen tir vas me l'aire.' " This is a song, written in exile, of love and of the pleasure of recalling the air and beauty and sunshine and people of Provence. Then, in another essay, "Troubadours—Their Sorts and Conditions," Pound reveals one of the reasons he responds so to this poem when he speaks of repossessing the world of the Provençal poets: "a man may walk the hill roads and river roads from Limoges and Charente to Dordogne and Narbonne and learn a little, or more than a little, of what the country meant to the wandering singers." He might be referring directly to the Vidal poem, the beginning of which Paul Blackburn translates thus:

> I suck deep in air come from Provence to here.
> All things from there so please me
> when I hear
> in dockside taverns
> travelers' gossip told
> I listen smiling,
> and for each word ask a hundred smiling words,
> all news is good
>
> for no man knows so sweet a country as
> from the Rhône down to Vence.
> If only I were locked between
> Durance and the sea!
> Such pure joy shines in the sun there.
>
> (p. 27)

And so one makes one's connections, among Vidal and Pound and Provence—connections such as remain afloat throughout the succession of cantos, attaching themselves to the affective centers here and there, un-

9. Ezra Pound, *Literary Essays* (Norfolk, Conn.: New Directions, 1954), pp. 95, 97, 113, *et passim*.

til we have the great outpouring in Canto 74, at the start of the Pisan group, of the poet's memories of Provence and much else. But these connections are not all provided in the poetry itself.

Still, the poems grow ever more absorbing as one becomes lost in the pleasures, and the ups and downs, of all that goes on in them. And yet—a second problem—their very richness and proliferation, the loading every rift with ore thrice over and then loading new rifts with more ore as the poem grows, militate against structural progression and completion. Clearly it is this overloading that constitutes the most serious problem of the very long would-be sequence; and with the overloading comes increasing reliance on the arbitrary, ultimately personal sensibility of the poet, no longer at the necessary remove from the poem's surface. Hence Pound's rhetorical self-indulgence, and the problems of clarity of reference at certain points, and the picking up of tonal currents at such great distances from one another that only the truly attentive and sensitized reader can respond—and then, sometimes, only at the risk of bringing external associations and sources of the kind we have to some degree just introduced (Pound's prose, Vidal's poetry) to support a reading that the poem itself may not sustain or simply to suggest an intellectual coherence rather than an affective one. The great short sequences of the early 1920s did not present these problems directly. On the other hand, their relatively open structure cleared the way for the more proliferating works of Pound and Williams and Olson and others—an inevitable development, hardly to be deplored because of the problems it would introduce. These problems parallel, in their "evolution," the ones presented by Joyce as he moved from *Dubliners* to *A Portrait of the Artist as a Young Man* to *Ulysses* to *Finnegans Wake*. For the general reader of poetry, there is no question that this evolution produces an ever more elusive and forbidding poetry, with no dependable points of contact, unless the reader has learned to experience poems as affective structures whose "meaning" inheres in a succession of interacting tonalities at varied pitches of intensity, and has also learned to be open to the kind of delayed echoes and reprises in new contexts we have been discussing. But the problems of overloading are not solved by having infinitely undemanding readers whose passive receptivity is as open to slack or willful passages as to superb ones. Difficulty in art is not a vice or a virtue but a *quality*, or affective dimension, like melodic echoing or stark diction. Pound's dependence on our holding multiple effects in mind—that is, in fluid solution, as he himself assumes the cantos are doing continuously—will always be a difficulty because the associations are idiomatically his to start with, though almost always derivable from the poems themselves.

The generations of modern poets born before 1900, and of their most gifted immediate successors, have had an aesthetic end in view despite their pushing against the boundaries of the closed forms they inherited. Their deliberate engagement with improvisational and random elements of poetic process has not been conceived as a resistance to the search for "difficult beauty." "Rigor of beauty is the quest"—with this sentence Williams begins the epigraph to his poetic preface in *Paterson;* and the joke in the quotation we find in the *Cantos* of Aubrey Beardsley's remark to Yeats ("Beauty is difficult, Yeats") is that anyone could think Yeats needed this information. But what has been called post-modernism tends to reject both the concept and the need. We find resistance to difficult beauty in the theory and practice of the Projectivist poets and their followers, in such highly discursive poets as John Ashbery and A. R. Ammons (writing in the wake of the later Stevens but indirectly of Pound as well), and in many others, British and American alike, who cultivate minimally intense styles, casually anecdotal, confiding, and prosaic.

The most influential Projectivist, Charles Olson, may stand as an example of the flattening, de-aestheticizing tendencies partly inspired by Pound himself. He once characterized Pound's poetics as an "ego-system," by which Pound talked as an equal or superior with minds of the past—only two of whom, Confucius and Dante, he considered "possibly" his betters. He thereby discovered, says Olson, an approach that "destroys historical time." This is "the methodology of the Cantos, viz, a space-field where, by inversion . . . he has turned time into what we now have, space & its live air." By contrast, according to Olson, Williams escaped Pound's confinement within his own ego but was trapped by a time-frame, "making his substance historical of one city (the Joyce deal)."[10] Olson thus reduces aesthetic questions to "methodology" and to the general "field" of a given poet's mind. He hovers on the edge of saying that structure is nothing more than zigzagging about within that "field." Olson's own poetry became less and less concerned with dynamic structure and increasingly discursive and random, and his career has encouraged lesser poets to write without regard to sound, rhythm, or tonal development. "Ego-system" or not, the exquisite music of the Kung Canto (13) has little to do with considering Confucius one's "better," but the undeveloped ear might hear only its prose-aphorisms and colloquial touches and never catch its rich lyric movement.

Pound's problem lies elsewhere. The modern poetic sequence meets Poe's old objection: that the "long poem" does not exist, for a given emotional impulse cannot be sustained beyond certain limits. The interaction among shorter units that might otherwise stand alone is the so-

10. Charles Olson, *Selected Writings* (New York: New Directions, 1966), pp. 81–83.

lution provided by the sequence. But the attempt by Pound—and others, but Pound especially—to go far beyond the length of the original short sequences of the early 1920s brings us back to the problem in a new way. In limiting his first sequence of cantos to a coherent group of sixteen, Pound brilliantly related a series of radiating affective centers. He balanced, for instance, the darkly fatalistic and archaically awesome tonalities of Canto 1, with its repossession of the *Nekuia* and its buoyantly sophisticated leavening toward the end, against the modern horror-visions of Cantos 14–16. More complexly, throughout the *Draft* he balanced a dazzling array of related and opposed tonalities—joy and desolation, erotic delight and suicidal despair, supernal visions and colloquial American humor—in changing mythical, historical, and personal contexts, while projecting his consciousness of the violent intrusion of war and politics and brutal historical realities on daily life. His full title, announcing the book as "the Beginning of a Poem of some Length," betrayed his exuberant ambitions but also belied the discovery this book, with *Mauberley, The Waste Land,* and the civil-war sequences, embodied: the sequence is *not* a long poem but a grouping of poems in dynamically coherent relationship. It is interesting that the 1930 volume *A Draft of XXX Cantos* is in effect an expansion of the first involving a reorientation of perspectives but very much dependent on the original balancings, which it seeks to reinforce. It does so beautifully, yet with a certain loss of impact because of the almost doubled length. With *Eleven New Cantos XXXI–XLI* (1934), however, which starts off with the Jefferson-Adams cantos and ends with the celebration of Mussolini, we are pretty much in a different universe of subjective reference despite the eches of the hell and war cantos in the first group and other equally important echoes of earlier motifs. Such echoes provide links among the canto-groups in each volume, and some of the poems providing the links are among Pound's most powerful—Cantos 36 and 39, for example, with their high lyricism. But the main dynamic reciprocities occur within rather than between the separate volumes, except for the relation of the *Pisan Cantos* to all that has been most telling in the earlier groupings: a selective culmination, if that be the right term, rather than an organic outgrowth.

Perhaps the contradiction we have been discussing, between the need for an aesthetic of interacting affect that has room for the sense of every possible mode of consciousness and feeling and the need for self-enclosed structures of crucial intensity at their most strategic points, rides all poetry. The balance Pound struck, imperfectly, may be as close as we can get to the heart's desire in an encompassing work that takes risks of commitment of every kind and that confronts intractable materials and intractable limits of form. The love of greatness of spirit and greatness of expression makes us cherish the *Cantos* whatever our objections. They gathered to a kind of greatness and then flew off in bright fragments. It is the afterbeat, in the depreciation of language, in talented droning, and in easy slackness and "experiments"—that is, in the trivializing of the discoveries of genius—that we are helpless to prevent though not to discount.

Chapter Ten:

William Carlos Williams's *Paterson*

1. Essentials of *Paterson*'s Structure

Of the great modern sequences, Williams's *Paterson* is certainly the most humanly available. It is also, however, so crowded with impressions, memories, incidents, and quick associations that, like the *Cantos*, it can sometimes seem all tumbling incoherence. And despite its stance of unpretentious ordinariness ("just another dog / among a lot of dogs"), it is far from common speech although it *involves* such speech:

> How strange you are, you idiot!
> So you think because the rose
> is red that you shall have the mastery?
> The rose is green and will bloom,
> overtopping you, green, livid
> green when you shall no more speak, or
> taste, or even be. My whole life
> has hung too long upon a partial victory.
>
> But, creature of the weather, I
> don't want to go any faster than
> I have to go to win.
> Music it for yourself.
>
> He picked a hairpin from the floor
> and stuck it in his ear, probing
> around inside—
>
> (30)[1]

1. *Paterson* (New York: New Directions, 1963), p. 30 in the current printing.

This incomplete passage (the beginning of *Paterson,* I,iii) will illustrate the combined simple colloquialism and poetically rigorous quality of a good deal of Williams's writing. The first line and the final three, and most of the middle stanza, have the sound of spontaneous speech; and nothing could be more commonplace, and comically inelegant, than the action the closing lines present so plainly. Yet everything in the first stanza after the opening line has the ring of a more formal lyrical tradition, and the second stanza's subtlety makes its casualness of tone a bit deceptive and riddling. The poet laughs at himself for thinking that the artist's genius for catching and isolating some aspect of experienced reality gives him "mastery." The rose is red—he may convey his sense of that vividly. But it is green, too, with the persistent indomitable greenness of nature that outlasts and overgrows poet and poem alike. At best, then, his art gains him a "partial victory" only. Nevertheless, even as just another "creature of the weather," he must do what he can: the decision of the second stanza. And so we get a momentary buffoonery, with the hairpin, and thereafter a series of effects that carry the poem lightly, idiosyncratically, along toward a tenderly mock-heroic treatment of the illusion of poetic mastery.

Before quoting the rest of the passage, however, we must note another aspect of the dynamics of the opening lines. It is an aspect easily overlooked—namely, Williams's subtle shifts of person to suggest a complex sense of the experiencing sensibility. The poem begins in the second person, then shifts to the first, then to the second again, and then (over the next thirty lines) to the third person—but always in reference to the poet himself. That is, an internal dialogue becomes a meditation, then a momentarily sarcastic rejoinder to the overly modest and logical tone that began the process, then a whimsically appreciative account of the mind's wandering preoccupations, starting with the hairpin stanza:

> The melting snow
> dripped from the cornice by his window
> 90 strokes a minute—
>
> He descried
> in the linoleum at his feet a woman's
> face, smelled his hands,
>
> strong of a lotion he had used
> not long since, lavender,
> rolled his thumb
>
> about the tip of his left index finger
> and watched it dip each time,
> like the head

of a cat licking its paw, heard the
faint filing sound it made: of
earth his ears are full, there is no sound

(30–31)

By this point in the poem, within its Picasso-like intersection of planes of sensibility, we have been treated to a number of shifting tones: the initial cry of affectionate self-rebuke; the sardonic question; the insistent and slightly appalled call to recognition of nature's self-renewing energy; the confession of limitations; the self-acceptance and counter-irony of "Music it for yourself"; the deliberately self-demeaning grossness of the hairpin moment (a continuation of the defiant tone just before, in its way); the mock-scientific trivial notation about the melting snow; the sudden little complex of erotic association (woman's face, smell of lavender lotion, attention to hands); the amusing connection of the fingers' movement with the action of a cat licking its paw; and the alertness to the sound of thumb rubbed against finger. And then the poem leaps into visionary affirmation of the richness of the realm of imagination and desire, happily streaming clouds of Blakean language and Joycean epiphanies and Stevensian eloquence and ending with a double, serious image of the poet's thoughts as "trees" of "heroic" desire from which his mind drinks:

: And his thoughts soared
to the magnificence of imagined delights
where he would probe

as into the pupil of an eye
as through a hoople of fire, and emerge
sheathed in a robe

streaming with light. What heroic
dawn of desire
is denied to his thoughts?

They are trees
from whose leaves streaming with rain
his mind drinks of desire :

(31)

The passage completes the opening movement of I,iii. (The colon at the end suggests that what is to follow bears a closely reciprocal relationship to the perspectives so far developed.) In the course of it the poet has proved the futility of art to himself, but then refuses to accept the defeatism his own eloquence is thrusting upon him. Instead, his lapse into uncontrolled reverie carries him into something like ecstatic anticipation. As a whole, the passage epitomizes Williams's basic method: keeping the poem in action in such a way as to undercut set or logical positions.

The method works ebulliently here. The issue raised in the first stanza is familiar, the subject of many a famous poem: Can art catch the gist of natural things and fix it within a humanizing aesthetic order? But Williams quickly breaks off any meditative centering by his ceaseless attention to the inner workings of his awareness and imagination. The stanzas could well stand as an independent poem. But in context they serve beautifully as counter-stress against the relatively impersonal I,i and the self-doubting depression of I,ii. At the same time, they usher into the sequence the poet's specific sense of the probing, chancy nature of his art, suggested by the double use of "probe" in the passage, once for a low-comedy effect and once exaltedly. His art is a matter of getting the set of things, of who and where we are, and watching the miracles occur, if they do—all logic and observation to the contrary, and our own ruthless recognitions of the downward set of existence to the contrary as well. No full-length pontificating, and no prolonged lugubrious maundering either.

All of which brings us to the nature of *Paterson* as a whole. It is most easily compared with the *Cantos*. Although it is only about one quarter their length, it too is a sequence of sequences, each three-part Book appearing as a separate volume before being combined with the others. (Each part is about the same length as Pound's longest cantos.) Whereas the *Cantos* marks a half-century or so of Pound's development, from his early thirties onward, *Paterson*'s five Books appeared over a relatively short twelve-year span, in 1946, 1948, 1949, 1951, and 1958 respectively—the period of Pound's *Pisan Cantos, Rock-Drill,* and *Thrones.* Both authors, we should remember, were in their sixties and seventies, writing with a vigor comparable to Hardy's in *Poems of 1912–13* and Yeats's in *Last Poems and Two Plays.* Both, too, of course, are further along in the modernistic stream than their great elders. *Paterson*'s improvisatory variations are like those in the *Cantos,* so that one may find, at any moment, a sudden shift of rhythmic base, or of tonal stream, or of voice or general style.

In II,iii, for instance, just one small portion of the section (two of its almost nineteen pages, 85–86) moves through eight such shifts:

First, two sentences from a literal weather report: "Sunny today, with the highest temperatures near 80 degrees . . ."

Then, a moment of erotic reverie:

Her belly her belly is like
a cloud a cloud
at evening

Then, an absurd sexual memory (comic "drama"):

He Me with my pants, coat and vest still on!
She And me still in my galoshes!

Then, a meditative passage, increasing in intensity as it goes from

—the descent follows the ascent—to wisdom
as to despair

to

From that base, unabashed, to regain
the sun kissed summits of love!

Then, five lines of the poet's self-observation as he watches his own thoughts and the way he mobilizes them poetically:

—obscurely
in to scribble . and a war won!

—saying over to himself a song written
previously . inclines to believe
he sees, in the structure, something
of interest:

Then, three highly romantic quatrains, beginning

On this most voluptuous night of the year
the term of the moon is yellow with no light

Then, back to the erotic dreaming, with the added terror of death:

Her belly . her belly is like a white cloud . a
white cloud at evening . before the shuddering night!

And finally, the beginning of the seven-page letter that closes Book II, hard-driving prose by a woman who challenges the poet's self-centeredness and describes "women's wretched position in society" and her own problems as a person struggling to be a professional writer. (This is the last example of a series that started as early as page 7.)

The technique here is perhaps even more collagist, in a purely verbal and affective sense, than Pound's. An apparently arbitrary succession of poetic notations creates its own associative and tonal ordering. Reference to literal weather suggests, in context here, the volatile, ceaseless shifting of one's internal weather. A dream of a woman's beauty floats up, and after it a memory of a ridiculous moment in lovemaking—whereupon are recalled two indissoluble links: that between high reverie and the body's occasional low comedy, and that between exaltation and despair. Then an insistent feeling asserts itself: the sense of a constant, waiting stillness of rapture, nourished by memory of past heights of delight. This moment of assertive awareness presents itself as part of a struggle to win artistic direction by the fullest openness. The direction now taken is toward the eroticism implicit in nature's beauty, and towards dreams of woman's beauty in itself. But a harsher female reality thwarts these dreams. It is embodied in the powerful, irrefutable complaint of the woman's letters challenging the male poet's whole way of life and thought.

In the same way, the whole of *Paterson* is collagist in organization, as are Pound's canto-groupings and many of the cantos themselves. Since, however, his numbered sections in each Book approximate the length of Pound's longest cantos, Williams has committed himself over and over to an extremely demanding long-distance effort—an intimate engagement with reality, memory, and fantasy that is all the more demanding because of the collagist method employed to keep the work moving and alive at every point. We are not claiming that *Paterson*'s every little shred and patch of colorful exclamation, prose quotation, or whatever, is necessarily superb—only that the work is actively structured in a demanding way. Pound's longer cantos, for instance the remarkable Canto 74 that starts off the Pisan group, present the same demand on the poet's energy and copiousness. But because the cantos are not of uniform length, whereas *Paterson*'s sections are relatively so, the occasional unusual proliferation of an individual canto has a special significance in the overall structure.

At any rate, with all their reciprocities and differences, these works by two friends illuminate each other. More demanding intellectually, and incomparable when in full flight, the *Cantos* outsoars *Paterson* without containing it. Williams seeks to locate the essential position of American sensibility. He does so, not grandiosely, but using himself, as man and poet, to gauge the field of operation. And like the Pound of the *Pisan Cantos,* he is holding off despair both personal and cultural. True, there are obvious, crucial distinctions. Imprisoned on a charge of treason and undergoing an exhausting physical and emotional ordeal in the Disciplinary Training Center near Pisa, Pound had to cope with depression and psychological instability. He drew on every resource of character, memory, and artistry, shoring them "against his ruins," and all these pressures and counter-efforts reveal themselves in specific cantos. The volume presents a poetry of psychological crisis.

The pressures in *Paterson,* on the other hand, as in the non-Pisan cantos, are the result of a slow accumulation over the years. They come to a head as an effort to master the condition of social and cultural crisis evidenced in sexual malaise, violence, and the loss of shared understanding, "a common language," in every phase of our lives. Yet clearly the two poets themselves speak a common language. Their exchanges of thought about poetry and other matters are part of *Paterson*'s fabric. Their preoccupations connect. Both are responding, as well, to their own looming old age and death, a pressure desperately at work in Pound's case and faced with an uncomplacent, informal directness in Williams's Book V. At the same time, their poetry is shot through with humor, elation, and ecstasy, and with a dedication to aesthetic values—Beauty—most explicitly stressed by Williams in his epigraph that begins "Rigor of beauty is the quest," in the "Beautiful Thing" passages of III,i and III,ii, in the "Two halfgrown girls hailing hallowed Easter" passage of I,ii, and especially in the whole drift of Book V. The "quest" for "rigor of beauty" dominates the self-perceptive passages of deepest internality, strewn through the sequence, of the sort we noted in discussing the dynamics of two pages of the third section of Book II.

A certain cyclothymic volatility seems common to all the great sequences, which often soar from despair to ecstasy through the transcendent elation of lyrical moments. The "Preface" to *Paterson* epitomizes such a movement, at first self-depreciative though persistent but, at the end, *singing.* Looking at the work in this light, we find a state of matured intelligence, at once vigorously alive and painfully dissatisfied but capable at any point of *letting go* into a transfigured realm. In Williams's language, this is the realm of the trees of thought, "from whose leaves . . . his mind drinks of desire." An intimate, confessional restlessness holds sway in most of *Paterson.* The restlessness has to do, centrally, with the difficult relations of the sexes, ranging from the fear of commitment of more cultivated persons—their inability to bridge sexual differences through mutual trust and responsiveness—to rape and brutality of many kinds. This preoccupation is, in *Paterson,* a driving element in a dream of beauty constantly tormenting us, as in the "Beautiful Thing" passages and other movements we shall be examining. But the sexual aspect of existence is not the whole of it for Williams. The infinite particulars of life and imagination are a challenge to an aesthetic of keenly active observation and response. As we are told in III,i:

> The province of the poem is the world.
> When the sun rises, it rises in the poem
> and when it sets darkness comes down
> and the poem is dark .
>
> and lamps are lit, cats prowl and men
> read, read—or mumble and stare
> at that which their small lights distinguish
> or obscure or their hands search out
>
> in the dark. The poem moves them or
> it does not move them. . . .
>
> (99–100)

By a somewhat elaborate process, this large perception is related, in the same section, to an even larger, more inclusive one that reveals the major energizing motifs of the sequence:

> The riddle of a man and a woman
>
> For what is there but love, that stares death
> in the eye, love, begetting marriage—
> not infamy, not death
>
> tho' love seems to beget
> only death in the old plays, only death, it is
> as tho' they wished death rather than to face
> infamy, the infamy of old cities .
>
> . . . a world of corrupt cities,
> nothing else, that death stares in the eye,
> lacking love: no palaces, no secluded gardens,
> no water among the stones; the stone rails
> of the balustrades, scooped out, running with
> clear water, no peace .
>
> The waters
> are dry. It is summer, it is . ended
>
> Sing me a song to make death tolerable, a song
> of a man and a woman: the riddle of a man
> and a woman
>
> What language could allay our thirsts,

what winds lift us, what floods bear us
                                        past defeats
but song but deathless song?

(106–7)

The major energizers we refer to are "the riddle of a man and a woman," the nature of the "corrupt cities," the need to "stare death in the eye," the overriding concern with finding a true "language," and the dream of beauty: "deathless song." The sequence is orchestrated by their interaction. Book I opens with the allegorically embodied city and the painful search for a language and song to transcend the sexual riddle, and it introduces the encompassing sense that "the province of the poem is the world." Book II engages with the corruption of the city-populace by the sheer defusing of its significant traditions, but also finds, joyfully, lingering traces of myth and ritual that once gave aesthetic point to life and death for the folk. (Thus, in the passage just quoted, "it is . ended" bears traces of the Crucifixion.) An aspect of the corruption against which these primitive traces of joy sometimes gleam is the indifferent inspirationalism of an outworn evangelical Christianity, on the model of Billy Graham's creed, that serves as apologist for the economic system. Book III is a blazing pursuit of beauty and passionate meaning amid squalor, violence, and the obliteration of cultural memory. Book IV presents a "pastoral" comedy of sexual crisis and then fixes on one figure, Mme Curie, who rose above that crisis—and above the world's indifference to its own creative possibilities, despite the corruptness of science and of medicine. Book V embodies the great human effort to cope with death and tragic knowledge through the vision of art. Each Book, as we have suggested, has its cyclothymic dynamics of structure, with the same sort of verbal collage and complexity of tonal connections as we observed, in miniature, in the two pages of II,iii.

In all this the poet, sometimes "Mr. Paterson" or "Dr. Paterson" (his work as a physician among the poor is important to the poem), sometimes "Faitoute" (by a kind of empathy with another figure), and sometimes "I" or "he," is a fragmented collection of fractional selves, appearing as a vehicle of the poem's dynamics in various ways. Perhaps his most endearing self-dispersal as an animated carrier of tones and affects comes early in I,i, in lines making a gentle transition between the allegorical and topographical purview at the start of Book I ("The Delineaments of the Giants") and the poem's later sensuous confessionalism:

A man like a city and a woman like a flower
—who are in love. Two women. Three women.

Innumerable women, each like a flower.

But

only one man—like a city.

(7)

Surely the most delicately couched gesture toward promiscuity, however bee-like rather than bull-like, in the history of poetry! In terms of *Paterson*'s structure, the poem here points to the importance of all the female figures that appear in its several Books. A city contains many women; so does a man's mind concerning itself with the right understanding between the sexes. From both sides of this double embodiment (city as man, man as city), the poem has as it were a love affair—whether guilt-ridden, helplessly admiring, beauty-shaken, or self-defeating—with each of the women it centers attention on. In Books I–II the first and last woman's voice heard comes in prose-passages quoted from the letters of the desperate writer called "C." and "Miss Cress," who reproaches the poet ever more bitterly for neglecting her. The neglect is not only personal (his refusal to become her close friend or, perhaps, lover) but professional. He has not helped her career as a writer as she would have wished—and indeed, the odds against any woman writer of limited means, as detailed in Miss Cress's merciless waves of complaint, are crushing. Short of devoting his life to her, there would be no way the poet could begin to solve her problems. Still, and rightly, he is tormented by her assumption that, having shown some real interest and given some real help, he bears a certain responsibility toward her whether or not he can act on it—and, by implication, to all others in the same predicament as hers.

A primitive solution, impossible for us, is presented with humor that scarcely conceals the intense seriousness of the issue it points up, in the description, near the end of I,i, of a *National Geographic* photograph of an African scene:

I remember
a *Geographic* picture, the 9 women
of some African chief semi-naked
astraddle a log, an official log to
be presumed, heads left:

Foremost
froze the young and latest,
erect, a proud queen, conscious of her power,
mud-caked, her monumental hair
slanted above the brows—violently frowning.

Behind her, packed tight up
in a descending scale of freshness
stiffened the others

and then . .
the last, the first wife,
present! supporting all the rest growing
up from her—whose careworn eyes
serious, menacing—but unabashed; breasts
sagging from hard use . .

Whereas the uppointed breasts
of that other, tense, charged with
pressures unrelieved .
and the rekindling they bespoke
was evident.

Not that the lightnings
do not stab at the mystery of a man
from both ends—and the middle, no matter
how much a chief he may be, rather the more
because of it, to destroy him at home

. . Womanlike, a vague smile,
unattached, floating like a pigeon
after a long flight to his cote.

(13–14)

Here is a society that speaks a common language—a shared culture whose sexual, familial, and human mutuality are undeniable. The problems are not *all* solved. As the poem observes slyly, "the lightnings" of female sensibility will harry the "chief" in any case—the Mona Lisa smile, "womanlike," whose mysterious import he will never quite fathom. Again and again we are reminded of the desperate case the poet is speaking to, regardless of any whimsy or exuberance along the way. As he says in I,ii, shortly before a reprise of the "lightnings" that "stab at the mystery of a man":

On the embankment a short,
compact cone (juniper)
that trembles frantically
in the indifferent gale: male—stands
rooted there .

The thought returns: Why have I not
but for imagined beauty where there is none

or none available, long since
put myself deliberately in the way of death?

(20)

Paterson I,ii, in fact, turns sharply to the predicament suggested in these lines. I,i has indicated broad concerns and basic positionings, and the second section begins in personal bewilderment:

There is no direction. Whither? I
cannot say. I cannot say
more than how. The how (the howl) only
is at my disposal (proposal) : watching—
colder than stone .

a bud forever green,
tight-curled, upon the pavement, perfect
in juice and substance but divorced, divorced
from its fellows, fallen low—

Divorce is
the sign of knowledge in our time,
divorce! divorce!

(18)

The knot of associations here proliferates quickly, from the initial confused "howl" to the bud, full of potentiality, that has fallen and is seen as a symbol of "divorce" because separated from the parent bush and other buds. "Divorced / from its fellows, fallen low," it embodies "unfledged desire, irresponsible, green, / colder to the hand than stone." The word "divorce" sets the poem shouting—

Divorce is
the sign of knowledge in our time,
divorce! divorce!

The image and the shouts ride side by side in the poem, without forced connection although the words used suggest the many modern lives that are detached from either tradition or communion. Soon a closeup emerges of another sort of green bud. Again the associations are unforced:

Two halfgrown girls hailing hallowed Easter,
(an inversion of all out-of-doors) weaving
about themselves, from under
the heavy air, whorls of thick translucencies

poured down, cleaving them away,
shut from the light: bare-
headed, their clear hair dangling—

Two—
        disparate among the pouring
waters of their hair in which nothing is
molten—

two, bound by an instinct to be the same:
ribbons, cut from a piece,
cerise pink, binding their hair: one—
a willow twig pulled from a low
leafless bush in full bud in her hand,
(or eels or a moon!)
holds it, the gathered spray,
upright in the air, the pouring air,
strokes the soft fur—

                    Ain't they beautiful!

(18–19)

The girls appear as a vision of pure erotic loveliness, tinged with a surreal undertone as in a Delvaux painting: innocent, darkly subject to cosmic sexual tides. The passage connects with language heard in the first section of the poem: "Innumerable women, each like a flower." It recalls paradisal moments in Pound's *Cantos:* nymphs and goddesses seen amid "translucencies" of light. "Hailing hallowed Easter," the girls are a profanely epiphanic vision. Yet the setting is here and now, a city street, and the words describing the girls echo those used for the Passaic Falls in I,i, precise yet allegorical. Here is a perfect instance of the poem's unfailing, volatile response to the presence of beauty, too often marred, in our midst. One girl has pulled a twig from a bush; the act of unconscious destructiveness matches the fallen state of the bud in that it embodies a "divorce." Nevertheless, she exclaims: "Ain't they beautiful!" She is saying this about the pussy willow buds, and the poem is saying it about her and her companion.

The girls' beauty is something to be reconciled with all the other preoccupations of I,ii. The key is a rich, bitter poignancy. We have quoted the section's opening lines. Their alienation and confusion ("There is no direction. Whither?") are set against the endless, tantalizing promise "tight-curled" within a green bud, or abundant in the roar of the river going over the falls, or femininely exciting in the apparition of the girls. All these modes of promise are constantly "challenging our waking"—the phrase becomes a short-lived refrain—and so is the oppressive awareness of neglected history that isolates the questioning mind. "He," the overriding awareness—being neither erudite nor at all confident that his thoughts are accurate—feels "alone / in a wind that does not move the others." They are "divorced" from the organic sense of a region's or a civilization's memory. Here, in Paterson, the Indian once was master, revolution occurred, waves of immigrants arrived—all forgotten, like the incidents of violence, suicide, tragic loss, and pointless daring that mark local history.

> : a body found next spring
> frozen in an ice-cake; or a body
> fished next day from the muddy swirl—
>
> both silent, uncommunicative
>
> (20–21)

The poem grows more confessional a few pages later, in a passage of rare gentleness and personal honesty that provides a key affect of the sequence. As we have been suggesting, the touchstones of emotive structure that count most in *Paterson* are usually the sexually charged ones. In the relations between men and women (and in IV,i a relationship between two women as well) are concentrated the virtues and despairs—the realities—of the culture. Such is the underlying assumption, the drive, of *Paterson,* which in a sense is a curve of tonalities that, with a few exceptions, are at their most intense and pervasive in the sexually suffused points of involvement.

In the passage in I,ii that we have called confessional, a man and woman are talking together, intimately, yet with a certain distance between them too. The atmosphere is completely civilized; they are in a state of almost-communion, but something, a failure of shared language, prevents lovemaking. The passage, like that describing the "halfgrown girls" in springtime, is full of metaphoric language reflecting the felt presence of the waters of the Passaic River, crashing over the falls. Although the man and woman are not literally seated beside the water, the language suggests that they are:

> we sit and talk
> I wish to be with you abed, we two
> as if the bed were the bed of a stream
> —I have much to say to you
>
> We sit and talk,
> quietly, with long lapses of silence
> and I am aware of the stream

that has no language, coursing
beneath the quiet heaven of
your eyes

which has no speech; to
go to bed with you, to pass beyond
the moment of meeting, while the
currents float still in mid-air, to
fall—
with you from the brink, before
the crash—

to seize the moment.

We sit and talk, sensing a little
the rushing impact of the giants'
violent torrent rolling over us, a
few moments.

If I should demand it, as
it has been demanded of others
and given too swiftly, and you should
consent. If you would consent

We sit and talk and the
silence speaks of the giants
who have died in the past and have
returned to those scenes unsatisfied
and who is not unsatisfied, the
silent, Singac the rock-shoulder
emerging from the rocks—and the giants
live again in your silence and
unacknowledged desire—

(24–25)

Unexpressed on the one hand, unacknowledged on the other, because the stream of reciprocity here "has no language" that will reconcile the physical and spiritual, the "desire" that needs a mutually assumed depth of context cannot be fulfilled by sex alone. There is no completed "story" here, but a yearning for a new kind of lovers' joy to compensate for the miserable past. "Giants"—striking rock-formations on the landscape, but also past generations and their most vital men and women, some of them legendary figures—haunt the moment of desired communion. The word "unsatisfied" seems to echo Yeats's "The Magi," where, "by Calvary's turbulence unsatisfied," archetypal figures of the Christian myth return continually to seek a new revelation on the "bestial floor." The "giants" here seem to bear the same relation to the would-be lovers, except that the living and the dead, and the anthropomorphic landscape, all share in the finely held moment of fearful anticipation so purely evoked. The figures of local legend to which the documentary prose-inserts introduce us—the minister's wife who fell from the rock-ledge to her death in the falls, the daredevil who made his career of leaping into the water, and others—are waiting upon the moment of truly meaningful love communion: a "fall" into love's shared depths of union, a "crash" from the brink that would redeem their failures.

That moment would reverse, as well, the destiny of woman implied in Miss Cress's letters. We should be overreading to see her, symbolically, as a tragic giantess out of the whole past of women's suffering, haunting male sensibility with guilty remorse. Yet her letters do suggest something of the sort. The eternal moment of incomplete rapport between the non-lovers is charged with all the emotional strains of *Paterson*. Necessarily unresolved, it is a moment of introspective realization acted out in reverie, not a confrontation in the objective world. It cannot, therefore, reverse or redeem the past. Further along, in section II,iii, the sequence resists the destructively pressing sense of failure of such moments by stressing the enormous volatility of the subjective life. The section first yields ground by dropping into fatalistic depression ("The descent beckons"); but later it counters this affect ("With evening, love wakens") and then moves into rich response to the world's marvels ("On this most voluptuous night"). *Paterson* has its allegorical side and copes with cultural crisis, but its main streams of feeling derive from the unpredictably varied emotional climate of private existence.

This unpredictably varied, or extremely volatile, or cyclothymic aspect of *Paterson*'s dynamics, and the poem's manysided preoccupation with female personality and the sexual relation generally (both in themselves and for what they betray concerning our whole social covenant), are the essential forces in the work's quality and unfolding. The other great preoccupations, a rigorously challenging search for "beauty"—the operation of the aesthetic principle in guiding basic life-values—and a critique of the economic interests dehumanizing our culture and stripping away the people's memories and traditions, are closely related to these primary energizing forces in the structure of the sequence. The point needs some stressing, since Williams did attempt a preliminary, if vague, mapping of *Paterson* in various statements conveniently reprinted under the heading "Author's Note" in later editions of *Paterson*. Actually, they reveal little of the work's organic structure, providing only a sort of conceptual scheme. The scheme, however, may well have enabled Williams to release the poem's real movement by laying out supposed boundaries hardly visible from within the work itself.

Williams's statements deal more with encompassment than with progression: "to find an image large enough to embody the whole knowable world about me," he says in the *Autobiography*. Although the titles he gives his first four books ("The Delineaments of the Giants," "Sunday in the Park," "The Library," and "The Run to the Sea") and the dedication to the untitled Book V ("To the Memory of Henri Toulouse-Lautrec, Painter") give one to think, only the fourth is suggestive of movement of some kind. But taken in conjunction with the "Author's Note," they do argue a sort of plan. Book I introduces the city of Paterson's mythic symbolism in the sequence. Book II is a closeup of the city park, of the people relaxing there of a Sunday, and of the observer's thoughts about what he sees and hears and about his own place in the scene. Book III tries to find, in the "library" of our heritage, even as fire, tornado, and flood destroy it (all this both literal and allegorical), a "language" or voice for the "modern replicas"—the city's inhabitants and their present customs. They are "replicas" in that they reproduce the past under new conditions, and also because the present is a *reply* to the past. Book IV selects significant memories from a lifetime and considers them in relation to time's inexorably obliterating flow. Book V, in Williams's words, reflects "many changes" and takes the whole work "into a new dimension." The poet has "come to understand . . . that many changes have occurred in me and the world" and has been "forced to recognize that there can be no end to such a story [as] I have envisioned." The "changes" are clear from the text of Book V. He has, a little belatedly, come to see the enormity of the Holocaust as well as the Bomb; and now, too, the altered perspectives of advanced age, doubtless made more acute (although the poem does not mention this) by the strokes he has suffered, have matured his sense of the reciprocities of beauty and tragic awareness. These changes have enhanced his understanding of how passionate intensity links opposites in the artistic scheme of things—how "the virgin and the whore" are one—and accounts for the Book's dedication to Toulouse-Lautrec, artist of Paris brothels and cabarets.

Within this very general context, hardly an elaborated plan, a swarming mass of feeling and alertness is at play, with its own gravamen and momentum and stops and turns and starts. Thus, Book II ("Sunday in the Park") lifts its head out of the subjective morass of Book I and tries to fix attention on the objective, empirical life—

> Outside
>
> outside myself
>
> there is a world . . .
>
> (43)

—by means of a Sunday walk in the park. The walk reveals the flagrant, pointless sexuality of a populace without imagination or remembered rituals of communion. Zest for catching raw reality struggles with revulsion—perhaps a reflex of something like voyeurism—and sometimes creates an illusion of prurient moralism:

> 3 colored girls, of age! stroll by
> —their color flagrant,
>   their voices vagrant
> their laughter wild, flagellant, dissociated
> from the fixed scene   .
>
> But the white girl, her head
> upon an arm, a butt between her fingers
> lies under the bush   .   .
>
> Semi-naked, facing her, a sunshade
> over his eyes,
> he talks with her
>
> —the jalopy half hid
> behind them in the trees—
> I bought a new bathing suit, just
>
> pants and a brassier   :
> the breasts and
> the pudenda covered—beneath
>
> the sun in frank vulgarity.
> Minds beaten thin
> by waste—among
>
> the working classes SOME sort
> of breakdown
> has occurred. Semi-roused
>
> they lie . . .
>
> (51)

At first, this passage may seem prudishly shocked and hostile. It is indeed shocked—a reflex of coming out of one's own mind into bright sunlight and observing human beings unselfconsciously in action. *There* is ordinary life, nakedly revealed, its sexuality flaunted yet not exactly cherished as a transforming enchantment or a source of rich communion: the "frank vulgarity" for which the great mysteries and possibilities are nothing special. Yet the scene is painted in bold and clear outline, and the poem keeps moving uphill, among the various groups of picnickers, to the summit of the hilly grounds. All these lives, bodies, postures are magnetic in themselves despite the always surprising insensitivity the people display.

Loiterers in groups straggle
over the bare rock-table—scratched by their
boot-nails more than the glacier scratched
them—walking indifferent through
each other's privacy .

—in any case,
the center of movement, the core of gaiety.

Here a young man, perhaps sixteen,
is sitting with his back to the rock among
some ferns playing a guitar, dead pan .

The rest are eating and drinking.

The big guy
in the black hat is too full to move .

but Mary
is up!
Come on! Wassa ma'? You got
broken leg?

It is this air!
the air of the Midi
and the old cultures intoxicates them:
present!

—lifts one arm holding the cymbals
of her thoughts, cocks her old head
and dances! raising her skirts:

La la la la!

(56–57)

At this point the crucial transition of the section takes place—"frank vulgarity" is redeemed by the awakening of atavistic Dionysian memories in the old woman's mind: "the cymbals of her thoughts." Subjective reveries and what the external world has to show come together here. The scene's grossness is only part of its meaning, the ground on which the "old cultures" of Europe, with their pre-Christian memories—partly evoked by the sunny atmosphere as of southern France (the "Midi")—reassert themselves. What now appears is aesthetic transcendence created out of life's sexual energies, beyond mere flagrancy ("laughter wild, flagellant, dissociated") and dull immodesty ("just / pants and a brassier"). It is burlesque—the dance of an aged, goatlike "nymph" inviting a satyr. And yet the old Italian woman is acting out an ancient fertility dance-ritual:

—the leg raised, verisimilitude .
even to the coarse contours of the leg, the
bovine touch! The leer, the cave of it,
the female of it facing the male, the satyr—
(Priapus!)
with that lonely implication, goatherd
and goat, fertility, the attack, drunk,
cleansed .

(58)

It is Williams's special genius to be able to project this variety of transcendence again and again in his work: a process of reorienting perceptions of the ugly and unpalatable. In II,i the process is revealed in slow-motion—through a series of impressionistic snapshots. They come between the purely subjective self-reminder, at the start, that the outside world really exists and the transforming vision of Mary's sexual dance, which satisfies the human need for artistic "mastery" over the ruck of life. Henceforth in the sequence the new perception is the key to the structuring of affects, while the wistful tonality ("I wish to be with you abed") and the tonalities of depressed longing and inhibited fastidiousness are subordinated, becoming almost a ground against which the major progression is boldly outlined—or a massing of resistance that it must overcome.

Paterson's emphasis, as a result, now shifts to an altered context of feeling: acceptance of life's grossness, as integral to beauty-creating ecstasy. The displacement is reinforced by a recollection that intrudes itself into the scene we have been discussing. This interruption comes between the description of Mary beginning to dance and the superimposed vision of the satyr-dance:

Remember

the peon in the lost
Eisenstein film drinking

from a wine-skin with the abandon
of a horse drinking

so that it slopped down his chin?
down his neck, dribbling

over his shirt-front and down
onto his pants—laughing, toothless?

Heavenly man!

(58)

Eisenstein's vision of the peon's inspired drinking isolates the man's "abandon" and animal spirits, his blissful gaiety that no decrepitude can down, so that his sheer slovenliness seems a necessary ingredient of his ecstasy. Williams's lines follow Eisenstein's lead enthusiastically, recreating the isolation of the Dionysian letting-go by their final adjectives, "laughing, toothless," and final epithet, "Heavenly man!" The scene is framed aesthetically by this isolating emphasis, just as her dance frames old Mary's vitality.

In these various passages of II,i, the sequence gets into action in a new way. First, it unsentimentally opens up humanity's true "common language." At the same time, it prepares us for the deeper plunge into the tragic and aesthetic depths of sexuality in the "Beautiful Thing" passages of III,i and III,ii. And it also prepares us for the complexly confessional burlesque-sexual "Idyl" in IV,i, the humane idealism in IV,ii of the movement centered on Mme Curie, and the opening out of the widest range of sensibility in Book V through the convergence of states of sexual, sacred, and aesthetic transport. Even the Holocaust is seen in the context introduced by Mary's dance, when we are told, in V,ii, that the "satyric dance" enacted by all art is our one means of coping with horror and catastrophe.

2. In Pursuit of "Beautiful Thing"

Turning now to the "Beautiful Thing" passages, especially in III,ii, we can see that their white-hot intensity, and the human predicament they try to deal with, escalate the pitch of feeling with which the sequence engages the central question of sexual communion. The anger and dismay of these passages add subtle dimensions of feeling, a body of love and despair, to the whole of *Paterson* and thereby give organic density to the finely sustained rhetoric of grief and transcendence in the crucial V,ii. Their combined eloquence and humane immediacy, of a peculiarly intimate volatility, come out of a wrenching contact with the complexities of suffering: Williams's experience as a doctor called to treat a young woman who has been gang-raped. The contact, and the raging awareness brought to life, make for a new turn in the aesthetic process to which *Paterson* is given—the absorption and transcending of the vile. The bedraggled victim becomes the object of a fanatical, reluctant love, not literally involving sexual desire for her but nevertheless a lover's passion. The poem's baffled intensity is a blazing engagement with the poverty, cruelty, and perverse treatment of beauty in a world that has lost its bearings. The "Beautiful Thing" (later the phrase will carry over to the elusive essence of communion concealed within the ordinary or defeated life) constantly addressed here is, first of all, the young woman the doctor has come to see. The description is at once literal and mythical, naturalistically clinical and romantic:

> —the small window with two panes,
> my eye level of the ground, the furnace odor .
>
> Persephone
> gone to hell, that hell could not keep with
> the advancing season of pity.

And:

> Shaken by your beauty .
> Shaken.
>
> —flat on your back, in a low bed (waiting)
> under the mud plashed windows among the scabrous
> dirt of the holy sheets .
>
> You showed me your legs, scarred (as a child)
> by the whip .
>
> (125–26)

Having established these simple axes, the movement gets under way in earnest—a passionate violin solo punctuated by rasping city noises:

> But you!
> —in your white lace dress
> "the dying swan"
> and high-heeled slippers—tall
> as you already were—
> till your head
> through fruitful exaggeration
> was reaching the sky and the
> prickles of its ecstasy
> Beautiful Thing!
> And the guys from Paterson
> beat up
> the guys from Newark and told
> them to stay the hell out
> of their territory and then
> socked you one
> across the nose
> Beautiful Thing

for good luck and emphasis
    cracking it
till I must believe that all
desired women have had each
    in the end
    a busted nose
and live afterward marked up
    Beautiful Thing
    for memory's sake
to be credible in their deeds

Then back to the party!
    and they maled
and femaled you jealously
    Beautiful Thing
as if to discover whence and
    by what miracle
there should escape, what?
still to be possessed, out of
    what part
    Beautiful Thing
should it look?
    or be extinguished—
Three days in the same dress
    up and down   .

        I can't be half gentle enough,
half tender enough
        toward you, toward you,
inarticulate, not half loving enough

(126–28)

This last passage is in essence the climax of *Paterson*. It was first published, in virtually identical form, as part of a poem called "Paterson: Episode 17" in *New Directions 2* (1937). That title will suggest an early conception of *Paterson* as an accumulation of affect-laden units—"episodes." But, equally indicative, the writing of the passage itself reflects a unifying vision of the poem's character. It is extremely concrete in detail, mixes street language with elegantly evocative, tender, and toughly witty language, and moves easily among levels of awareness—direct perception, allegorical vision, wry generalization, and maddened personal feeling. These contradictory elements combine with equally contradictory sexual feelings and confusions—a sort of promiscuity of erotically linked tonalities of desire, resentment, and moral outrage and fascination all at once at the rapist postures and idiom of brute *machismo*, and gentleness toward the victim. It is interesting that when he came to do *Paterson,* Williams already had his climactic affect waiting in the wings.

In the movement of the whole sequence, the tremendously touching, self-revealing, and often comical confrontations in dialogue form of IV,i confirm the earlier subjective discoveries, which culminate in the "Beautiful Thing" passages, in a broadly tolerant way. These confrontations are "pastoral" dialogues in which two older persons, a wealthy lesbian in advanced middle age ("Corydon") and a man who is similar to the poet ("Paterson"), have separate relationships with a young practical nurse and masseuse ("Phyllis"). The two sets of relationship do not interact directly, but they both represent distortions of love possibility. Phyllis will give herself to no one, although with Corydon she is professionally attentive and friendly in a toughly self-reliant way and with Paterson she has a sensitively responsive and even passionate attachment despite refusing actual intercourse. Williams's work here has great charm and energy. One can only admire his ear for American speech, his humorous understanding of Corydon's poetic and spiritual aspirations and of Phyllis's stubborn integrity, and his sense of the absurd—half harking back to Chekhov and Mark Twain, half ahead of its time. He was probing the ways that awkwardness, humiliation, and misdirection connect with depths of human feeling—an elusive connection amid distrust, starved and bruised hopes, and murky though desperately intent purpose. There, in that context, lurks the "Beautiful Thing," could we but touch its essence. It is harder, of course, to glimpse it in one's own psyche than in the circumstances of the raped black girl of the previous Book, a deprived and perhaps subnormal person for whom it is all too easy to feel a sentimental compassion. Clearly, IV,i is a sort of counterbalance to the transcendence gained earlier through overcoming disgust and condescension. Here it is not another sort of being but one's own supposedly cultivated sensibility that is sweaty with sexual embarrassment and something like grossness, so that perspectives furnished by old Mary and by "Persephone" merge with the poet's.

These comments may seem too solemn for the hearty, bawdy, hardboiled side of IV,i. In the dynamics of *Paterson,* the succession of brief scenes here provides dramatic relief and an independent vivacity, while the section is nevertheless shot through with tonalities familiar from earlier movements. The confessional strain continues, for one thing, in the speech directly attributed to "The Poet" and in the scenes between "Dr. Paterson" and Phyllis. For another, the poetry of Corydon gives Williams a wide range for self-parody. Corydon is supposed to be an amateur poet, unsure of herself and, though her writing is similar to Williams's own in its concerns, half-successful at best. She is certainly capable of an appealing kind of bad poetry:

If I am virtuous
condemn me
If my life is felicitous
condemn me
The world is
iniquitous

(160)

But also, through her, the sequence provides a lovely sort of whimsy that parallels certain effects in the very first passage we quoted, the opening movement of I,iii—but without being constrained by a more rigorous discipline or exalted poetic style than Corydon possesses:

You dreamy
    Communist
where are you
    going?

To world's end
    Via?
Chemistry
    Oh oh oh oh

That will
    really
be the end
    you

dreamy Communist
    won't it?
Together
    together

(160)

Corydon's style reflects Williams's sensibility without his developed art:

. . drives the gulls up in a cloud
Um . no more woods and fields. Therefore
present, forever present

or:

The gulls, vortices of despair, circle and give
voice to their wild responses until the thing

is gone . then, ravening, having scattered
to survive, close again upon the focus,
the bare stones, three harbor stones . . .

(161)

The comic intrusion of "um" suggests a poet losing track of her poem for a moment, or just thinking about whether it is going well as she reads. Otherwise, passages like the last two echo the style of much of the rest of *Paterson* despite deliberately contrived lapses of diction or rhythm. Taken as a whole, IV,i, "An Idyl," presents its poets—Paterson, Corydon, and The Poet—modestly and ironically. It deliberately plants them in unglamorous contexts, teasing them somewhat and showing them as victims of our general crisis of love and trust. In this sense the section is cooler and less oracular than the "Beautiful Thing" movement. It foreshadows the elegiac restraint and acceptance of the aging process with which Book IV will end—not because IV,i is at all restrained itself, or elegiac, but because of its mellowed recognition that "I," the implied poet-self behind all the other poet-figures, is but a part of the human comedy.

In another sense, this self-parodying reduction of the poet-figure is a preparation for the surprising turn in IV,ii. This section opens with a tender address by the poet to his fourteen-year-old son, whom he has taken to a lecture on atomic fission—that development which holds so many possibilities for good and evil. Much of the rest of the section is devoted to Mme Curie, with strong emphasis on her quiet simplicity and humble origins and on the fact that she was pregnant while conducting her fruitful experiments: just one human being among the many, apparently—apart from the "radiant gist" of her genius. In her way she is ideal woman, marvelously fulfilled and creative at every level, her life harmonious with both the scientific and the democratic ethos of the age. Her courage is that of explorers and voyagers who take risks, and open dangerous possibilities in the interest of knowledge and power. No other figure in *Paterson* is regarded so admiringly, for her embodiment of the creative spirit is the true meeting-ground of the sexes:

—with ponderous belly, full
of thought! stirring the cauldrons
. in the old shed used
by the medical students for dissections.

(177)

and:

Ah Madam!
this is order, perfect and controlled
on which empires, alas, are built

But there may issue, a contaminant,
some other metal radioactive
a dissonance, unless the table lie,
may cure the cancer . must
lie in that ash . Helium plus, plus
what? Never mind, but plus . a
woman, a small Polish baby-nurse
unable .

Woman is the weaker vessel, but
the mind is neutral, a bead linking
continents, brow and toe

and will at best take out
its spate in mathematics
replacing murder

Sappho vs Elektra!

(179–80)

There is a curious, ennobling recognition saturating the Mme Curie section. *She,* the configuration of values and impressions presented to us in IV,ii, reveals a sense of human existence at its best. It is a "best" naturally acquainted with pain, a condition associated with womanly life when it is free of ignorance and violence and can develop its true human energies. Williams is original, and unPoundian, to speak of "Sappho vs Elektra!" rather than "Propertius vs Achilles!" And the feminine emphasis is sustained in his treatment of Mme Curie's daily life and work. The poem breathes loving nostalgia for her quality of work and existence: totally dedicated, even Spartan, and yet suffused with the more agreeable associations of Paris. The evocations, even of the city's atmosphere, are essentially female. And they parallel, in a happier, civilized context, the earlier presentations of old Mary and "Beautiful Thing." That is, they suggest the fulfillment of other women's thwarted possibilities that one can often glimpse behind a love of finery and a vitality in search of joyous release. The connections blaze up in these lines, especially, of the Curie section:

Paris, a fifth floor room, bread
milk and chocolate, a few
apples and coal to be carried,
des briquettes, their special smell,
at dawn: Paris .

> the soft coal smell, as she
> leaned upon the window before de-
> parting, for work .
>
> —a furnace, a cavity aching
> toward fission; a hollow,
> a woman waiting to be filled
>
> —a luminosity of elements, the
> current leaping!
> Pitchblende from Austria, the
> valence of Uranium inexplicably
> increased. Curie, the man, gave up
> his work to buttress her.
>
> But she is pregnant!
> Poor Joseph,
>
> the Italians say.
>
> Glory to God in the highest
> and on earth, peace, goodwill to
> men!
>
> Believe it or not.
>
> (175–76)

The playful ending here adds a touch of folk-skepticism that sounds like Phyllis's wisecracks in IV,i and connects with old Mary of Book II. Marie Curie was not, after all, the Virgin Mary *rediviva.* And all "miracle" is in question. Yet she does remain as close to miracle as we are likely to get. The movement of the passage dances around all this—at first through the charmed mood of her mornings, almost in the tone of an *aubade;* then in an allegorical exclamation, at once sexual and scientific in phrasing, of considerable intensity; and at last in a mixture of bawdy talk and Christian outcry that connects Mme Curie with the Madonna despite surrounding laughter. Together with the lines we quoted earlier from the Curie section, this dance of associations (Paris morning, smell of *briquettes,* furnace, womb, fission, miraculous birth or discovery, subordination of the man, bawdy joking, sacred and profane thought) prepares us well for Book V, with its many female-centered passages. These include the reverie in V,i on "the Virgin and the Whore," to which the lines on Mme Curie's double "luminosity" bear a direct relationship. In V,ii, the pitch is established at the very beginning by a translation of a famous poem of lesbian desire by Sappho, an extraordinary poem of trembling arousal; later in the section, in a very different style but with fully engaging force, we are given the poet's vision of the common Muse on the city streets—an elusive, momentary vision that links his passionate concerns with the world's realities. And in V,iii our attention is called to a Brueghel *Nativity* in a manner recalling and compressing all these volatile perspectives.

Not to labor the matter, the centering on Mme Curie has prepared us for the widened orientation of Book V: in terms of simple subject-matter, the swing of *Paterson* from a basically local perspective to one involved with Europe and its traditions. Her entrance into the poem marks an increased emphasis rather than an absolute shift of attention. We could hardly forget the old Italian woman, Mary, dancing in the "Midi" atmosphere of the park, or the vision of "Persephone" in the slum basement; such links with European myth and reality are very much a part of the entranced imagination, bitter moments, and wry irony of the earlier Books. But Mme Curie, the Polish-French scientist, a woman presented as a possible model for the poet's son in choosing a career, plays a remarkable, pivotal role in the affective life of the sequence. She is its sole heroic presence, an alternative persona for the poet's ideal conception of his own role. Especially, the tonalities swirling about her connect two centers of encompassing transport in *Paterson.* These are the world of ignorant vulnerability of the "Beautiful Thing" section (Mme Curie herself has come at the "radiant gist" that is the undiscovered possibility in the raped girl's stupefied life) and the world of heroically beautiful effort amidst tragic and sophisticated knowledge toward which the poem reaches out in the climactic part of V,ii:

a letter from a friend
saying:
For the last
three nights
I have slept like a baby
without
liquor or dope of any sort!
we know
that a stasis
from a chrysalis
has stretched its wings
like a bull
or a Minotaur
or Beethoven
in the scherzo
from the Fifth Symphony
stomped
his heavy feet

I saw love
mounted naked on a horse
on a swan
the tail of a fish
the bloodthirsty conger eel
and laughed
recalling the Jew
in the pit
among his fellows
when the indifferent chap
with the machine gun
was spraying the heap
he had not yet been hit
but smiled
comforting his companions
comforting
his companions
Dreams possess me
and the dance
of my thoughts
involving animals
the blameless beasts

(223–24)

The intensity of this passage escalates from the friend's report of temporary escape from a state of misery relieved only by drinking and drugs, to the resounding images so charged with love, terror, and the power of music, and then to the memory (from a documentary film, perhaps) of the doomed, smiling Jew in the pit. The Jew, "comforting his companions," takes on a stature like Mme Curie's despite the fact that his poetic existence is limited to only eleven lines. This is the sole reference, in this sequence that began to appear the year after World War II, to Nazi genocide (as opposed to *none* in Pound and Eliot). Yet the war must have been a powerful factor in the wider range of Book V we have noted. In any case, the neurosis implied in the friend's letter, the associations of death with power and beauty in the visionary images, and the Holocaust closeup of helpless people being machine-gunned present unbearable pressures that connect with all the notes of violence and distortion of personality in the preceding Books. Then the quiet self-characterization in the final five lines provides an aesthetic distancing from all the passion, very much like the self-discounting at the end of "Gerontion" or of "Meditations in Time of Civil War."

This similarity to the practice of other poets is not unusual. The introduction of a note of modesty after a passage has reached certain heights is common in poetry. At the same time, what Williams does here represents the far greater stress on art and the role of the artist in Book V than before. The lines beginning "Dreams possess me" are the poet's apology for the vision that has just issued forth and, possibly, for the hubris of identifying that vision with Beethoven's and the heroic Jew's—and, in lines just preceding the foregoing passage, with the visions of Leonardo, Dürer, Bosch, Freud, Picasso, and others. Beginning with the figure of Mme Curie, he has begun to conflate supreme commitment and vision in any aspect of life with the greatest art. But beyond this, all of Book V is a contemplation of the poet's sense of his art and its relation to his memories and preoccupations and to the fact that he has grown old.

In old age
    the mind
        casts off
     rebelliously
    an eagle
from its crag

     —the angle of a forehead
         or far less
     makes him remember when he thought
         he had forgot

            —remember

         confidently
  only a moment, only for a fleeting moment—
         with a smile of recognition . . .

(207)

Thus the start of V,i, after which we are carried on a wondrous associative journey involving direct observation (spring returning, with "the song of the fox sparrow / reawakening the world / of Paterson"), the complex relations of innocence and corruption (evoked by memory of a play of Lorca's concerning a pubescent bride and her aged bridegroom and developed in terms of the Unicorn tapestries and a supposed sight of a unicorn by Audubon in the woods "northward of Kentucky"), memories of Paris, the superior endurance of the realities of artistic vision to those of unconsidered, transient experience, and the relation of these drifting preoccupations to the challenge of death:

The Unicorn
has no match
or mate . the artist
has no peer .
Death
has no peer:
wandering in the woods,
a field crowded with small flowers
in which the wounded beast lies down to rest .

We shall not get to the bottom:
death is a hole
in which we are all buried
Gentile and Jew.

The flower dies down
and rots away .
But there is a hole
in the bottom of the bag.

It is the imagination
which cannot be fathomed.
It is through this hole
we escape . .

So through art alone, male and female, a field of
flowers, a tapestry, spring flowers unequaled
in loveliness.

Through this hole
at the bottom of the cavern
of death, the imagination
escapes intact.

he bears a collar round his neck
hid in the bristling hair.

(211–12)

This imagery and way of thinking seek, not altogether successfully, to introduce an entirely new tonal range into the sequence: a music at once death-entranced and buoyant, though full of sadness too. The effort here is to open on a ringing chivalric note, and then follow it with quatrains whose melodic base is a fatalistic chant of death and dissolution. Unfortunately, the varied refrain is like something one might hear in a cheery folksong. "There is a hole / in the bottom of the bag"—one can almost hear children shouting these words around a campfire. But the passage also has an overlay of the marveling, pondering mind observing the images rising in the poem itself. The tone recalls the start of Book V—"In old age / the mind / casts off / rebelliously." Here the "rebellion" is against absolute acceptance of death; although "Death / has no peer," it is asserted that "the artist / has no peer" either. The figure now shifts, from personified Death to the squalor of "a hole / in which we are all buried / Gentile and Jew"—perhaps anticipating the Jew and his companions in "the pit" who are seen in the next section—and the "hole" figure gets enough variation (the context is different in each quatrain) so that it suggests both grave and womb. We escape death through the "hole" of imagination. The poem is presenting the artist, and others who use imagination with the utmost commitment, as "having the mastery" after all. But the imagery is drastically elemental and recalls the language that was used for Mme Curie in the throes of creative labor: "a furnace, a cavity aching / toward fission; a hollow, / a woman waiting to be filled."

This is no easy assertion or bit of complacency. The language is severely impersonal in both instances. And when the poet is preparing, in V,ii, to speak with more personal fervor than anywhere else in *Paterson* except the "Beautiful Thing" passages, he maintains the impersonality at first by translating Sappho rather than diving into his own immediate imagination. Here too (although Williams's translation falters sadly) the conception is at the extreme pitch of feeling.

> Sweat pours out; a trembling hunts
> me down. I grow paler
> than dry grass and lack little
> of dying.
>
> (217)

When the poet begins to speak more personally, a few pages later, expressing his longing for a soul's companion, a Muse of the ordinary, he does not follow Sappho's example but is wittily colloquial for the most part. The style is open, full of rhythmic breaks that alter the emphasis when a thought is completed. Yet the poet is describing his deepest need, for the Athene to his Odysseus, and the loneliness and despair underlying the passage are not quite masked by its sheer charm. It is as if, between the tremors of Sappho's desire and the apocalyptic vision involving the Holocaust, the poet is constrained to mute his sense of personal deprivation. Still, the description and the event presented are very much in context. First of all, there is the assumption that all greatness of spirit and action is implicit in the givens of ordinary life. Secondly, the modest anonymity of the woman's appearance recalls Mme Curie again. Third, there is the terror of divinity in the fact that she is a momentary apparition. "Grey eyes looked / straight before her"—as if to suggest the presence of "grey-eyed Athene." And, again like Mme Curie, she is superior to sexual differentiation without concealing her femaleness. At the same time she is the "lonely and / intelligent woman" who would meet the poet's own loneliness and intelligence. "She stopped / / me in my tracks," he says. The expression is perfectly colloquial but also evokes a thunderclap of revelation. A great event, an epiphany, is being handled lightly. Williams is, in a manner, avoiding hubris. A brief sample:

Her
    hips were narrow, her
                                        legs
thin and straight. She stopped

me in my tracks—until I saw
her
    disappear in the crowd.

An inconspicuous decoration
made of sombre cloth, meant
I think to be a flower, was
pinned flat to her
                        right

breast—any woman might have
done the same to
say she was a woman and warn
us of her mood. Otherwise

she was dressed in male attire,
as much as to say to hell
with you. Her
                expression was
serious, her
                feet were small.

And she was gone!

(219–20)

The Brueghel Nativity-painting described in V,iii is the last of the sexually charged centers of attention in *Paterson.* The description notes all the sacred and secular elements combined in the painting—the richly clad Magi, the poor peasants, the soldiers looking slyly at Joseph, the "greybeard" whose young wife holds the naked Baby "(as from an / illustrated catalogue / in colors)" on her lap. All the tonal streams we have been discussing come together in the presentation and implications of this painting. Book V completes the transcendence that was only partially reached in IV and shows all of human passion and catastrophe and faith encompassed by the artist's drive to perceive accurately and to achieve the enduring resonance of the most daring acts of imagination. Not that the poem makes this claim for itself, but it undertakes the same task as Brueghel did and strikes a similar equilibrium.

We have followed the major affective movements of *Paterson* as they reveal themselves: a pursuit of apperceptions of the particular that will open up the "language" available to us through our insight into the character of our sexual communion. The burden of an entire culture is determined by the state of this communion and the degree to which it finds aesthetic expression. Rape and violence on the one hand, longing and visions on the other, and passionate engagement with women's experience of love and sex are the urgent elements of awareness through most of the sequence, contrary to the vague schematic outline Williams early laid out.

Nevertheless, a fair number of *Paterson*'s passages of interest are only indirectly involved with this decisive cluster of driving moments of tonality. Important examples include the evangelist's harangue in II,ii and the satirical lyric "America the golden" in the same section—both deployed ironically in the course of the poem's wilful brooding over where the crisis of the civilization truly lies. There are some seventy distinct passages, extremely varied in their focus, that make up the larger affective units in the organic body of *Paterson*. The work is also studded with almost a hundred prose-entries: quotations from personal letters, historical and journalistic documents, often with a pointed economic relevance along half-Poundian, half-Marxian lines, accounts of brutal or highly eccentric behavior, descriptions of primitive ritual and unusual sexual behavior, aesthetic notations, and factual tables. Sometimes, because of these insertions and protrusions, the poem seems to have no formal ground or frame but to resemble cave-paintings in which the conformation of stone along the walls is incorporated in the painting. The prose-entries feed an idiosyncratically objective context into the poem, and the poem in its turn serves as a context for them. The whole work is a persistent, passionate attempt to assimilate the scattered tonalities of a fragmented modern American culture into the emotional dynamics we have been observing.

The question of place, of the importance of Paterson as the locale of the poem, needs to be seen in this same perspective. Paterson and its environs are certainly an important literal and allegorical ground for the sequence. Yet—again—the poem's full subjective import cannot be limited to the one locale, for the simple reason that it depends above all on the overriding, linked issues of sexual interaction and artistic imagination. Moreover, both these issues become increasingly imbued with a tragic private sense and world view, especially when the sequence rises to its heights in Book V. But all the five Books, in any case, plunge again and again into internal states having nothing to do with locale. Even Book I, which most immediately reflects an original, fresh intention of giving mythic body to Paterson and of following a scheme, concludes with a drop into depression that is countered by the changing energies of a highly varied emotional life whose cycles have their own rhythms. And, at the end of the sequence, what takes over decisively is the state of utter openness projected in V,iii in the passage beginning "The (self) direction has been changed." Although crucially placed, these sections, I,iii and V,iii, are only two examples of the ultimately subjective, non-thematic dynamics of *Paterson*.

Part Four

Neo-Regionalism and Epic Memory

Chapter Eleven:

British and Irish Models

1. "Neo-Regionalism" and Modern Poetry

We closed the preceding chapter with comments on "place" in *Paterson.* Our main point there was that, while *Paterson* is saturated with regional preoccupations, they do not determine its structural dynamics. But at the same time they do count. Williams delved lustily into the history, topography, literary and journalistic treatment, and general studies—whatever he could find—of Paterson and the Passaic River and the whole area of northeastern New Jersey that is the poem's basic locale. In I,iii he quotes Pound as saying to him: "Your interest is in the bloody loam but what / I'm after is the finished product." If Pound meant that Williams was not concerned with making a work of poetic art, his thought was a gross simplification. But Williams did get close to his raw materials: the physical realities of a place, and the look and voices of the people. These materials, and his effort to find a "language" for them, provide the magnetic field within which he worked.

His initial allegorical presentation of the city of Paterson as a sleeping giant establishes this field. The characters wandering through the poem are mostly derived from local experience, the image of the Passaic River flows through the poem, and the crash and rocky heights of the falls enter its metaphorical life. Both contexts of imagery—river and falls—are essentials of the poem's organic life and form a surround for its sense of a possible common language. This sense is itself related to a basic question involving place: What is the shared, felt meaning of this locality? The haunting sense of a world of buried memory has much to do with the epic dimension of a poetic sequence. Ancient epics carried the essence of a culture and its ideals in their heroes' natures and tasks and confrontations with supernatural beings. Oral or written, these works were, in this function at least, sacred texts with an irresistible momentum of doom and mission. Modern sequences inherit their aura of urgent destiny from ancient sources.

We are dealing here with a need, or instinct, to affirm a kind of immortality by creative renewal: the verbal hallowing of the lives, manners, and speech of countless generations. No profound argument is called for to show that Whitman acted on this need. It may, at first, seem to be forcing things to say the same of Emily Dickinson, yet her poems' fey power derives from their bitter combat with the thought of death and of the disappearance of identity. In the course of her struggle, she brings heroically into the open the secret sensibilities of women in a confined area of existence—privileged, articulate, yet artificially hemmed in. She plays stormily with ambiguous equations between literal death and the "death" that is moral and emotional disaster, and she pits the sacred, virtually forbidden mysteries of recovered personal feeling against the assumptions of religious orthodoxy. Far more than Mme Curie, she reveals the "Beautiful Thing" or "radiant gist" lurking beneath the repression and the blocked communion. Her idiom of revelation is an Amherst lady's, its passionate precision wrung from native modes of thought and speech. Literal place is present more subtly in her work than in that of most other poets, but it comes through undeniably—in her *Bay Psalm Book* versification, her Yankee turns of dialect and humor, her religious wryness, and her relentless psychic self-mortification.

In their simple physical detail, the sequences of Hardy, Yeats, and Eliot are more obvious forerunners of Williams's neo-regionalism. The dead wife in Hardy's *Poems of 1912–13* is virtually a spirit of place who has pined away because she was taken from her beloved Cornwall after her marriage. The poem's disturbing intensity depends mainly on the husband's remorse after her death, a remorse heightened by the fact that even now she is buried in an alien place. He imagines that she is trying to draw him back to her native haunts; and finally, in the anguished state he has been brought to, he actually does seek them out. Yeats's civil-war sequences project scenes of Ireland in bloody chaos, and of his own displacement within it, even while he seeks to superimpose an idealized vision, compounded of past aristocratic virtues, on the uncontrollable reality around him. Eliot roots *Four Quartets* in particular, little known places whose continuing daily reality he seeks to connect with a past still alive in it, still capable of being summoned up and of defeating time as we usually understand it. His effort bestows, on the generations making up a local "life of significant soil," that virtual immortality of creative renewal to which we have referred.

The effort of superimposition in these works, combined as it is with an uncompromising acknowledgment of everything that, empirically speaking, must frustrate the effort, by its nature demands artistry of the highest order. Edgar Lee Masters's *Spoon River Anthology* (1915) antedates the sequences of Yeats and Eliot and shares their sense, and Hardy's, of the massed memories that make up a "significant soil." But his necropolitan monologues cannot rival the quality of the other poets' imagination. The neo-regionalism of which we speak involves powerful recovery of the deepest memories of a region or nation that is radically out of touch with them, and therefore out of touch with the richer experiences and values underlying its own history. Neo-regionalism may employ local color and realistic character portrayal as Masters does, but it is something more embattled and demanding. Its heroism resides in its refusal to yield up cultural memory to oblivion. More positively, the imaginative effort has to do with reaching a state of awareness that reaffirms a transcendent identity—the sense of long continuity between the significant past and the freshest involvement in the present moment, and of one's own place in this continuity.

Among American poems, Hart Crane's *The Bridge* (1930) and Charles Olson's much more recent *The Maximus Poems* (1960, with later proliferations) certainly meet these terms in what they attempt. We shall turn to these American works in the next chapter, but first some observations about neo-regionalism in British and Irish sequences may prove useful. Such works would certainly include Hardy's masterpiece, *Poems of 1912–13,* which still towers above all other modern sequences of the British Isles save Yeats's (and Eliot's, if we consider him—as we doubtless cannot fully do—English rather than American). The earned lyrical intensity and cumulative psychological discovery of Hardy's sequence gives it special authority despite its lapses. Later British and Irish neo-regionalist sequences should perhaps be thought of as beginning with the largely unacknowledged influence of Hugh MacDiarmid (Christopher Grieve), whose *A Drunk Man Looks at the Thistle* (1926), written in literary Scots, outstrips most of its successors in sheer music and vitality. These successors would at least include David Jones's *The Anathémata* (1952), Basil Bunting's *Briggflatts* (1965), Geoffrey Hill's *Mercian Hymns* (1971), John Montague's *The Rough Field* (1972), and Thomas Kinsella's *Notes from the Land of the Dead* (1973).

Typically, these works have a special political dimension. Their nostalgia for older, indigenous ways can be militantly reminiscent of separatist rhetoric. They stress regional diction and history in opposition to the culture of the "Southrons," as Bunting calls them—the dominant English classes. As with Williams's *Paterson,* these sequences have an organic context of "place" that is a genuine source of energy within their associative field. It determines neither their structural dynamics nor their final poetic quality, yet it does play a role in those values. And we should remember that we are not dealing with a merely provincial poetry. The relation between Pound's method and Williams's—that is, the interaction between Pound's polylingual, cross-culturally preoccupied emphasis and Williams's nativist emphasis—may be taken as a model of the internal tensions within any one of the sequences. Modern poetry has moved simultaneously into both emphases.

2. Hugh MacDiarmid's *A Drunk Man Looks at the Thistle*

MacDiarmid's *A Drunk Man Looks at the Thistle* precisely reveals such an interaction of nativism and cosmopolitanism. Two of its most obvious characteristics are its Scots diction and its references—by direct allusion, imitation, or parody—to a wide variety of non-Scottish sources from Alexander Blok to Else Lasker-Schüler and T. S. Eliot, and from Spengler to Dostoievsky, to say nothing of the whole tradition of English poetry. On the one hand, MacDiarmid characterized his effort as deeply nationalistic, "deriving its unity from its preoccupation with the distinctive elements in Scottish psychology which depend for their effective expression upon the hitherto unrealized potentialities of Braid Scots." And despite the satire of the poem against the sentimental literary Scots tradition and distortion of the memory of Burns's achievement in so much Scottish writing, MacDiarmid avowed his dedication to "the movement for the revival of Braid Scots which is being promoted by the Burns Federation, and to the Scottish Literary Renaissance movement which is seeking to re-establish the distinctively Scottish contribution in the literature of Europe." On the other hand, the poem is international and includes translations from several languages—"from the Russian, French, and German"—and indeed the reference to the "literature of Europe" implies an international dimension.[1]

These characterizations were published by MacDiarmid shortly before the poem appeared and were clearly not felt to be self-contradictory. Intrinsically they were certainly not so. One of the best lyric poems of the sequence, for instance, is the poem "Love." MacDiarmid's note tells us it was "suggested by the French of Edmond Rocher." Perfectly natural—a Scots poem based on a French one but gaining enormously from its conversion into what appears to be colloquial regional speech but is only partly so. The language is not ordinary Scots vernacular but "synthetic Scots"—a mixture with older literary Scots and, in the original

1. A more specifically political internationalism may be seen in MacDiarmid's "First Hymn to Lenin" (1930) and "Second Hymn to Lenin" (1932)—both written in Scots, however.

text, adapted in its spelling and some other usage to English conventions. The nationalist gesture is essentially nothing more than writing genuine poetry in one's own language, or at least within conventions that combine the spoken language with a native literary tradition, under circumstances that demand a modern, sophisticated adaptation of those conventions.

> A luvin' wumman is a licht
> That shows a man his waefu' plicht,
> Bleezin' steady on ilka bane,
> Wrigglin' sinnen an' twinin' vein,
> Or fleerin' quick an' gane again,
> And the mair scunnersome the sicht
> The mair for love and licht he's fain
> Till clear and chitterin' and nesh
> Move a' the miseries o' his flesh.
>
> [A loving woman is a light
> That shows a man his woeful plight,
> Blazing steady in every bone,
> Wriggling sinew and twining vein,
> Or flaring quick and gone again,
> And the more sickening the sight
> The more for love and light he's fain
> Till clear and shivering and fresh
> Move all the miseries of his flesh.][2]

Take this one poem by itself, and it projects the inadequacy a man feels to grasp or match the qualities of a woman's love—her physical and spiritual mingling with him that illuminates their differences; and in the poem all the light comes from her although it also can flare up and then disappear. The element of revulsion enters the poem surprisingly in the sixth line; the more sickening ("scunnersome") the light she throws on his inner nature, the more he pursues further illumination, until he exists in a very ecstasy of misery. The unbroken movement of the final four lines, with their delayed rhymes, contributes mightily to the shift of emphasis and increased complexity and intensity of the ending.

In the context of the whole sequence the intensity of climactic realization here is related to the poem's atmosphere of drunken exuberance mingled with melancholy, and to its lucid candor. There is an implied dramatic situation: one not unlike that of "Tam O'Shanter." The poet has

2. Our text of reference is Hugh MacDiarmid, *Collected Poems* (New York: Macmillan, 1962). We have "translated" the quotations as the simplest way of glossing them in context.

been drinking and wanders out into the night. Exhausted, emotionally volatile, and tipsily conversing with a hillside thistle, he explodes with a wide range of thoughts and dreams and memories and proclamations—variously lusty and painful and fanciful and sardonic, and on every subject from love to politics to literature to religion. Thus the sequence can present, side by side, many poems in different moods and styles as moments in a drunken monologue. The technique is close to Tennyson's in *Maud: A Monodrama* in providing a rationale for these juxtapositions, but there is no tight-knit "story" here. The "situation" is a point of convergence, in part comic, for the several tonalities, and the "drunk man" less a character than an associative device. So too the thistle is a shifting symbol, now betokening the bleakness of stark reality, now Scotland and her history, now raw maleness, now one's impersonal self, and now the struggle between one's bodily and spiritual natures (stalk and blossom). The list of possibilities is inexhaustible. His irresponsible condition allows the "drunk man" to see the thistle however he chooses—just as maundering senility justifies the drift of vulnerable and piercing images and memories in "Gerontion," and just as the need to "shore" the "fragments" haunting the mind "against my ruin" justifies the drift in *The Waste Land* and in the *Pisan Cantos.*

In its original publication *A Drunk Man Looks at the Thistle* ran its 2685 lines of poetry continuously, without breaks to indicate that it was made up of 59 separate poems. The only external indication that one had reached a new poem came from shifts of form, for the poems were not separately numbered or titled. Whitman did the same thing with early editions of *Song of Myself,* but at last divided and numbered the poems (1881 edition). When MacDiarmid's *Collected Poems* was about to be published in the United States, in 1962, his editor suggested that American readers would be greatly aided by having titles for the separate poems, especially since very few knew Scots at all. MacDiarmid quickly accepted the suggestion, providing titles with such alacrity that it seemed he had thought of them from the start despite his saucy original "Author's Note," from which we quote a small portion:

> Drunkenness has a logic of its own with which, even in these decadent days, I believe a sufficient minority of my countrymen remain *au fait.* I would, however, take the liberty of counselling the others, who have no personal experience or sympathetic imagination to guide them, to be chary of attaching any exaggerated importance . . . to such inadvertent reflections of their own sober minds as they may from time to time—as in a distorting mirror—detect in these pages . . . It would have been only further misleading these good folks, therefore, if I had (as, arbitrarily enough at best, I might have done) divided my poem

> into sections or in other ways supplied any of those "hand-rails" which raise false hopes in the ingenuous minds of readers . . .[3]

The position is an attractive one. A poem that is a pure flow of lyrical affects, feeding into one another, exploding into contrasting states, and interacting organically, should perhaps not allow itself to suffer interruption along the way. Insofar as it moves dynamically, the poem's structure should reveal itself in the fullness of time. Yet with a long work made up of clearly separate, independent pieces whose very differences create its dynamics, the position becomes dubious. The separate parts—usually, in truth, separate poems—do exist. Insofar as the work is a sequence, their function as radiating centers becomes clearer through demarcation, which, simply by accenting the fact that they are far from comprising a single, seamless unit, emphatically helps the poet to overcome Poe's objections to the long poem.

An ideally sensitized reader would find the point at issue irrelevant. Such a reader would, as a matter of course, take note of breaks in tonality and identify the affective divisions. But that very response corresponds to what a poet does in numbering divisions, giving them titles, or using some other typographical indication. A poem's flow is full of interruptions—shifts of tonal color, context, and intensity, to say nothing of rhythmic variations—that must be discerned if its quality is to make itself felt. This would seem to be why most sequences have their clearly noted sections. In his late seventies MacDiarmid was nevertheless persuaded to revert to his original unbroken arrangement, and so the reader may compare the *Collected Poems* version with the 1971 redaction edited by John C. Weston.[4] For convenience here, we have chosen to refer to titles in *Collected Poems* rather than to line-numbers (the format in the 1971 edition).[5]

The two opening poems, "Sic Transit Gloria Scotia" and "A Vision of Myself," illustrate the problem just discussed. Poems of 120 and 48 lines respectively, both are written in iambic pentameter quatrains with ballad rhyme—a redundant form over such a long stretch in the first poem alone, and even more so if repeated without variation for twelve stanzas more. Their tone is in certain respects constant throughout both poems: an energetically opinionated holding forth, moderately racy and witty,

3. In *A Drunk Man Looks at the Thistle* (Edinburgh: Blackwood, 1926).
4. New York: Macmillan, 1962; Amherst: University of Massachusetts Press, 1971.
5. For like reasons, we retain the original dialectical spelling, rather than the standard Scots spelling developed by a later generation than MacDiarmid's, which the poet was persuaded to accept for the 1971 edition. The earlier usage helped make the work more available to the general reader of English poetry, and the process of accommodation reflected the floating irony and conscious nationalist insistence rather beautifully.

with an occasional more intimately personal note. The following quatrains are representative:

> You canna gang to a Burns supper even
> Wi'oot some wizened scrunt o' a knock-knee
> Chinee turns roon to say "Him Haggis—velly goot!"
> And ten to wan the piper is a Cockney.
>
> No' wan in fifty kens a wurd Burns wrote
> But misapplied is a'body's property,
> And gin there was his like alive the day
> They'd be the last a kennin' haund to gi'e . . .
>
> [You cannot go to a Burns supper even
> Without some wizened stump of a knock-knee
> Chinaman turns round to say, "Him Haggis—velly goot!"
> And ten to one the piper is a Cockney.
>
> Not one in fifty knows a word Burns wrote
> But misapplied he's everyone's property,
> And if there was his like alive today
> The last to reach an understanding hand they'd be.]
>
> (from "Sic Transit Gloria Scotia")

> I lauch to see my crazy little brain
> —And ither folks'—tak'n itsel' seriously,
> And in a sudden lowe o' fun my saul
> Blinks dozent as the owl I ken't to be.
>
> I'll ha'e nae hauf-way hoose, but aye be whaur
> Extremes meet—it's the only way I ken
> To dodge the crust conceit o' bein' richt
> That damns the vast majority o' men.
>
> [I laugh to see my crazy little brain
> —And other folks'—taking itself seriously,
> And in a sudden blaze of fun my soul
> Blinks dully as the owl I know it to be.
>
> I'll have no halfway house, but always would be where
> Extremes meet—the one way I'm sure I can
> Dodge the curst conceit of being right
> That damns the vast majority of men.]
>
> (from "A Vision of Myself")

One might be hard put, without the poet's typographical help, to say confidently that there were two poems here. The mere fact of separation, even without the titles, is illuminating in the *Collected Poems* version; one comes far more quickly to the slight yet indicative modulation in tone in the second poem—indeed, it becomes obvious instead of being barely discernible at the tail end of a long harangue. The first poem, "Sic Transit Gloria Scotia," is dominated by a publicly directed dismay at contemporary Scotland's cultural disgrace and self-betrayal. The second poem, "A Vision of Myself," pursues a defiant, self-analyzing, dreamily imaginative individualism and openness. The harangue has reoriented itself. It is still bitterly jocular but points the way to the concentrated lyricism and rich visionary excitement of the third poem, "Poet's Pub."

The question of format, then, is of importance. Running both poems together without a break makes for an artistic disproportion. The subtler individualism of the dozen quatrains at the end must balance an extended denunciation (in 30 quatrains) of Scotland's susceptibility to sentimental exploitation of her language and her traditions (including the cheapening of Burns's poetry). And because the rhetoric of "Sic Transit Gloria Scotia" is so expansive, one needs a high absorptive capacity not to feel saturated by the time the modulations of "A Vision of Myself" arrive—themselves not a masterpiece of condensation. Only a Scot, immersed in the heady atmosphere of nationalist cultural politics, would possess that capacity to the full. For the rest of us, the price of sustained attention is flagging sympathy. We can respond to the fine indignation with a certain empathy, discounting its parochial fervor, but will still find it all somewhat labored. The sense of a highly outraged noble sensibility is diminished by the long diatribe and the joke about the "wizened scrunt o' a knock-knee / Chinee." But the poem begins winningly, and the first flashes of satire are lively wit:

> I amna' fou' sae muckle as tired—deid dune.
> It's gey and hard wark coupin' gless for gless
> Wi' Cruivie and Gilsanquhar and the like,
> And I'm no' juist as bauld as aince I wes.
>
> The elbuck fankles in the coorse o' time,
> The sheckle's no' sae souple, and the thrapple
> Grows deef and dour: nae langer up and doun
> Gleg as a squirrel speils the Adam's apple.
>
> Forbye, the stuffie's no' the real MacKay,
> The sun's sel' aince, as sune as ye began it,
> Riz in your vera saul: but what keeks in
> Noo is in truth the vilest "saxpenny planet."
>
> And as the worth's gane doun the cost has risen.
> Yin canna thow the cockles o' yin's hert

Wi'oot ha'en' cauld feet noo, jalousin' what
The wife'll say (I dinna blame her fur't).

It's robbin' Peter to pey Paul at least. . . .
And a' that's Scotch aboot it is the name,
Like a' thing else ca'd Scottish nooadays
—A' destitute o' speerit juist the same.

[I'm not so much drunk as done in—dead tired.
It's pretty hard work coping glass for glass
With Cruivie and Gilsanquhar and the like,
And I'm just not the man that once I was.

The elbow stiffens in the course of time,
The wrist's not so supple, and the throat
Grows dry and dull; no longer up and down
Spry as a squirrel bobs the Adam's apple.

Besides, the stuff's not the real McCoy.
The sun's own self once, as soon as you began it,
Rose in your very soul; but what peeks in
Nowadays is in truth the vilest "sixpenny planet."

And as the worth's gone down the cost has risen.
You cannot thaw the cockles of your heart
Without having cold feet now, thinking of what
The wife will say (not that I blame her for it).

It's robbing Peter to pay Paul at least. . . .
And all that's Scotch about it is the name,
Like anything else called Scottish nowadays
—All destitute of spirit just the same.]

Putting matters simply, this is good talk, splendid talk, colorful and confiding, but it does go on. When the second poem arrives at last, we are politely relieved. It brings with it a new, welcome dimension of sheer emotional volatility: "Heichts o' the lift and benmaist deeps o' sea" ("Heights of the heavens and inmost depths of the sea"). And here the drunk-man role enables a reeling shift, from the gadfly stings of the canny local intellectual to the rhetoric of suffering:

And in the toon that I belang tae
—What tho'ts Montrose or Nazareth?—
Helplessly the folk continue
To lead their livin' death!

[And in the town that I belong to
—What though it's Montrose or Nazareth?—
Helplessly the folk continue
To lead their living death!]

The exuberant talk continues, and the sardonic critique along with it. But the sense has been added of an erratically "extreme" temperament that can sink into absolute gloom. Although the preceding poem had anticipated the linking here of Christ and the modern poet—"A greater Christ, a greater Burns may come"—the parody of Shelley leavened the comparison and perhaps implied, facetiously, that the true poets are, willy-nilly, revolutionary saviours *manqués*. But the lines just quoted suggest a personal despair reinforcing the ironies, and an affinity with the spirit of *The Waste Land.* They prepare us for the decisive affective change in the third poem, adapted, MacDiarmid's note tells us, from the Russian of Alexander Blok. "Poet's Pub" presents a scene at once sordid, eerie, and exalted. A filthy pub full of rowdy drunks is invaded by a beautiful lady out of Elfland:

But ilka evenin' fey and fremt
(Is it a dream nae wauk'nin' proves?)
As to a trystin'-place undreamt,
A silken leddy darkly moves.

[*But every evening fey and strange*
(Is it a dream no waking proves?)
As to a trysting-place undreamt,
A silken lady darkly moves.]

The "leddy" from Elfland has clear affinities with the archetypal divinity whose nature and mythical origins and appearance in poetry and song of earlier centuries Robert Graves so ingeniously examines in *The White Goddess.* So do the ages-old "lass" of "O Wha's the Bride" and the other dread, irresistible manifestations of female power and mystery in the sequence. We shall turn to these other poems shortly. "Poet's Pub" anticipates them, and it is interesting to note that it moves between the quotidian and the visionary in its treatment of woman. Only one mildly humorous effect interrupts its intense accumulation of contrasting tones of squalor and epiphany. We are told that, although "ilka evenin' " the poet conjures up the "leddy" from his whiskey glass, his wife knows nothing of this dream that lifts him out of his parochial "captivity"—"Jean ettles nocht o' this, puir lass."

Such homely confidences—in this instance the husband hiding his intoxicated secret passion from his good wife—are important in the body of the sequence. Jean is the touchstone, throughout, of earthy common sense and healthy-mindedness. Male exuberance and a poet's wild, transforming imagination are not quite of her world, whose reality is as irreducible as it is, after all, desirable. The "drunk man" loves his wife, and his politics, and his "coupin' gless for gless" with boon companions, and even the hillside thistles. In the progression of the sequence, revulsion wars with hearty unsqueamishness, and tenderness for Jean with Chaucerian frankness, and all these complexes of feeling with mystical, aestheticizing reverie.

"Poet's Pub" is the first of the adaptations from the poetry of Continental writers scattered throughout the work. "Love," which we quoted earlier on, is another. Usually these poems are italicized, a device that points up their role as concentrated moments of high lyrical intensity, but MacDiarmid is inconsistent in this usage. "Love" is not italicized, nor is the brilliant ballad of his own composition, "O Wha's the Bride?"—a poem that secularizes and redirects the mystery of the Virgin, making it at once demonic and generously erotic. The exchange in the third stanza, and the promise in the two concluding quatrains, give us the furthest reach of paganized imagination—retrieved from Christian tradition—in the sequence. (The italics used for the "bride's" replies emphasize the mystery, of course, in a way that setting the whole poem either in roman or italics would not have done.)

O wha's the bride that cairries the bunch
O' thistles blinterin' white?
Her cuckold bridegroom little dreids
What he sall ken this nicht.

For closer than gudeman can come
And closer to'r than hersel',
Wha didna need her maidenheid
Has wrocht his purpose fell.

O wha's been here afore me, lass,
And hoo did he get in?
—*A man that deed or was I born*
This evil thing has din.

And left, as it were on a corpse,
Your maidenhead to me?
—*Nae lass, gudeman, sin' Time began*
'S hed ony mair to gi'e.

But I can gi'e ye kindness, lad,
And a pair o' willin' hands,
And you sall ha'e my breists like stars,
My limbs like willow wands.

And on my lips ye'll heed nae mair,
And in my hair forget,
The seed o' a' the men that in
My virgin womb ha'e met. . . .

[O who's the bride that carries the bunch
Of thistles glittering white?
Her cuckold bridegroom little dreams
What he shall know this night.

For closer than husband can come,
Closer to her than herself,
Who did not need her maidenhead
Has wrought his purpose fell.

O who's been here before me, lass,
And how did he get in?
—A man that died ere I was born
This evil thing has done.

And left, as it were on a corpse,
Your maidenhead to me?
—No lass, husband, since Time began
'S had any more to give.

But I can give you kindness, lad,
And a pair of willing hands,
And you shall have my breasts like stars,
My limbs like willow wands.

And on my lips you'll heed no more,
And in my hair forget,
The seed of all the men that in
My virgin womb have met.]

This is MacDiarmid's best-known poem, its chthonic charm compounded of mystifying foolery and erotically visionary terror—Gothic effects, in a sense, yet exquisitely lyrical and, in context, a poem of strangely liberated sexual exultation. It appears toward the end of a cluster of poems engaged with love and sex that take over the sequence for an extended period about a third of the way through. These poems pick up notes introduced very early, in "Poet's Pub," in the self-deprecatory "The Looking Glass," and in the fear-laden "The Unknown Goddess," all of which strongly suggest a a complex psychic state. That is, "Poet's Pub" presents a dark and guilty dream of a faery beloved that betrays "puir Jean"; "The Looking Glass" sees the male self as unsavory; and "The Unknown Goddess" (another adaptation from Blok's Russian) is a pure thrill of dread: "*A licht I canna thole is in the lift*" ("*A light I cannot bear is in the sky*") and

The ends o' space are bricht: at last—oh swift!
While terror clings to me—an unkent face!

[*The ends of space are bright: at last—oh swift!*
While terror clings to me—an unknown face!]

To these poems (and others in like vein), the group that includes "O Wha's the Bride" adds a more direct contemplation of the meaning of sexual love. Often the contemplation is both earthy and psychologically keen, as in "The Psycho-Somatic Quandary," "Love," "In the Last Analysis," "The Spur of Love," "The Feminine Principle," and "The Light of Life." But "O Wha's the Bride" combines the subtle and often bitter realism of these poems (and their affinities with Lawrence's writing) with the supernatural dimensions of the first group, and with a kind of transport of joy despite all the negative notes. Its bearing seems to be toward creation of ecstasy out of the elements of pain; and perhaps it is significant that the poem "Man's Cruel Plight," which presents a half-playful yet serious absorption in the preoccupations of "Masoch and Sade," precedes "O Wha's the Bride" by only a half-dozen poems.

We can, in fact, say that the psychosexual resonances of this sequence are pervasive and deep. They are, most subtly, aspects of a genuine poetry of place, for they embody the interactions among lust and guilt, a morbid romanticism, stereotyped assumptions about the different attitudes of the sexes, and a powerful dream of a state of freedom beyond all these trapped perspectives of Scottish sensibility. The work gets its major emotional force from this vital element. The wit and passion of the more discursive stretches (which take over more and more in the second half of the sequence) are admirable but overdone. These expansive portions threaten to sink the sequence as a sequence, for they impose a factitious continuity of disputation and tend to drown out the centers of affective power. Yet the very language in which they are written, and their reflection of national concerns and habits of thought, and even their intimacy with the supernatural, give them a profound nativism despite the lengthiness.

MacDiarmid, finally, had a genius for squirreling away the writings of other poets and absorbing them—by allusion, parody, quotation, and translation—into his own work. Here the Scots dimension is especially interesting as instrumental in a nativist cosmopolitanism. Via MacDiarmid, Blok and Lasker-Schüler become as it were Scottish poets, and their work becomes part of the Scottish canon—an extraordinary energy assimilates their sensibilities to his. It makes *A Drunk Man Looks at the Thistle* the liveliest and most humanly appealing (apart from Yeats's work) of the British sequences, despite the pressure to convert a confusion of tonalities into a triumph of discourse. The final poem, "Yet Ha'e I Silence Left," points up the artistic problem and the appeal of the work together, and shows how conscious MacDiarmid was of his tendency toward garrulousness. Like Pound in the *Cantos*, he dreams of a redeeming recourse to stillness—the isolation of what Yeats called "the solitary soul"—to free him from the inward turbulence that drives the sequence every which way. At the end, he is able to use humor to avoid full solemnity while intensifying the feeling of dread mystery through contrast with Jean's wifely wit.

> Yet ha'e I Silence left, the croon o' a'.
>
> No' her, wha on the hills langsyne I saw
> Liftin' a foreheid o' perpetual snaw.
>
> No' her, wha in the how-dumb-deid o' nicht
> Kyths, like Eternity in Time's despite.
>
> No' her, withooten shape, wha's name is Daith,
> No' Him, unkennable abies to faith
>
> —God whom, gin e'er He saw a Man, 'ud be
> E'en mair dumfooner'd at the sicht than he.
>
> —But Him, whom nocht in man or Deity,
> Or Daith or Dreid or Laneliness can touch,
> *Wha's deed owre often and has seen owre much.*
>
> O I ha'e Silence left,
>
> —"And weel ye micht,"
> Sae Jean'll say, "efter sic a nicht!"
>
> [Yet have I Silence left, the crown of all.
>
> Not her, who on the hills long since I saw
> Lifting a forehead of perpetual snow.
>
> Not her, who in the deepest dead of night
> Shows forth, like Eternity in Time's despite.
>
> Not her, without shape, whose name is Death,
> Not Him, unknowable except to faith
>
> —God who,[6] if ever He saw a man, would be
> Even more dumbfounded at the sight than he.
>
> —But Him, whom nought in man or Deity,
> Or Death or Dread or Loneliness can touch,
> *Who's died too often and has seen too much.*

6. The "whom" in the original seems a typographical error.

O I have Silence left,

—"And well you might,"
So Jean'll say, "after such a night!"]

3. Thomas Kinsella and a Landscape with Ancestral Figures; Basil Bunting's *Briggflatts*

Modern poetry, in the British Isles especially, is ever more a poetry of the recapture of lost worlds. As we have seen in *A Drunk Man Looks at the Thistle,* this entails a nation's or region's deep history and speech and mystical visions; but also it can involve sunken memories of families and the primal impressions of early childhood. A poet who, like Kinsella, engages ably and bravely with these worlds reaches past surface charm and nostalgia to something like discovery. He is coping with the intractable in this engagement. As Kinsella puts it in his poem "Ritual of Departure," he must confront a

> Landscape with ancestral figures . . . names
> Settling and intermixing on the earth,
> The seed in slow retreat, through time and blood,
> Into bestial silence.
> Faces sharpen and grow blank,
> With eyes for nothing. . . .

Kinsella has burrowed into his familial and psychological past most persistently, and with far greater freedom from the rhetoric of cultural conservatism than most other poets. He is an elegist with a bitter, grieving, melodious tongue, whose most effective poems unfold in a manner difficult to illustrate in brief quotations. They tend to open quietly, perhaps humorously, in the midst of reverie or of a commonplace situation, and then to move by easy modulations into their full realization. The poem "His Father's Hands," in *Peppercanister Poems,* is an example. It begins with a domestic closeup of the poet and his father drinking together and arguing about something. Their gestures are comically intimate: "I drank firmly / and set the glass down between us firmly," while "His finger prodded and prodded, / marring his point." A quick association recalls the childhood memory of the poet's grandfather, a cobbler, pressing tobacco into his pipe and cutting new leather. Details multiply lovingly and vividly; and suddenly we are with the grandfather, now very old, playing the fiddle with "his bow hand scarcely moving" and "whispering with the tune." The poem breaks into song, combining the words of the old tune with the poet's interpolations:

with breaking heart
whene'er I hear
in privacy, across a blocked void,

the wind that shakes the barley . . .

A sweet yet piercing rapport across the generations has been evoked, and next we hear Grandfather's voice telling the boy something about family history. ("Your family, Thomas, met with and helped / many of the Croppies in hiding from the Yeos.") It is, as it were, the personal history of Ireland's common people: the battles, the escapes, the hangings, the migrations, the available occupations and trades—all the crucial experience that gives a long body and range of personality to the multitudes whom ordinary history renders anonymous and unimportant. Near the end of the poem an impersonal, terrified vision of a menacing landscape in an evil-driven land gives a hard, darkly gleaming turn to the language. The poem thus casts an ironic and tragic mood over the final references to Grandfather's "blocked gentleness" and the child's memories.

"His Father's Hands" is but one poem in a complex sequence called *One* (1974), which as a whole is a nightmarish plunge into the "ghost companionship" of past worlds. But it does touch on most of Kinsella's preoccupations, as unfolded in the three reciprocal sequences *Notes from the Land of the Dead* (1973), *One*, and *A Technical Supplement* (1976).[7] Three poems in the first of these, "Ancestor," "A Hand of Solo," and "Tear," are notable for their fusion of intense revulsion and love within the precise rendering of childhood drama. *A Technical Supplement* is charged with horror. Its literal slaughterhouse closeups constitute a hideous indictment of human bestiality; all of history, including the Holocaust and a hopeless sense of our destiny, is embodied in these closeups, and the internalization of the horror produces subtly morbid realizations in the poetry. The pressure on human endurance, which Kinsella had explored earlier in his marriage-sequence *Wormwood* (1966), is seen, in these three later sequences, as inseparable from clear awareness and from the brute nature of social reality. Working-class life (including involvement with the daily events of the slaughterhouse) and struggle, psychic agony, and the deeply private backwards exploration of family memories converge in the bitterly elegiac hardness of Kinsella's work.

Such poetry of lost worlds and human loss, strongly rooted in the common life but not limited thereby in its sensibility, seems a natural

7. In *Poems 1956–1973* (Winston-Salem, N. C.: Wake Forest University Press, 1979), pp. 129–73, and *Peppercanister Poems* (Wake Forest University Press, 1979), pp. 51–72 and 73–96 respectively. The sequence *Wormwood* is in *Poems 1956–1973*, pp. 65–73.

tendency in sequences written in the British Isles. MacDiarmid and Kinsella provide a Scottish and an Irish example, and the best known of the strictly English sequences of lost worlds is Basil Bunting's *Briggflatts*. Less richly varied and copiously energetic than *A Drunk Man Looks at the Thistle*, and less undeviatingly introspective and elusively associative in feeling than Kinsella's sequences, its appearance late in the poet's career (1966) seemed a confirmation of Pound's methods by a writer both overwhelmed by the master's magnetic genius and yet somehow true to his own very different, if slighter, talent.

Briggflatts came to us in Ezra's long wake. The affinities and derivations are clear. *Briggflatts* even has its hell-canto: Part III, with the necessary scatological smell. In general, the sequence presents the usual Poundian mixture of tonalities that has influenced so many other modern works—the affirmations amidst drastic alienation, the rhetorical outbursts against the evils of modern urban culture, the moments of keenest observation or dancing fantasy or passionate memory, the flashes of ironic or exalted insight superimposing the real or mythical past upon the present instant, and the devout aestheticism that yet does not blur an uncompromisingly Anglo-Saxon gloom and even fatalism about life as it is.

Yet within the Poundian shell Bunting made a small, pure, very English creation of his own: a living stream of verse that makes its way from beginning to end, disappearing from sight at times when the urge to bluster and convince takes over or the voices of other poets—not only Pound but a chorus of others from Tennyson to Auden—break in with a kind of static. Various currents of tonality make up the essential movement, combining to form an ultimately elegiac poem. They include lyrically celebratory moments and memories of young love and of the hard-working but wholesome life of artisans—memories darkened by the poet's sense that he has betrayed their purity by abandoning the provincial world of his youth for the fleshpots of urban and cosmopolitan spiritual corruption. The world thus forsaken is the world of Northumbria and its Anglo-Saxon past, evoked by direct historical reference and by the fairly frequent use of a starkly alliterative, compressed line. The evocation of a primal native culture, with its axe-swinging warriors, earthy basic language, and raw contact with resistant nature, is another important current of tonality, one that supports the heavy depression with which the sequence contends and to which it eventually accommodates itself. It reinforces the idealized memory of the stonemason's rigorous yet enchanted realm that lies at the heart of the poet's nostalgia for his youthful past. The stonemason's tools and skills were exercised against the same resistant nature with which primitive folk contended. To have left his tutelage, and the love of his daughter, is identified with turning one's back on the regional life and language and history that are the very wellspring of one's most compelling meanings.

Part I of *Briggflatts* is the heart of the sequence and its most successful section. Its dozen thirteen-line stanzas, each ending in a rhymed couplet, combine Anglo-Saxon with modern versification in flexible units allowing for complex development and for many shifts of feeling and intensity. The exquisite opening stanza is at once sheer song, charming comedy, and ominous vision. It is saturated with a sense of bucolic locality and with the mixture of elation and near-lugubriousness characteristic of the whole work:

> Brag, sweet tenor bull,
> descant on Rawthey's madrigal,
> each pebble its part
> for the fells' late spring.
> Dance tiptoe, bull,
> black against may.
> Ridiculous and lovely
> chase hurdling shadows
> morning into noon.
> May on the bull's hide
> and through the dale
> furrows fill with may,
> paving the slowworm's way.[8]

The ancient, familiar psychic and literary linking of vital sexuality with the death-principle is obvious here yet held at a distance by the sheer delightful buffoonery of the first two sentences, though they too have their shadows. We have been charmed into the dominant tonal realm of the sequence, where currents of buoyancy and power and decay and fatalism constantly mingle in varying proportions. Then, in the second stanza, this mingling of opposites (spring and fertility, intractable reality and death—the bull enjoined to "dance tiptoe" and "the slowworm's way") continues in a grimmer key. Notice, for instance, how much more emphatic its final couplet is than the first stanza's close.

> A mason times his mallet
> to a lark's twitter,
> listening while the marble rests,
> lays his rule
> at a letter's edge,
> fingertips checking,
> till the stone spells a name
> naming none,
> a man abolished.

8. Our text of reference is Basil Bunting, *Collected Poems* (Oxford: Oxford University Press, 1978).

Painful lark, labouring to rise!
The solemn mallet says:
In the grave's slot
he lies. We rot.

The middle stanzas (5–9) of this opening poem center on the very young, just pubescent lovers. The fifth stanza especially evokes the homespun passion and magic of that remembered time, placing it within its context of a laboring world of plain folk whose everyday life vividly reflects their long history of tough perseverance. We see the "children" in their astringent Eden, lying together in the horse-drawn cart the mason uses to fetch marble for his craft:

Stocking to stocking, jersey to jersey,
head to a hard arm,
they kiss under the rain,
bruised by their marble bed.
In Garsdale, dawn;
at Hawes, tea from the can.
Rain stops, sacks
steam in the sun, they sit up.
Copper-wire moustache,
sea-reflecting eyes
and Baltic plainsong speech
declare: By such rocks
men killed Bloodaxe.

Apart from the fey lyricism of the poem's opening lines, Bunting's great achievement in this first movement of *Briggflatts* is his recovery of a lost world of reality: its decisive sensuous detail, the body of its physical presence. Thus, two stanzas further on, the journeyers are home again.

Rain rinses the road,
the bull streams and laments.
Sour rye porridge from the hob
with cream and black tea,
meat, crust and crumb.
Her parents in bed
the children dry their clothes.
He has untied the tape
of her striped flannel drawers
before the range. Naked
on the pricked rag mat
his fingers comb
thatch of his manhood's home.

The nostalgia here may be too pungent, the recalled pubescent sexuality at once sentimental and a little brackish; but the atmosphere that saturates this stanza and the one quoted just before has the authority of indelibly significant memory. The authority is reinforced by another, more impersonal sort of memory imbedded in the older regional language deployed throughout this section: "Their becks ring on limestone," "fellside bleat," "fog on fells"—and with it the heavy stresses and echoes of an ancient poetry. Neither the historical nor the personal past can be restored: "No hope of going back." Even the recovery in words is painfully difficult, like the mason's work: "It is easier to die than to remember." And yet the opening poem *has* remembered, in the face of a debilitating depression that rides almost every stanza and controls the whole sequence except in certain limited respects.

The four remaining sections of *Briggflatts* cope with the work's prevailing depressive perspective in various ways. One way is the exaltation of disciplined workmanship with intractable materials: "No worn tool / whittles stone." The axe swung by fighting forebears, and their speech that was a sharp, rock-splitting weapon in its own right, were tools for yet different types of stonemasonry. So the work presents a staunch ideal, despite the pervasive imagery of cultural and private defeat. In Part II, the inevitable downfall of the axe-wielders is contemplated in terms suggesting a historical betrayal; there is a certain parallel here to the Poundian view of the triumph of fraudulence in the modern world:

> Loaded with mail of linked lies,
> what weapon can the king lift to fight
> when chance-met enemies employ sly
> sword and shoulder-piercing pike,
> pressed into the mire,
> trampled and hewn till a knife
> —in whose hand?—severs tight
> neck cords? Axe rusts. . . .

In the same section, music and myth are drawn upon as sources of morale for the poem. The defeated king's struggle, in the lines just quoted, finds a curious counterpart in men's efforts to encompass natural process in the organic structures of their own creation. The musical theme is developed in four subtly unfolding sexains that take us from a simple, amiable equation—

> Starfish, poinsettia on a half-tide crag,
> a galliard by Byrd—

to the more complex, deliberately unpleasant proposition that a

> rat, grey, rummaging
> behind the compost heap has daring
> to thread, lithe and alert, Schoenberg's maze.

The mythical motif is introduced in the concluding, and climactic, stanza of the section. This stanza reinforces the impression created in Part I of the sequence that the most intense notes of affirmation in *Briggflatts* will have a nearly pornographic glow of erotic transport. Part II ends with lines recalling the myth of Pasiphaë, who

> heard the bull-god's feet
> scattering sand,
> breathed byre-stink, yet stood
> with expectant hand
> to guide his seed to its soil;
> nor did flesh flinch
> distended by the brute
> nor loaded spirit sink
> till it had gloried in unlike creation.

The pedigree for such a passage is: out of Ovid, by Ezra Pound, with a definitive rankness contributed by Basil Bunting.

Apart from this climactic finale, Part II has a rather wandering movement. It begins with language of utter desolation, in a recognizable, even trite, poetic mode. The alienated poet walks through London streets disheartened by the things that disheartened Blake, Wordsworth, and Eliot before him. (One difference, however, is the "rankness" just mentioned: the special turn of sexual obsessiveness that colors his sense of—and participation in—metropolitan squalor. Then the poem turns away from the city, with its available "sluts" and consequent opportunities for the poet to grow "sick, self-maimed, self-hating.") Suddenly a language of self-questioning is introduced that lifts the poem's sensibility out of the romantically autobiographical ruck. Instead of the plaints of an Eliot *manqué*—

> He lies with one to long for another,
> sick, self-maimed, self-hating,
> obstinate, mating
> beauty with squalor to beget lines still-born—

the question is posed whether one can somehow contrive to redirect one's destiny:

What twist can counter the force
that holds back
woods I roll?

According to Mallarmé's "*UN COUP DE DÉS*," which lurks behind the passage in which these lines are imbedded, no trick of the dice will work to change things. No direct answer is given to the hopeless question in Bunting's poem. Instead, we find ourselves, without transition, in the midst of a fantasy-voyage in cold northern seas, its context that of the Anglo-Saxon "Wanderer." The shift of poetic focus here culminates in the doom-laden intoning endemic to such verse. And after this shift we are borne, once more abruptly and arbitrarily, toward the appealingly sensual south: escape to Italy and lush self-indulgence, with the diction reminiscent of Browning's "The Englishman in Italy." But this mood, too, alters. It is too relaxed and free for Anglo-Saxon severity to bear, a desertion of the rigors of chisel and mallet and rime-cold seas and the ever-presence of stark deprivation and uneaseful death: "Wind, sun, sea upbraid / justly an unconvinced deserter."

So Italy—dream-Italy—is left behind. With its "white marble stained like a urinal" and its innumerable teeming dead, it is no dream-world really anyway, and certainly no genuine salvation. The poem must return to the true ground of its being, the realm of old Bloodaxe and his descendants, whose stone is no soft marble but as hard as death. It must face directly into the challenge, despite near-hopelessness and a knowledge of helpless corruption, of rocklike native realities. Those realities include the literal landscape of one's birth and early life and heritage, and they demand the cultivation of renewed indigenous art despite cosmopolitan seductions. They call, too, for repossession, sweatily and experientially, of certain mythical events—a repossession like Ovid's but at greater risk than his because crucial personal commitment is at stake. The events chosen (the myth of Pasiphae and the bull) are conceived as projections of agonizing, probably destructive elementary choices entailing an oddly hard-pressed ecstasy, one that would seem more like its opposite to an imagination less in love with pain as a proof of identity.

Be that as it may, from here on the sequence strikes various balances in the long struggle with a profoundly depressive subjective state. Part III is the "hell-canto" of *Briggflatts,* visualizing modern man as reduced to dung-vending in the marketplace. Only nature's cleansing rhythms can purge away the vision of sheer foulness presiding over this section. Parts IV and V, after the drop of III into the abyss of total revulsion, settle into something like release through acceptance of life's meanness and through living within one's emotional means. Part IV finds sources of energy in the candor and death-preoccupation of Cymric poetry. The poet accepts poverty, a new and very earthy love, and the plainest satisfactions, along with his irrevocable separation from the stonemason's world. Nevertheless, separation has broken the heart permanently:

> Stars disperse. We too,
> further from neighbours
> now the year ages.

These lines, ending the section, make a transition to the very lyrical Part V, which approaches Part I in its formal character. There is no rhetoric to mar the melody of this movement, which begins with winter-images rather than the vernal ones that open the poem. The sequence has returned home, finding sufficient calm, and an entrancement with the barren landscape, that together make a special music of sheer perception, even when what is perceived is chill and barren:

> Light lifts from the water.
> Frost has put rowan down,
> a russet blotch of bracken
> tousled about the trunk.
> Bleached sky. Cirrus
> reflects sun that has left
> nothing to badger eyes.

Because it is in part a reprise of the earlier sections, and because it also moves into a wider, cosmic frame of reference, the fifth section is somewhat overextended. Otherwise, however, its precise conversion of feeling into a distanced, impersonal language of impressionistic nature-description is as effective as it is surprising at this critical point in the sequence. Bunting chose to end with neither a bang nor a whimper. Rather, the movement is straightforward and elegantly controlled; its reverberations are at once joyous and cool, until a few notes of loss and regret chime in at the very last. Of course, the "Coda" that follows plunges the work into the sea of primordial despondency again. So be it.

Briggflatts is Bunting's one developed effort in the direction of a major modern poem. It was to a very slight degree foreshadowed in his longish poem "Villon," written in 1925 when he was twenty-five, which contained the following quatrains that would have been his major claim to attention were it not for the sequence he wrote forty years later:

> Remember, imbeciles and wits,
> sots and ascetics, fair and foul,

young girls with little tender tits,
that DEATH is written over all.

Worn hides that scarcely clothe the soul
they are so rotten, old and thin,
or firm and soft and warm and full—
fellmonger Death gets every skin.

All that is piteous, all that's fair,
all that is fat and scant of breath,
Elisha's baldness, Helen's hair,
is Death's collateral. . .

And General Grant and General Lee,
Patti and Florence Nightingale,
Like Tyro and Antiope
drift among ghosts in Hell,

know nothing, are nothing, save a fume
driving across a mind
preoccupied with this: our doom
is, to be sifted by the wind.

These acidly elegiac quatrains form a small movement that beautifully catches Villon's poignancy and verve. It both echoes the fifteenth-century master's pathos and Pound's ways of projecting the vitality of such poetic music for his own contemporaries. It is not hard to see that the spirit of these quatrains is still retained in *Briggflatts*, with its heavy morbidity and its contrasting melodic buoyancy in the opening passage and here and there elsewhere. But the later work is anchored in precise observation, together with a complex visual and aural repossession of the lost personal past and the partial compensations of fantasy and practical compromise that make the work "An Autobiography" (Bunting's subtitle). The modern confessional sequence, with its strong component of such repossession and the necessary "neo-regionalism" that provides so much of its ground and body, took renewed energy from the *Pisan Cantos* and allowed each poet of sufficient artistry to find his or her one crucial work of self-encompassment. (We would suggest that, for Kinsella, this work is still in the making: a distillation and recombining related to his poems that we have touched on, but independent of them too.)

4. David Jones's *The Anathémata*, Geoffrey Hill's *Mercian Hymns*, and John Montague's *The Rough Field*

The effort to summon up a lost way of life through sensuous evocation, bringing to bear the whole sense of self of the remembering poet, is as serious and moving an artistic enterprise as can be imagined—provided the literal details of history and memory do not swamp the aesthetic structure. Pound's ambiguous yet resonant pronouncement is the poetic issue:

> nothing matters but the quality
> of the affection—
> in the end—that has carved the trace in the mind
> dove sta memoria
>
> (Canto 76)

In a purely personal sense, we may take this passage as a resolution of self-questionings having to do with love and remorse over neglected relationships, half-apologetic but finally no more so than the "Pull down thy vanity" passage in Canto 81. Poetically, however, we take "the quality of the affection" to be the pure quality or heft of a felt realization, as a measure of the artistic conviction and authority of a work. Pound's very phrasing implies a process of aesthetic conversion: "carved the trace in the mind / dove sta memoria." The wonderful compression here not only balances but unifies the two concerns, the one private and the other impersonal because focused entirely on the mobilizing of language into poetic structure. The "carving" metaphor establishes the reciprocity between what remains in the self's inmost depths and the detached transformations of art. The reciprocity may be a paradox logically; but in art, and certainly in the kind of poetry we are discussing, it is the indispensable basis both of a necessary volatility and of self-transcendence.

David Jones's *The Anathémata* does not have quite this sort of balance or fusion. Its warm baths of ethnological wonderment and linguistic bemusement lave us in waves of association that are, in their fashion, alive with richly immediate thought. But we find nothing in this work like the passages of personal memory in *Briggflatts* that infuse our senses with their evocation of how a way of life once felt. Using free but recurrent rhythms and an often riddling diction, *The Anathémata* maintains its forward movement over the long haul without quite reaching heights of intensity. Its passion lies in its delight at the convergence of cultures and the centuries-long echoing of languages and dialects, together with an elegiac dismay at the loss of faith and its rituals over the ages. At the same time, a kindly human feeling accompanies the poem's erudite forays, reflected in its colloquial and cockney moments but even more, perhaps, in its longing for a complete grasp of all the particular meanings that people have attached to places and images and ceremonial communion: all knowledge and all experience is our province, could we only hold fast to it all.

And when
where, how or ever again?
. . . or again?
Not ever again?
never?
After the conflagrations
in the times of forgetting?
in the loops between?
before the prides
and after the happy falls?

Spes!
answer me!!
How right you are—
blindfold's best!
But, where d'you think the flukes of y'r hook'll hold
next—from the *feel* of things?
(p. 93)[9]

The major tones of *The Anathémata* bring together a questioning incantation punctuated by exclamatory insistences, a language of physical energy, and a sense of awe allied to the poet's Roman Catholicism and the phrasing of the Roman Mass. Ceaselessly, the poem chants the mysteries of the time-layers ("deposits") underlying British consciousness: the continuing presence of Celtic, Teutonic, and Roman strains, conditioned by European and ecclesiastical history. Meanwhile, page by page, we find a massive accompaniment of footnotes—one of the most unabashed instances of the sort that poetry affords. Read unsympathetically, these footnotes weigh the poem down hopelessly. They keep reminding us, however, of the linguistic, historical, mythical, and religious research—*loving* research—with which every rift has been loaded. One sympathetic and sensitive reader, Peter Levi, has suggested that the notes are pure ore:

> David Jones' poetry . . . is marked by an extreme, even a difficult precision. The *Anathémata* certainly needs several readings, and one can grasp the precise relations of words only after mastering the material in the notes. Fortunately the notes are most interesting even if there were no text, so this is not a painful process. Any poet who needs to be precise and particular over an enormous range of language as well as facts must necessarily make such a demand.[10]

9. Our text of reference is David Jones, *The Anathémata* (New York: Chilmark Press, n.d.; and Viking, 1965).
10. Peter Levi, S. J., "The Poetry of David Jones," *Agenda*, V (Spring–Summer 1967), 85.

For Mr. Levi (at that time Peter Levi, S. J.), the theological, Biblical, and archetypal absorption of the notes, and their archaeological and linguistic concerns generally, were so intriguing as to make them an independent source of satisfaction and therefore, perforce, an added value for the poem. The sense of this emphasis is partly reflected in Mr. Levi's assertion that the role of "a prince" (that is, a figure of power and responsibility in history) in Jones's poetry "cannot be understood without reference to the voluntary sufferings of Christ."[11] Indeed, one great attraction of the poem is its immersion in an imagined revelation of a world so dense with mythic power it makes old faiths and ways more real than our present empirical lives. The glamour of the effort is attractive even to secularist and untraditionalist readers, although such readers would be more likely than not to resist a certain cloying half-archness in the style at certain points. They may see the Joycean infusions and shifting line-breaks, together with the running spryness of thought and image, primarily as devices—however ingenious and effective—for countering the inertness of the implied discourse. And of course the battery of footnotes, while it has the fascination of dictionaries and encyclopedias, has its drawbacks as well in the realm of poetic dynamics. Even when we grant the lively, often playful, and fully engaged spirit at work giving plastic form to an intractable mass and thereby itself trying "to master the material in the notes," the riddling questions and the often obvious efforts to charge the poem's batteries with Joycean and colloquial effects themselves take on a certain redundancy. A characteristic passage:

> And was it the Lord Poseidon got him
> on the Lady of Tyre
> queen of the sea-marts
> or was his dam in far Colchis abed?
> did an Argo's Grogram sire him?
> Certain he's part of the olden timbers: watch out for
> the run o' the grain on him—look how his ancient knars are
> salted and the wounds of the bitter sea on him.
> He's drained it again.
>
> (96–97)

We shall return again, briefly, to *The Anathémata*. Despite its many shifts of focus, rhythmic patterning, and tones, it does not quite—in its capacity as a 243-page poetic exploration of the archetypally subjective history of England, amply annotated—avoid being a discursive work, at least in the Lucretian sense (and allowing for the aspect of free formal

11. Ibid., p. 84.

improvisation familiar to modern poetry). It might be considered a poetic exposition in eight chapters, using certain characteristics of the modern sequence without quite having its essential emphatic discontinuities and confessional dimension. One can see why this poem, with its genuine virtuosity and high achievement, has nevertheless not inspired much direct imitation. It is not merely a matter of its originality and idiosyncrasy, or even its sheer bulk and heavy learning. Charles Olson, in the United States, worked with equally unwieldy materials and without Jones's dependable technical skill at the middle level. But his affective units are usually more distinct in *The Maximus Poems,* and so are the moments of rapturous or angry song and of personal feeling and experience; sometime these moments project an engagement with the intractable that threatens to get out of hand through sheer intensity. As a sequence (we do not speak of its special qualities as a work *sui generis,* and as a storehouse of implied values of the sort that deeply engage truer sequences as well) it is too laden with its layers of treasured cultural memory—the *anathémata* of its title, as we shall see further along. The pleasures of a rich connoisseurship of erudition, with a strong bent toward religious devotions combined with distaste for the vulgar ignorance of a rootless, pragmatic age, provide it with a thick insulation of knowledgeable complacency despite its humane awareness. And the sting of the personal is missing—Pound in the cage remembering Provence, Bunting ruing his necessary betrayal of his lost mentor and lost love. The "quality of the affection" certainly matters greatly, and staunchly, here, but in a more distanced way than in our model sequences, without their emphasis on its decisive role in lyrical structure.

Appearing in 1951, *The Anathémata* marked the postwar crisis in British poetry that accounts for the rise of a wild or rogue tendency in figures like Peter Redgrove and Ted Hughes. Their reaction to the perfecting, by someone like Jones, of traditionalist poetic sensibility embedded in the "experimentalism" of an earlier generation was to return to primitivistic sources: picking up, as it were, from Lawrence. That on the whole they have not prevailed in England can be seen in most current British verse, and in the praise given Geoffrey Hill's *Mercian Hymns.*[12] This volume is in one sense a miniature *Anathémata,* with comparable allusive trappings and with a certain number of lexical and historical end-notes that might almost have been cullings from Jones's leftovers, so that we might all too readily lose sight of its simple poetic qualities. It is made up of prose-poems that often engage ear and imagination lyrically—thus, the endings of Hymns XXVII and XXVIII:

12. London: André Deutsch, 1971.

> After that shadowy, thrashing midsummer hail-storm, Earth lay for a while, the ghost-bride of livid Thor, butcher of strawberries, and the shire-tree dripped red in the arena of its uprooting

and:

> . . . A solitary axeblow that is the echo of a lost sound.
>
> Tumult recedes as though into the long rain. Groves of legendary holly; silverdark the ridged gleam.

Such moments, however faintly derivative, are far from being discursive and are dynamically spotlighted in their framed isolation. *Mercian Hymns* suffers from self-conscious, recherché knowingness and allusiveness, yet it does manage to sustain lyrical integrity—not as fully as the finer moments in *Briggflatts* but enough to keep the sequence of highly concentrated poems in motion. Like Bunting more than Jones, Hill balances the cruelty of past worlds against their nourishing values. The crassness of state power, past and present, inspires him to satire. Nevertheless, his sequence laments the dwindling away of the shared memory of a people's tribal communion and organic reciprocity with the land, somehow preferring old social wrongs to the alienating, less personal forms of modern power. (This, at any rate, is the basic gravamen, although concealed somewhat by the pervasive irony.) For instance, the three poems to each of which Hill, in his "List of Hymns" that follows the sequence, gives the title "Opus Anglicanum" (Hymns XXIII–XXV) suggest a devolution of communal aesthetic standards over the centuries. They take us from the high humanism of English embroidery and English Romanesque sculpture in the Middle Ages to the antihumanism of nineteenth-century capitalist production: that is, from the labor that was also art and embodied transcendence over life's hardships and death itself to labor that was sheer exploitation. Thus, the difference between the opening of Hymn XXIII—

> In tapestries, in dreams, they gathered, as it was enacted, the return, the re-entry of transcendence into this sublunary world. *Opus Anglicanum,* their stringent mystery riddled by needles: the silver veining, the gold leaf, voluted grape-vine, masterworks of treacherous thread—

and the full bitter outburst of Hymn XXV, the most striking of the hymns:

> Brooding on the eightieth letter of *Fors Clavigera,* I speak this in memory of my grandmother, whose childhood and prime womanhood were spent in the nailer's darg.

> The nailshop stood back of the cottage, by the fold. It reeked stale mineral sweat. Sparks had furred its low roof. In dawn-light the troughed water floated a damson-bloom of dust—
>
> not to be shaken by posthumous clamour. It is one thing to celebrate the "quick forge," another to cradle a face hare-lipped by the searing wire.
>
> Brooding on the eightieth letter of *Fors Clavigera*, I speak this in memory of my grandmother, whose childhood and prime womanhood were spent in the nailer's darg.

Perhaps the phrase "the nailer's darg" impedes the poem somewhat; Hill is constrained to explain it in an endnote citing the *O.E.D.* definition ("a day's work, the task of a day"), which also notes Ruskin's repeated use of the word in *Fors Clavigera* and elsewhere. But "darg," in context, is at least associated with the "nailer's workshop" although we could never derive its literal meaning from the fact, and it does connote things dark, hard, and bleak. And the poem is a rare surfacing in the sequence of its strongest current: a mingled anger, humiliation, and guilt at the insult to humanity felt especially in working-class family memory but in general history and personal recollections as well. As with Kinsella and Bunting, a profound emotional loyalty to lower-class origins betrays itself at a crucial point in the sequence. The wound in innumerable British and Irish poems, the memory that unlocks the poet's sense of his duty to free his people symbolically from the intolerable historical offense done them by poverty and class oppression, is exposed in this poem. It has a close kinship with the pervasive tones of a whole body of poetry: tones of disillusionment, of strenuous effort to undo the brutal past by reconceiving it and giving the anonymous victims a virtual new life and identity thereby. This effort receives special stress in the unusually forthright Hymn XXV, with its stark, identical first and last lines.

Four other major streams of tonality cut through *Mercian Hymns*, which consists of thirty poems, each made up of from one to four extended lines of the sort we have been quoting. The streams often merge in eccentric and revealing combinations. One such stream has the aura of barbaric dignity associated with the most ancient past. Another has the bad smell of power in whatever context: the ruthlessness of Offa, the ancient Mercian ruler, but also that of modern rulers and in fact of anything—process itself—that makes for the ruin and dismantling of cultural memory. Thus, in Hymn XII we see men digging thoughtlessly into a soil that contains tokens of every kind from the forgotten life of the past, including ancient coins they retrieve. Wealth, dissociated from a living culture, is seen as a kind of excrement that all of us "accrue."

> Their spades grafted through the variably-resistant soil. They clove to the hoard. They ransacked epiphanies, vertebrae of the chimera, armour of wild bees' larvae. They struck the fire-dragon's faceted skin.
>
> The men were paid to caulk water-pipes. They brewed and pissed amid splendour; their latrine seethed its estuary through nettles. They are scattered to your collations, moldywarp.
>
> It is autumn. Chestnut-boughs clash their inflamed leaves. The garden festers for attention: telluric cultures enriched with shards, corms, nodules, the sunk solids of gravity. I have accrued a golden and stinking blaze.

A fourth stream of tonality projects or reflects fantasies of power and greatness in the poet himself, both as child and as adult. Here we have a good deal of sticky confessionalism, somewhat resembling Bunting's but derivative from many other writers including the early Auden, Randall Jarrell, and Robert Lowell. A fifth stream evokes the essential continuities, literal and mythical, of the region: earth-smell, flora and fauna, the continuing tribal sense. All the quotations we have given from *Mercian Hymns* (except Hymn XXV) are colored by this stream to a greater or lesser degree.

Hill avoids the bland expansiveness of Jones and the Poundian haranguing of Bunting partly by the sheer discipline of compression. In addition to his poems' brevity, each of Hill's lines (itself compressed even when it runs longer than usual) constitutes an independent associative unit. This is so even when its relationship to the other units in a poem is intellectually and emotionally obvious; but ordinarily we have association by juxtaposition of unlike affective units. Moreover, the associative leap called for is a wrenching one—a torque-effect that twists the poem abruptly into a new positioning, as the final sentence of Hymn XII, just quoted, does rather melodramatically.

The five streams of affect we have noted find their emotional source and their deepest channels in the sense of a region: its continuities from the furthest past, its encompassment of the private selves of its inhabitants together with the fauna and flora and composition of the soil, and the flow of power throughout its history—a history that even the most exploited inhabitant shares and, by way of the curious workings of cultural memory, remembers as the history of *our* (not *their*) power and achievements. So, in *Mercian Hymns,* the Anglo-Saxon past is linked to the present moment and the ancient King Offa with his successors to power over the centuries and with the "remembering" inhabitant (the poet) who rules over a kingdom of repossession and imagination in his private sensibility. Hill's introduction, at the head of his section of notes called "Acknowledgments," explains:

> The historical King Offa reigned over Mercia (and the greater part of England south of the Humber) in the years AD 757–796. During early medieval times he was already becoming a creature of legend. The Offa who figures in this sequence might perhaps most usefully be regarded as the presiding genius of the West Midlands, his dominion enduring from the middle of the eighth century until the middle of the twentieth (and possibly beyond). The indication of such a timespan will, I trust, explain and to some extent justify a number of anachronisms.

The poet's self-identification as an Offa of the modern "Mercian" imagination, the carrier of British cultural memory and its embodiment in his literal self, is made clear early on in the sequence, most strikingly in Hymn IV:

> I was invested in mother-earth, the crypt of roots and endings. Child's-play. I abode there, bided my time: where the mole
>
> shouldered the clogged wheel, his gold solidus; where dry-dust badgers thronged the Roman flues, the long-unlooked-for mansions of our tribe.

In the next Hymns the identification is given its specifically modern, personal grounding: "Dreamy, smug-faced, sick on outings—I who was taken to be a king of some kind, a prodigy, a maimed one" (Hymn V); "The princes of Mercia were badger and raven. Thrall to their freedom, I dug and hoarded" (Hymn VI); "Ceolred was his friend and remained so, even after the day of the lost fighter: a biplane, already obsolete and irreplaceable, two inches of heavy snub silver. Ceolred let it spin through a hole in the classroom-floorboards," and later "he lured Ceolred, who was sniggering with fright, down to the old quarries, and flayed him," after which "he journeyed for hours, calm and alone, in his private derelict sandlorry named *Albion*" (Hymn VII).

In these quotations (assembled, confessedly, with incomplete regard to their contexts) we may see examples of how Hill has dealt with the intractable problem of aligning the sense of a private, sensitive self with the long body of English memory and English power. Hymn IV comes at this identification of personality with everything that is "significant soil" subtly and precisely; while suggesting the special preoccupations of the solitary child whose portrait is revealed in succeeding poems, it converts the sense of personality into something like floating potentiality, a function of human receptiveness. The piling-up of detail may suggest—despite the tone of reticent self-revelation—Whitman's method and exuberance, and even a touch of Browning as well. The later poems spell out the portrayal with varying originality and success. The tone of Hymn

V, for example, recalls Auden's more dismal school-memories, and Hymn VI starts with a dash of Dylan Thomas. Hymn VII, shifting to the third person, falls into a fashionable mode of sadistic reverie, yet does project an introspective child's fantasy of power. A wide realm of play on possible dimensions of the world of the sequence, in which the poles are Offa at the height of his reign in that barbaric old world and "Offa" the dreaming modern intelligence, at once totally inward and assimilating the whole past and present of existence, real and imagined, into itself, characterizes the movement of *Mercian Hymns*. There is room in the course of all this play for the suggestion of the enmeshment of region in nation and nation in world, especially in the contexts of terror and tragedy. Thus, Hymn XVIII:

> At Pavia, a visitation of some sorrow. Boethius' dungeon. He shut his eyes, gave rise to a tower out of the earth. He willed the instruments of violence to break upon meditation. Iron buckles gagged; flesh leaked rennet over them; the men stooped, disentangled the body.
>
> He wiped his lips and hands. He strolled back to the car, with discreet souvenirs for consolation and philosophy. He set in motion the furtherance of his journey. To watch the Tiber foaming out much blood.

One might view Hill's sequence as an expansion, in its own eccentric way, of Pound's "The Return." He summons up worlds of the past, and virtually calls out, in a poem like Hymn XVIII:

> Haie! Haie!
> These were the swift to harry;
> These the keen-scented;
> These were the souls of blood.

Hill, of course, adds dimensions of facetiousness and of complex identifications, from the start. Yet he begins, in Hymn I, with a wryly mixed bag of incantatory appositives to invoke the full-blooded past, revealing a similar intent to Pound's despite his half-comic surface and only minimally interesting rhythmic movement:

> King of the perennial holly-groves, the riven sandstone: overlord of the M5: architect of the historic rampart and ditch, the citadel at Tamworth, the summer hermitage in Holy Cross: guardian of the Welsh Bridge and the Iron Bridge: contractor to the desirable new estates: saltmaster: money-changer: commissioner for oaths: martyrologist: the friend of Charlemagne.
>
> "I liked that," said Offa, "sing it again."

The falling away of Offa's presence at the end of the sequence, too, resembles the "tentative" advance and "slow" and "pallid" fading out of Pound's "leash-men": "And it seemed, while we waited, he began to walk towards us he vanished." The suggestion is that a powerfully desiring imagination has brought a complex vision into focus under pressure of a compulsion to realize a state of organic oneness—but for a short time only, so that all that is left behind are a few extant tokens of a once vivid past: "he left behind coins, for his lodging, and traces of red mud." The "quality of the affection" here is, finally, a sort of self-conscious ruefulness, almost embarrassment, at being the helpless carrier of burdensome memory and loss, alleviated only by self-ironies and by the lyrical richness of one's most stirring images.

In a sense, *Mercian Hymns* is a recasting of *The Anathémata* in more vividly poetic form—a highly concentrated reorientation of Jones's work—just as much as it is an expansion of "The Return." More so, because the specific subject matters of the two British poets are so close to one another, and the kinds of consciousness of local memory and the overlays of successive cultures and languages provide the same problem of synthesis and identity. Structurally, the Poundian element is dominant, however; that is, the visionary process becomes more and more vigorously central and then dies away—a model for MacDiarmid as well, of course. Incidentally, too, one would imagine Jones's Preface to *The Anathémata* must have given Geoffrey Hill real guidance, especially in passages like

> We are, in our society of today, very far removed from those culture-phases where the poet was explicitly and by profession the custodian, rememberer, embodier and voice of the mythus, etc., of some contained group of families, or of a tribe, nation, people, cult. But we can, perhaps, diagnose something that appears as a constant in poetry by the following consideration:
>
> When rulers seek to impose a new order upon any such group belonging to one or other of those more primitive culture-phases, it is necessary for those rulers to take into account the influence of the poets as recalling something loved and as embodying an ethos inimical to the imposition of that new order. . . . Poetry is to be diagnosed as "dangerous" because it evokes and recalls, is a kind of *anamnesis* of, i.e. is an effective recalling of, something loved. [We would add: *or* passionately hated.]

and:

> this artist [James Joyce], while pre-eminently "contemporary" and indeed "of the future," was also of all artists the most of site and place. And as for "the past," as for "history," it was from the ancestral mound

that he fetched his best garlands and Clio ran with him a lot of the way—if under the name of Brigit. So that although most authentically the bard of the shapeless cosmopolis and of the megalopolitan diaspora, he could say

> "Come ant daunce wyt me
> In Irelaunde."

Thus David Jones in 1951, presenting the essential premises of that neo-regionalism whose aim is repossession of the living past as it extends into the present moment and animates a people's identity and passionate meanings: "the quality of the affection," in Pound's words. As with the development of the sequence, the largely unspoken consensus of the poets reflected in these quotations defines a whole tendency and its burning motivations almost completely overlooked in the critical literature. Geoffrey Hill's partially gnomic or riddling style, with humorous turns that at times are private or precious or peculiarly scholarly, are foreshadowed in many of the incidentals of *The Anathémata.* Jones's title itself is illustrative. His Preface explains:

> I knew that in antiquity the Greek word *anathema* (spelt with an epsilon) meant (firstly) something holy but that in the N. T. it is restricted to the opposite sense. While this duality exactly fitted my requirements, the English word "anathemas," because referring only to that opposite sense, was of no use to me. I recalled, however, that there was the other English plural, "anathemata," meaning devoted things, and used by some English writers down the centuries, thus preserving in our language the ancient and beneficent meaning . . .
>
> Subsequent to deciding upon this title, I noted that in a reference to St John Chrysostom it was said that he described the word as "things . . . laid up from other things." And again that in Homer it refers only to delightful things and to ornaments. And further, that it is a word having certain affinities with *agalma,* meaning what is glorious . . .
>
> So I mean by my title as much as it can be made to mean, or can evoke or suggest, however obliquely: the blessed things that have taken on what is cursed and the profane things that somehow are redeemed: the delights and also the "ornaments" . . .

The self-delighted serious whimsy of this explanation indulges itself at far greater length than our ellipses-ridden quotation indicates. (The first ellipsis in the second paragraph is Jones's own.) There is an odd sense in which Hill's writing is a sort of condensed expansiveness of kindred character; that is, it engages in the same variety of self-indulgence—though with a certain air of reticence—that half-makes of otherwise serious poetry a British don's holding-forth at high table. Jones's various epigraphs, subtitles, and incidental elegances are amusingly superimposed at comparable expense. His riddling initial page, with its elusive projection of an archetypal culture-bearer, its footnoted Latin and English allusions (both referring to the Prayer of Consecration in the Roman Mass), its torque-shiftings of emphasis, and its piled up images and concrete appositives, clearly presages Hill's method:

> We already and first of all discern him making this thing other. His groping syntax, if we attend, already shapes:
>
> ADSCRIPTAM, RATAM, RATIONABILEM . . . and by pre-application and for *them*, under modes and patterns altogether theirs, the holy and venerable hands lift up an efficacious sign.
>
> These, at the sagging end and chapter's close, standing humbly before the tables spread, in the apsidal houses, who intend life:
>
> between the sterile ornaments
> under the pasteboard baldachins
> as, in the young-time, in the sap-years:
> between the living floriations
> under the leaping arches.
>
> (Ossific, trussed with ferric rods, the failing numina of column and entablature, the genii of spire and triforium, like great rivals met when all is done, nod recognition across the cramped repeats of their dead selves.)

The parenthetical passage that ends this quotation could well serve as one of Hill's extended lines, as indeed could the combined prose-and-verse passage that precedes it. The difference lies in Hill's greater focus on isolated poignancies of sensibility and memory, each presented in concentrated form with very little explanation built into the poem itself. We never sense, in *Mercian Hymns*, the very fullest deployment of dynamic shiftings. But with the *Anathémata*, despite intriguing surface variations, the *whole* emphasis is on a continuum—the mythical history of England unfolded in terms of the "deposits" of her rituals and culture and symbolic preoccupations. It is a masterly tour de force, a poetic repossession that is ultimately, however, masked discourse despite its teasing mysteries and parodic virtuosity.

At another extreme we may note the Irish poet John Montague's *The Rough Field 1961–1971.*[13] Like Kinsella, and to a greater extent than Bunting, Montague makes extensive use of his own private experience

13. Dublin: Dolmen Press, 1972.

and predicaments in this sequence that is centered on the tragic past and present of Northern Ireland. He writes out of the memories of a Catholic family in County Tyrone, and the eight sections of *The Rough Field,* subdivided into 45 poems and an epilogue, devote at least half their space to highly personal memories that carry much of the burden of the sequence. That burden has to do with loss of caste and general humiliation, the demoralization and dispersion of families, the violation of cultural pride by the condescending imposition of English speech and values, and the brutal struggles exploiting religious differences that have so violently divided the Irish people. It has to do, as well, with nostalgic memories edged with bitterness and with a receding world represented only by a dwindling number of elderly survivors: "shards of a lost culture." An overwhelming despondency dominates the volume, intensified by a sense of political hopelessness and an admittedly partly irrational dismay at the coming on of "liberated" attitudes, mechanization and commercialism, and relative affluence. For Montague shares with Jones and other poets of the British Isles, whether of the left or right politically, and on whatever side of the English-Irish question, the feeling of irrevocable loss as distinctive local pasts are forgotten:

> Only a sentimentalist would wish
> to see such degradation again:
> heavy tasks from spring to harvest;
> the sack-cloth pilgrimages under rain
>
> to repair the slabbery gaps of winter
> with the labourer hibernating
> in his cottage for half the year
> to greet the indignity of the Hiring Fair.
>
> Fewer hands, bigger markets, larger farms.
> Yet something mourns. The iron-ribbed
> lamp flitting through the yard at dark,
> the hissing froth, and fodder scented warmth
>
> of a wood stalled byre, or leather thong
> of flail curling in a barn, were part
> of a world where action had been wrung
> through painstaking years to ritual.
>
> ("Driving South")

What lends energy to the sequence and militates against sheer lugubriousness is the intensity of its closeups, a Hardyesque glow around the particulars of memory. There is a kind of "story" in *The Rough Field* that serves as a center of reference without dominating the structure. It has to do with the fragmentation of the Montague family under the pressure of poverty and political fear, so that John was born in Brooklyn, New York, where his father worked in a token-booth in the subway. Later he was sent back to Tyrone to live with his family and go to school there, in the townland of Garvaghey ("*Garbh acaidh*, a rough field," as the epigraph informs us). Still later, as a grown man, he returns to Garvaghey for a prolonged visit: the literal starting-point of the sequence, whose untitled first poem, in the section named "Home Again," describes the bus-ride from Belfast to "a gaunt farmhouse on this busy road." From here the section opens out associatively: recollections of how it felt in childhood to hear "the cock crow in the dark"; contemplation of a photograph of the poet's grandfather, "a rustic gentleman" whose "succession was broken," and the family's prosperity as well, by the course of Irish history during the war years and after; elegiac memory of a fiddle-playing uncle, rather wild, who disgraced himself and migrated to Brooklyn, where he became a speakeasy-owner "and died of it," having long left his "rural art" behind with his decaying fiddle; and a catalogue of old people who, "like dolmens round my childhood," are fixed vividly and archetypally in his mind forever—kindly old Jamie MacCrystal, whose cottage was robbed as soon as he died; Maggie Owens, with her animals ("Even in her bedroom a she-goat cried"), who was "reputed a witch"; the blind Nialls; Mary Moore, "a by-word for fierceness" in her "crumbling gatehouse"; "Wild Billy Harbison" the loyalist:

> Ancient Ireland, indeed! I was reared by her bedside,
> The rune and the chant, evil eye and averted head,
> Fomorian fierceness of family and local feud.
> Gaunt figures of fear and of friendliness,
> For years they trespassed on my dreams,
> Until once, in a standing circle of stones,
> I felt their shadows pass
>
> Into that dark permanence of ancient forms.

We have focused at some length on this opening section because its method of reconstructing the sense of the past as it impinges on present consciousness is radically different from Jones's and Hill's. *The Rough Field* is squarely based on immediacies of experience in a much more open and detailed fashion than *Mercian Hymns* and with the kind of confessional centering that is totally absent from *The Anathémata*. Montague's method requires a far more direct style, and far greater risks of exposing one's flaws of personality, than the English poets' masked or allegorical devices. Its points of focus merge with those of folk awareness.

The method is most inclusively represented, perhaps, in Section IV, "A Severed Head," whose title is based on a barbaric incident described in one of its epigraphs. The first of the six poems in this section begins pastorally; the poet is driving cattle along a mountain road just as he did when a boy, and the familiar scene unfolds with a kind of double vision. Some changes have taken place: a well-remembered farmhouse is now a trifle less bleak, with its radio aerial and its small added comforts; and, invisibly except to those who once knew them, the old inhabitants already named earlier in the sequence have disappeared from their deserted cabins. The tone is affectionate, and charged with muted loss in this poem of delicately modulated perception:

> . . . Uncurling
> Fern, white scut of *canavan*,
> Spars of bleached bog fir jutting
> From heather, make a landscape
> So light in wash it must be learnt
> Day after day, in shifting detail,
> Out to the pale Sperrins.
> "I like to look across," said
> Barney Horisk, leaning on his *slean*,
> "and think of all the people
> who have bin."
> Like shards
> Of a lost culture, the slopes
> Are strewn with cabins, deserted . . .

Succeeding poems in this section move into the wider memory of Ireland's ruinous history back to the time of Elizabeth: associations of places and names that make "The whole landscape a manuscript / We had lost the skill to read," drastic scenes of warfare and defeat, the drift of shared feeling aroused by a fiddler's playing that produces images in his hearers' minds of "anonymous suffering"—the "tribal pain" of "burnt houses, pillaged farms, / a province in flames." The fifth poem of the section, "A Grafted Tongue," returns the sequence to its initial personal, confessionally oriented direction: literally, the poet's problems as a stutterer, but in the context of the imposition of an alien culture and language. The poem begins with a bracketed psychological image, as if to explain its inner bearings before getting truly under way:

> (Dumb,
> bloodied, the severed
> head now chokes to
> speak another tongue:—

As in
a long suppressed dream,
some stuttering garb-
led ordeal of my own)

An Irish
child weeps at school
repeating its English.
After each mistake

The master
gouges another mark
on the tally stick
hung about its neck

It ends:

To grow
a second tongue, as
harsh a humiliation
as twice to be born.

Decades later
that child's grandchild's
speech stumbles over lost
syllables of an old order.

This is vulnerable writing, but it carries us deep into felt history. Its painful impact remains despite the distancing playfulness, and delight in names, of the poem that completes the section and ends rather joyfully:

And what of stone-age Sess Kill Green
Tullycorker and Tullyglush?
Names twining braid Scots and Irish,
Like Fall Brae springing native
As a whitethorn bush?

A high, stony place—bogstreams,
Not milk and honey—but our own:
From the Glen of the Hazels
To the Golden Stone may be
The longest journey
I have ever gone.

No doubt that last sentence has something portentous about it, but we shall nevertheless allow it to float lightly like a gay streamer behind the tragic misery of the rest of the section. The next two sections are highly personal, the fifth centered on Montague's father, the sixth on childhood memories and convivial evenings with fellow countrymen in Garvaghey. Again, the most personal poems are the ones that also have the widest reach in their historical and political connotations. Thus, the opening poem of the fifth section ("The Fault") starts with a painful recollection of civil war and a wounded policeman bleeding on the floor of the family's kitchen and of the inevitable but shameful fear that motivated so much of their behavior:

> my mother hobbling to the shed
> to burn the Free State uniforms
> her two brothers had thrown off
> (frugal, she saved the buttons):
> my father, home from the boat at Cobh,
> staring in pale anger at a Redmond
> Commemoration stamp
>   or tearing to
> flitters the polite Masscard sent
> by a Catholic policeman. But what if
> you have no country to set before Christ,
> only a broken province? . . .
>
>   ("Stele for a Northern Republican")

And in the sixth section ("A Good Night") the second poem, "The Fight," is a poem of childhood betrayal by a friend who destroyed the swallow's eggs John had run to fetch him to see—"the bitter paradox / Of betraying love to harm."

The Rough Field, finally, is somewhat overloaded, for Sections VII and VIII (especially the ten poems of the latter) pack a great deal of comment on the effects of "progress" and the various meanings, often ironic ones, of a largely unresolved revolution, and Section IX ("A New Siege") is an extended commentary in rattling dimeter on history. Only Section X ("The Wild Dog Rose"), a beautiful, many-sided poem about an old countrywoman, matches the power and intense identifications of the first six sections. The fierce-looking *Cailleach* (old woman or hag) who had once filled him with terror has become the poet's friend in this poem. She tells him a dreadful story about a drunk who tried to rape her and their terrible struggle before she somehow "broke his grip" and he fell asleep. The poem is a fit culmination for what we might call the true, inner sequence that remains despite Montague's piling on too much material. In this inner sequence we have a curve from the depressive return to the protagonist's sources at the start to the elegiac purity of the second section ("The Leaping Fire," whose four poems celebrate a saintly and sacrificial aunt), to an increasingly deeper tragic sense culminating in the fifth and sixth sections, and then (beginning in Section VI with "The Fight") a shift to emphasis on betrayal and beastliness in history and in the more intimate private life. Along the way it becomes clear that Montague owes debts to two masters especially—MacDiarmid and Williams (whose documentary "interruptions" in *Paterson* he imitates)—as well as to the "plain" yet accomplished style and realism of an Irish predecessor, Austin Clarke.

Chapter Twelve

Two American Models: Hart Crane and Charles Olson

1. Crane's "Voyages" and *The Bridge*

Williams's *Paterson* established American neo-regionalism so decisively that it is difficult to remember its many forerunners. This is especially true because so many sequences involving place in the senses we have mentioned were so powerfully influenced—that is, given a strong incentive to come into being—by the nature of *Paterson*. It opened up the prospect of exploring a particular locality, one's own, in every conceivable dimension including that of family memory. Thus, even Robert Lowell's *Life Studies* may be considered neo-regional. Although its concentrated confessional character has led us to discuss it in another chapter, dealing with "the poetry of psychological pressure," *Life Studies* grounds itself in the history of New England and of the Lowell family through the generations and illustrates once more the indispensability of streams of private intensities of feeling in a fully orchestrated neo-regional sequence. Charles Olson, the subject of the next section of the present chapter, fixed on Gloucester, Massachusetts, and himself as its embodying spokesman, in a manner that would have been unlikely had Williams not done his earlier spadework and mythmaking with Paterson, New Jersey. And even the British and Irish poets we have just been discussing (with the obvious exception of MacDiarmid) owe a great deal to the delayed influence of Williams, an influence much resisted in the British Isles for a very long time although it paralleled the powerful impact of Joyce.

The pitifully short-lived Hart Crane (1899–1932), sixteen years Williams's junior, must be considered a precocious predecessor by virtue of his sequences "Voyages" and *The Bridge*, both conceived in 1923 or a bit earlier and published in 1926 and 1930 respectively.[1] Crane's highly uneven work engages all too obviously with the problem of the intractable, but he did have a touch of greatness. Both his sequences are efforts toward visionary transcendence over the limitations of the empirical moment, but "Voyages"—shorter, more inwardly centered, and intellectually less ambitious—was easier for him to handle. A group of only six poems, beginning in fascinated terror and ending in willed affirmation, its movement is confined to two basic ranges of tension. The first is the pull between fear and enthrallment in the face of nature's impersonal power, which includes adult love and passion. The second, really an aspect of the same tension but erotically centered, is the disparity between the dream of ecstatic transcendence and the hard knowledge of an ultimate emptiness.

The Bridge, however, is Crane's sequence of place. Three times the length of *The Waste Land*, his intended major sequence contains sixteen poems: a proem ("To Brooklyn Bridge") set off separately, and the others in eight numbered sections. It presents itself as a specifically American epic with neo-regionalist emphases. To some extent it attempts to recast *Song of Myself* in new terms, using the language of private sensibility more extensively but anchoring its perspectives in national landmarks and symbols. These, in order but with some omissions, are Brooklyn Bridge, Columbus aboard the *Santa Maria*, New York Harbor, Rip Van Winkle, early explorers, the transcontinental railroads, the Mississippi, the American Indian, heroic prairie women, Whitman, the Wright Brothers, Dickinson, Isadora Duncan, Poe, and even the New York subway—the last a horror-journey somehow matched by the imagined leap into brilliant transcendence by way of an explodingly luminous Bridge-imagery in "Atlantis," the closing poem.

In the broadest sense, *The Bridge* tries to connect the kind of cosmopolitanism found in Pound and Eliot with a vision of America's past and present, and somehow to make these preoccupations also serve an introspective exploration like that of "Voyages." This extremely difficult, unwieldy enterprise certainly played an important role in preparing the ground for such a later work as *Paterson*. The titles of the proem and of the successive sections are indicative: "To Brooklyn Bridge," "Ave Maria," "Powhatan's Daughter," "Cutty Sark," "Cape Hatteras," "Three Songs," "Quaker Hill," "The Tunnel," and "Atlantis." "Powhatan's Daughter"

1. Text references are to *The Complete Poems and Selected Letters and Prose of Hart Crane*, ed. Brom Weber (New York: Anchor Books, 1966).

consists of five poems, and "Three Songs" of three more, all centered on American places, history, or legend. "Cutty Sark" and "Atlantis" both focus on the New York harbor-scene, and "The Tunnel" on the New York subway system. The other titles are self-explanatory.

Crane, however, does not *discover* American life and memories as Williams does. His sequence is not saturated with the intimately local sense of an experientially absorbed past that is one of *Paterson*'s triumphs. His use of American history and locales is more mechanically programmatic and rhetorical than Williams's, and so is his use of our myths and "fabulae." The latter term is Horace Gregory's, in his introduction to Williams's 1925 prose-work *In the American Grain,* a germinal little book which, as Gregory observes, provides important documentary and imaginative sources picked up by Crane and others. It was reviewed sympathetically by D. H. Lawrence and is close in spirit to Lawrence's *Studies in Classic American Literature* (1923). The American mystique of these two books has clearly infected that of *The Bridge*. But Crane does not so much explore its traditional symbols as he uses them to demonstrate the poem's nativist credentials—planting Columbus in Section I, Rip Van Winkle and prairie Indians and the Mississippi in II, the Wright Brothers in III, and so on.

But there are times, as in "The Harbor Dawn," when the poetry is lyrical and private in its tonalities. And there are moments when the contrived summoning up of mythical history turns into personal, autobiographically based intensity, as in part of "Van Winkle." More subtly, too, there are moments of immersion in the subjective awareness of some projected figure—for instance, the subway vision of Poe in "The Tunnel." At these junctures, the real face of the poem, brooding and fearful and bold all at once, emerges tantalizingly and then disappears. It bears a certain resemblance to the frightened, suicidal, daredevil sensibility of "Voyages," We must observe, indeed, that much of the best writing in *The Bridge* seems more appropriate for an expanded "Voyages" than for an epic celebration, trumpeting clichés themselves eroded by familiar oratory and textbooks and totally contradicted by the occasional truly poetic notes suggesting a finer music going on in a different key entirely. Section I, "Ave Maria," is a prime example. It begins the poem proper (as opposed to the brilliant proem, which we shall cite shortly) with that most schoolroomish, ur-American motif: the noble, devout, wonder-struck thoughts of Columbus on his homeward voyage from the new world he has just found. The self-indulgent verbosity of "Ave Maria" makes it unreadable, except in the occasional flashing image that liberates a tonality from its supposed context: "Invisible valves of the sea,—locks, tendons / Crested and creeping"; "This turning rondure whole, this crescent ring / Sun-cusped and zoned with modulated fire": "—Rush down the plenitude"; "Into thy steep savannahs, burning blue"; "white toil of heaven's cordons."

In a poem of some 93 lines, these few phrases stand out like strange jewels against the turgid norm, represented by the more typical lines of Columbus's prayer (57–62):

> O Thou who sleepest on Thyself, apart
> Like ocean athwart lanes of death and birth,
> And all the eddying breath between dost search
> Cruelly with love thy parable of man,—
> Inquisitor! incognizable Word
> Of Eden and the enchained Sepulchre.

This passage illustrates Crane's rhetorical copiousness and his love for Miltonic echoes; it also encourages us to turn our attention quickly elsewhere. Contrariwise, the atypical phrases we noted are extraordinary metaphoric forays centered on the sea's secret, impenetrable depths, untamable energy and multiplicity, and dangerous, thrilling, unencompassable vistas. These phrases, completely magnetic, project a sense of precarious engagement with irresistible reality at the keenest possible pitch, as do the best parts of "Voyages."

It is certainly to Crane's credit that he attempted the intractable task of repeating Whitman's triumph in terms of a new age—the fusion of affirmative national vision, in highly orotund language, with the purest lyricism, often vulnerable and confessional in its private reaches. But our age is less hospitable than former ones to rhetorical flight for its own sake; nor does it have the daily familiarity with such flights provided by politics and the pulpit in Lincoln's day, a familiarity that might have trained Hart Crane's ear to the differences between great, disciplined rhetoric and rodomontade. In addition, Crane's purely lyrical gifts lacked the scope and intensity that permitted Whitman to write great poetry despite his well-known flaws as a stylist. It is interesting that Crane, who admired Eliot's earlier work greatly and called himself an imitator of Eliot, criticized *The Waste Land* as "dead" and set out to write a less "negative" poetry than Eliot's—but in this conception of his task he was going against his own grain.

Instead of a counter-poem to *The Waste Land*, Crane's sequence is in certain ways an imitation. *The Bridge* presents a similar coping with a depressive state, linked with a comparable sense of a sadly fragmented existence, by affirming a vision of redemption against the odds. It follows *The Waste Land* formally, too, in juxtaposing a variety of tones and of kinds of verse. But *The Waste Land*, though so much shorter, is far more encompassingly resonant. Also, Eliot's reliance on a traditionally

Christian frame of "positive" reference enables a remarkable palimpsest-effect: modern sensibility atop ancient mysteries, as it were. The effect is richly thickened by his allusions to other religious traditions as well, to give depth and density to a perspective that might otherwise seem too contrived. By these means Eliot successfully exploits the gray area between humble faith and withdrawn despair and the rosy area between devout faith and exuberant delight. Except in "To Brooklyn Bridge" and in scattered passages, *The Bridge* does not approach Eliot's achievement. Leaving questions of ultimate talent aside for the moment, we can see that one very important reason is that Eliot does not split the essential subjectivity of his associative method off from any more public and declamatory method. His voices may shift, dizzying alterations of perspective and tonality may whip us here and there, but the method of building the poem organically, affect by affect, is not violated along the way.

It seems likely that the "right" solution for Crane would have been to use the structure of "Voyages" as the basis of the more ambitious *The Bridge,* subordinating the epic tonalities to the dominant subjectivity of the former sequence. This would have brought him closer to Eliot's truthfulness to psychic process and to the demands of his own ear and idiosyncrasies of style. Except when it slips into a curious false inspirationalism (a forcing of "vision"), Crane's method in "Voyages" is like Eliot's in *The Waste Land* in its associative purity and quick metaphoric leaps. But it does have its own distinctive character: the mingled eroticism, terror, and exuberant energy; the greater formal regularity; and the more openly *singing* quality—partly a result of the fact that, unlike Eliot, Crane does not intrude new rhythms and dramatic tonalities within a section of his sequence. Also, he often employs a rather stately, periodic construction (however wildly images and connotations may be thrashing about inside their Latinate confines) and a diction learned from Milton. Thus, the ending of "Voyages" VI:

> Beyond siroccos harvesting
> The solstice thunders, crept away,
> Like a cliff swinging or a sail
> Flung into April's inmost day—
>
> Creation's blithe and petalled word
> To the lounged goddess when she rose
> Conceding dialogue with eyes
> That smile unsearchable repose—
>
> Still fervid covenant, Belle Isle,
> —Unfolded floating dais before

Which rainbows twine continual hair—
Belle Isle, white echo of the oar!

The imaged Word, it is, that holds
Hushed willows anchored in its glow.
It is the unbetrayable reply
Whose accent no farewell can know.

To some extent this final passage of "Voyages" presents forced vision of the sort just alluded to. But it does present the periodic structure and Miltonic phrasing as well, and in a highly strategic place. The goddess's imagined final reassurance, despite everything "betrayable" in love's vulnerability, promises "unsearchable repose." The echo from Milton's Final Chorus in *Samson Agonistes* is a striking adaptation of diction and context:

All is best, though we oft doubt
What th'unsearchable dispose
Of Highest Wisdom brings about,
And ever best found in the close.
Oft he seems to hide his face,
But unexpectedly returns,
And to his faithful champion hath in place
Bore witness gloriously; whence Gaza mourns,
And all that band them to resist
His uncontrollable intent.
His servants he, with new acquist
Of true experience from this great event,
With peace and consolation hath dismissed,
And calm of mind, all passion spent.

The sympathetic reader will find in Crane's closing lines a touching affinity with Milton's. His final stanza, especially, is close in tone to the first four lines and the concluding line of the Chorus. His "unsearchable repose" echoes Milton's "unsearchable dispose," an echo Crane later reinforces by the phrase "unbetrayable reply" in his penultimate line. This reinforcing is itself an echo of Milton's Chorus, where the phrase "uncontrollable intent" magnificently recalls and buttresses "unsearchable dispose." The atmosphere of trust and reconciliation with fatality is comparable in the two passages, although neither glosses over tragic truths. In *Samson Agonistes* "Gaza mourns" the hero's real death; in "Voyages" the poet knows the "accent" of "farewell" and can only dream it is unknown elsewhere, "beyond siroccos," in the "Belle Isle" of lovers' dreams where "lounged" Aphrodite smiles. Of course, Milton's Chorus has a rocklike strength. It is based on his genius for blending narrative, exposition, and the language of devotion with the felt spirit of his sacred text, well within the culturally shared wisdom of his age. Crane's ear allows him to borrow certain of Milton's resonances, but he cannot take for granted any sort of culturally shared context of ingrained and militant faith. He needs, therefore, to furnish new images of revelation ("like a cliff swinging or a sail / Flung into April's inmost day") and of the oneness of deity and sacred text ("Creation's blithe and petalled word," "Belle Isle, white echo of the oar!"). These images depend for their conviction, tentative and optative as it is, on freshness, surprise, and intensity.

The ending of "Voyages," like "To Brooklyn Bridge," demonstrates that Crane might well have been able to fuse heroic and epic notes successfully with the more sensitive private lyricism of most of his best work. Easier said than done, no doubt—which is probably why he wrote two sequences in the first place. Nevertheless, there is much in *The Bridge* to indicate the structure that might have been. We have alluded to the brilliance of the prefatory poem, which realizes a subtle balance of heroic and inwardly negative tonalities—a remarkable affective duel in which the darker side prevails despite the remote, grandly glittering attractions of what is being overtly proclaimed. On one side we have the projection of an image of industrial triumph that, apparently, embodies a transcendence far more dependable than any given us by nature, fantasy, or popular art. These latter deceptive alternatives are implied, successively, by the exquisite freedom of the gull's flight, the "apparitional" sailboats that "cross / some page of figures" office-workers are toiling over, and the "cinemas / panoramic sleights" that lure multitudes who hope to see, this time, some miraculous key to life. In contrast, the dazzling celebration of the bridge's glories might be thought of as unqualified trumpeting of American destiny and human promise—were it not for the many counter-notes that enter with the "bedlamite" in the fourth stanza. The alienated, dark view of the celebrator's self in the next-to-last stanza is particularly telling:

> Under thy shadow by the piers I waited;
> Only in darkness is thy shadow clear.
> The city's fiery parcels all undone,
> Already snow submerges an iron year . . .

This pivotal quatrain undoes any possible domination of the poem by a mood of glamourizing celebration. It might be taken as an expression of humility—a distant echo, perhaps, of Eliot's "Come in under the shadow of this red rock" in *The Waste Land* and of comparable shadow-images in "The Hollow Men." But it also emphasizes a state of psychological and, by a subtle resonance, social desolation: "Already snow submerges an iron year." This tonal break prepares us for the closing stanza's shift from awestruck, elated vision to wistful petition:

> O sleepless as the river under thee,
> Vaulting the sea, the prairies' dreaming sod,
> Unto us lowliest sometime sweep, descend,
> And of the curveship lend a myth to God.

Only in this one poem, where the dream of the bridge's transcendence is so continually linked with the need to obliterate a painful present and therefore suffers internal resistance in the very texture of the language, does the celebration ring true. It has a sufficiently astringent context. The first three quatrains wryly note the restricted state of daily existence ("the chained bay waters"), the way that visions of freedom "forsake our eyes," and the expectation of a revelatory "flashing scene / Never disclosed" that forever haunts and deceives the multitudes of movie-goers. And even the celebration itself grants that the "guerdon" the bridge offers is "obscure as that heaven of the Jews." Its promised "accolade" will merely be "anonymity time cannot rise": irrevocable obliteration, a freedom from the burden of self-awareness and of constant guilt. This is the negative gravamen of the apparently buoyant line: "Vibrant reprieve and pardon thou dost show."

Apart from its deeper affinities with Eliot's work and from the indications in "To Brooklyn Bridge," there is much evidence throughout *The Bridge* that it might have become a sequence whose main organic body resembled that of "Voyages"—that is, was much more centered on successive states of feeling and awareness and much less on public symbols and values. In fact, as we have suggested, the two works could well have been integrated had Crane been able to master the problems involved. We have noted a few points at which *The Bridge* is close to the spirit of "Voyages": for instance, "The Harbor Dawn," first poem of the "Powhatan's Daughter" group constituting Part II of the sequence. In this poem the setting, atmosphere, and general tenor exactly fit the situation of magnetic, unstable love, with some intrusive factor of brute impersonality in the outside world, of "Voyages." This is true as well of "Cutty Sark" (Part III) and of "Southern Cross," the first of the three-poem group "Three Songs" comprising Part V. All these poems share the yearning and excitement, if not the audacity, of the best of "Voyages." On the negative side, both sequences, especially *The Bridge,* suffer from their lapses into bombast in strategic places. This is surprising in the author of such small masterpieces, beautiful yet mordantly clear-eyed, as "Passage," "Repose of Rivers," and—in the sequences—"Voyages" V and VI ("Meticulous, past midnight in clear rime" and "Where icy and bright dungeons lift") and the stanzas of "The Tunnel" (Part VII of *The Bridge*) beginning "Whose head is swinging from the swollen strap." One can only attribute the lapses to a certain panicky forcing in order to show completed works under what the poet felt to be pressing, competitive circumstances of his career.

Leaving this speculation aside (and with it the supposedly climactic final section, "Atlantis," with its disastrous and unintentionally self-parodying outbursts), we can see that *The Bridge* has the makings of an American epic couched in primarily subjective, associative units. Although there is something crudely mechanical in Crane's method of planting his American materials at various points along the way, the stream of sensibility does move through the work as well. It follows its own dynamics in a sort of underground movement beneath the thematic surface. "To Brooklyn Bridge," we have seen, gets things started with brilliant authority. It is neo-regionalist in its fashion, but also a unique fusion of heroically celebratory verse with introspective lyricism. We have seen, too, how "Ave Maria" has a few true lines amid the dross Crane manufactures to buoy up the symbolic icon of Columbus as the first heroic figure of the American myth. After this unfortunately faked set piece the next section, "Powhatan's Daughter," cumbersomely bunches five poems together as Part II. These are essentially unconnected with "Ave Maria" except in the far-fetched sense that we are all voyagers trying to keep our bearings.

"Powhatan's Daughter" begins with two highly personal poems, "The Harbor Dawn" and "Van Winkle." In the first—one of Crane's best—two lovers awake in a room overlooking New York's harbor. Their world of sleep and dreams and passion is projected onto the sounds and atmosphere of the scene outside their room, and at the same time they are touched into a new state by all that is impersonal out there. (The poem parallels the more ruthlessly chilling "Voyages" V and should be a companion-piece in that sequence—or rather, in the combined sequence that might have been.) Then "Van Winkle," a largely autobiographical piece, summons up the poet as a child, already alienated from a harsh father and indifferent mother. "The Harbor Dawn" is dense with the full richness of a floating, fantasy-saturated moment. "Van Winkle" is at once touching and—whimsically, cheerfully—self-distancing. From the depressive penultimate stanza of the proem to the images of impenetrable glory in the few best lines of "Ave Maria" to the memories of real, thwarted love in a confusing, fascinating world in the opening poems of "Powhatan's Daughter," a sort of psychological chaining has carried us into intimate recesses of a volatile psyche.

Rip Van Winkle, of course, is part of American mythical history, and

Crane's poem suggests the deep sleep of unawareness within which new cultural forms emerge. The poet manages to include references to Pizarro, Cortez, Priscilla Alden, and John Smith as well as to Washington Irving's creation—virtually an entire elementary school curriculum—and indeed the worst stanzas make the sentimental point clear:

> Times earlier, when you hurried off to school,
> —It is the same hour though a later day—
> You walked with Pizarro in a copybook,
> And Cortes rode up, reining tautly in—
> Firmly as coffee grips the taste,—and away!
>
> There was Priscilla's cheek close in the wind,
> And Captain Smith, all beard and certainty,
> And Rip Van Winkle bowing by the way,—
> "Is this Sleepy Hollow, friend—?" . . .

Even the best stanzas are somewhat infected by the imposed facile nostalgia, and perhaps have a too-easy bitter sadness as a result. But this observation may be unfair:

> So memory, that strikes a rhyme out of a box,
> Or splits a random smell of flowers through glass—
> Is it the whip stripped from the lilac tree
> One day in spring my father took to me,
> Or is it the Sabbatical, unconscious smile
> My mother almost brought me once from church
> And once only, as I recall—?
>
> It flickered through the snow screen, blindly
> It forsook her at the doorway, it was gone
> Before I had left the window. It
> Did not return with the kiss in the hall.

This passage has its Proustian twist despite the self-pity, and one sees the confessional need that produced it but could not find original language for it. The first line echoes Eliot's "Rhapsody on a Windy Night" and "Portrait of a Lady," the sixth *The Waste Land,* the last a moment in "Gerontion." One can only think that Crane deprived his poem of the kind of original force it might have had, and *needed,* by compromising its process with his forced thematic allusions. The pure associative stream of the inner, truer poem demands an entirely different level of artistry. Still, the stream is not utterly clogged and blocked. We are left with its aura of sadness, and its pressure to find reassurance or a promise of healing love that will reconstruct some lost primal bliss, as the most enduring emotional resonance in the best sections of *The Bridge.* And the same pressure elicits a fiercer response of angry vision in "The Tunnel."

"Voyages," too, is driven by this pressure. For the most part, however, it faces directly into the hard and bitter aspects of its awareness. It begins with a warning to heedless children (anyone's innermost self, ultimately, among them) against venturing too far into the sea. Playing on the shore with death's very tokens—empty shells, dried seaweed, dried sticks—they imagine a power over nature they do not possess. The notion of the life-cycle as our primal "voyage" is embedded in the image of the children's sinewy, active bodies as "spry cordage"—that is, as ship's rigging—while the sea is personified as female. ("Fresh ruffles of the surf," "caresses / Too lichen-faithful from too wide a breast," and "The bottom of the sea is cruel"—all these phrasings suggest the feminine. The last two give the suggestion of destructive passion, without absolutely specifying the personification, and the second poem completes the identification of the sea as voluptuous female.) The simplicity of this opening poem does not conceal its prophetic grandeur and sense of pre-ordained doom—tonalities that would have lent authenticity to the epic pretensions of *The Bridge* if only "Voyages" could have been integrated with it.

> Above the fresh ruffles of the surf
> Bright striped urchins flay each other with sand.
> They have contrived a conquest for shell shucks,
> And their fingers crumble fragments of baked weed
> Gaily digging and scattering.
>
> And in answer to their treble interjections
> The sun beats lightning on the waves,
> The waves fold thunder on the sand;
> And could they hear me, I would tell them:
>
> O brilliant kids, frisk with your dog,
> Fondle your shells and sticks, bleached
> By time and the elements; but there is a line
> You must not cross nor ever trust beyond it
> Spry cordage of your bodies to caresses
> Too lichen-faithful from too wide a breast.
> The bottom of the sea is cruel.

The terror of the second stanza (where we are in the realm of thinly disguised, majestically threatening gods) is modulated but not removed by the tender concern of the third. The emphasis is so strongly on the pitiless elements and the vulnerable children in this first poem that the feminine dimension of the imagery is subordinated. The beginning of the next poem changes all that, spelling out precisely this dimension, and with it the whole irresistible force of sexuality against which it was futile to issue any warnings:

—And yet this great wink of eternity,
Of rimless floods, unfettered leewardings,
Samite sheeted and processioned where
Her undinal vast belly moonward bends,
Laughing the wrapt inflections of our love;

Take this Sea, whose diapason knells
On scrolls of silver snowy sentences,
The sceptred terror of whose sessions rends
As her demeanors motion well or ill,
All but the pieties of lovers' hands.

We are flooded with Gargantuan eroticism in the first stanza; moreover, the bottom of the sea may be cruel, but the stanza's closing line seems to have forgotten the first poem's tutelage. Things are reversed in the second stanza, though, as the sea's cruelty is seen as stern justice—our deserved doom—yet a justice that shows mercy toward one kind of unconscious suppliant, recognizable through "the pieties of lovers' hands." The stanza hovers on the verge of sentimentality, yet introduces exquisite cosmic imagery of the sea as a beautiful passionate woman moving through fated dance-turns while "her turning shoulders wind the hours."

And onward, as bells off San Salvador
Salute the crocus lustres of the stars,
In these poinsettia meadows of her tides,—
Adagios of islands, O my Prodigal,
Complete the dark confessions her veins spell.

Crane, in his essay "General Aims and Theories," cites the image of "adagios of islands" in this passage as an instance of the "logic of metaphor" in poetry. "The reference," he writes, "is to the motion of a boat through islands clustered thickly, the rhythm of the motion, etc. And it seems a much more direct and creative statement than any more logical employment of words such as 'coasting slowly through the islands,' besides ushering in a whole world of music." The association in the poet's mind is perfectly self-evident, although no boat is mentioned in this poem. The basic association of "adagio" is with a slow movement in music or in ballet. One might, indeed, feel a boat's movement as an adagio among islands, and here that feeling has been projected onto the islands themselves—as if they were dancing to slow music or were somehow part of the music themselves. And further, they are an aspect of the personified Sea, an all-powerful object of cosmic desire. Suddenly, though, in speaking of the prodigally endowed Sea, another being is addressed as "my Prodigal," no doubt the partner, in the first stanza, who shares "the wrapt inflections of our love." There are two "Prodigals," the Sea and the human lover, and from here on in the sequence there will be a certain deliberate confusion between them. The Sea will be at times a living mistress, and the lover at times seem all watery depth, entered as light penetrates the sea ("Voyages" III).

The associative shifts involved in all this open up a world of floating sensuality, love-fear, and visionary vagueness. The imagery is so lovely, the rhythm so caught up in a state of death-accepting exaltation, that the sudden assertion of mystical transcendence in the penultimate stanza is almost convincing—especially as, in a shift like that in "To Brooklyn Bridge," assertion is changed to prayer in the final stanza. The closing image, "The seal's wide spindrift gaze toward paradise," embodies the final set of the poem: a yearning toward the truth of the vision that has just been dreamed through before us. But in itself it is far more than the embodiment of an abstraction. The simple picture is itself more than that: the seal looking out past the spindrift, the suggestion that its gaze is itself a kind of spindrift, the thought that paradise might perhaps be out there—and all outside the ken of human awareness. The power of this image is like that of the whole of stanzas 1 and 3: so engaging that we could hardly care what the supernal aspirations are that pad the poem's prayer to the seasons and the waters of the deep.

But "Voyages" III and IV press further into unqualified assertion of transcendence, so much so that they destroy the fine balance of II. Meanwhile, however, they do reinforce the suggestions in II of a "real" lover's presence. The impassioned language of sexual union in III and the forced language of delirious transformation in IV seem to insist that the descriptions of love have an immediate human subject, however ambiguously interchangeable at times with the sea and fatality. "Voyages" V, immensely compelling, makes the shift of emphasis explicit while giving the sequence a subtly ambiguous mystery. This poem restores the strength of the sequence through its reassertion of the hard, impersonal principle of existence, going beyond the warning notes of "Voyages" I into a realm of ultimate dismay. The lovers are disturbed out of sleep by the "merciless white blade" of the moonlit bay estuary outside their window late at night, "past midnight in clear rime."

> —As if too brittle or too clear to touch!
> The cables of our sleep so swiftly filed,

Already hang, shred ends from remembered stars.
One frozen trackless smile . . . What words
Can strangle this deaf moonlight? For we

Are overtaken. . . .

The strangeness and authority of evocation in this poem are remarkable. "This deaf moonlight" has welded the bay waters into a blade of intransigent reality cutting through dreams that linked the lovers to the stars. The blade is a "tidal wedge" forcing the lovers apart too. It is as though a symbol of inevitable separation had created itself as a beautiful, relentless, terrifying destroyer. The lovers' ensuing conversation is one-sided and ambiguous. One of them speaks of the unfathomable, unique character of their love, but in words that apply equally well to that blade, or wedge, of ruinous light. The other replies silently, in thought only:

. . . "There's

Nothing like this in the world," you say,
Knowing I cannot touch your hand and look
Too, into that godless cleft of sky
Where nothing turns but dead sands flashing.

"—And never to quite understand!" No,
In all the argosy of your bright hair I dreamed
Nothing so flagless as this piracy.

The passage has the chill of deep recognition. We are almost tempted to stress the fact that, empirically, this is a homosexual love-poem, with an imagery of guilt and isolation rising from the poet's ambivalent psyche. But the poem gives no overt clue, and phrases like "godless cleft" and "flagless piracy" resonate beyond such specifics. The cluster of associations is not self-analytical or subject to rational explanation. "That godless cleft of sky" echoes the last line of the opening stanza—

Meticulous, past midnight in clear rime,
Infrangible and lonely, smooth as though cast
Together in one merciless white blade—
The bay estuaries fleck the hard sky limits.

One cannot love, *and* experience such utter, non-human, brilliantly revealed meaninglessness, at one and the same moment. In the semi-dialogue, the "you"—the lover addressed—feels this and tries to distract the "I," who is absorbed in the bright vision of ultimate emptiness: the surface of the moon, that dead satellite. If we may digress for an instant, a similar image takes over at the climactic ending of Crane's "Passage":

". . . from the Ptolemies / Sand troughed us in a glittering abyss." A remote echo, perhaps, of "Ozymandias," the image is a characteristically desolate one in Crane, revealing dead—that is, impersonal—energy at the heart of an illusory beauty, power, and light. The ending of "Repose of Rivers," too, is comparable: "I heard wind flaking sapphire, like this summer, / And willows could not hold more steady sound." In "Voyages," the first poem, with its sun that "beats lightning on the waves" and its waves that "fold thunder on the sand," prepares us for the climactic imagery of Poem V. An interesting variant, incidentally, occurs in "North Labrador," relevant here because of the way sexual feeling is linked, by apparent contrast, to a "cold-hushed" world of "leaning ice":

> "Has no one come here to win you,
> Or left you with the faintest blush
> Upon your glittering breasts?
> Have you no memories, O Darkly Bright?"

There is no answer in "North Labrador" except the pointless passing of time in a place where there is neither life nor seasonal change—a place that is more a cosmic condition than a locality, the same condition focused on in "Voyages" V with infinitely greater intensity. There the half-dialogue of the lovers is a balancing of ardent expression of gratification against that terror at brilliant nothingness; and there is a suggestion that the brilliant nothingness is within us as well. That is one reason we shall "never . . . quite understand" our love-transports. The silent lover thinks that the vision of the "frozen trackless smile" linking the unknown barren reaches of the cosmos with the room of love is the ultimate "piracy," a seizure of his whole self greater than that perpetrated by any human predator of our feelings and desires. "Your bright hair" has a double direction: the lover's hair, but also the bright moonlit estuary perceived as the Sea's golden hair. "Argosy" may suggest its literal meaning—in this context as a rather romantic term for a fleet of invading pirate ships—but it seems likely that Crane meant to suggest the Argonauts and the Golden Fleece stolen from Colchis despite the fact that the word "argosy" is not derived from "Argo." The associations of mortally perilous adventure, passion, and high-handed, piratical behavior in the legend of Jason and Medea are powerfully relevant to the emotional situation of this poem.

This is the climactic moment of the sequence. The final stanza of "Voyages" V presents a gentling of the stark, wild realizations into which we have been taken. It is a reconciliation with fatality, an acceptance of parting and of impersonality (seen here in its psychological dimension as the realm of death and the dead) that is addressed equally to the lover, the sea, and the speaker's own self. Crane makes full use in this poem of the fact that a poem is not a literal communication but a structure of affects. "Voyages" V *uses* dialogue, *suggests* a scene and a relationship, *envisions* a dead, impenetrable universe—but is a construct of tonalities and in no sense a narrative or argument. The final set of the poem holds all its elements in a shifted, quieter balance. It picks up the ambiguous, touching suggestiveness of the ending of II: "The seal's wide spindrift gaze toward Paradise." Only now the tensions and active character of V as a whole are still present in the softened language:

> But now
> Draw in your head, alone and too tall here.
> Your eyes already in the slant of drifting foam;
> Your breath sealed by the ghosts I do not know:
> Draw in your head and sleep the long way home.

"Sleep the long way home."—This reconciliation with the disastrous is echoed in the first quatrain of "Voyages" VI, but with an emphasis on vibrant awareness of immersion in an alien state of being (like Conrad's "in the destructive element immerse"). The suggestion here of absolute alert, individual human sensibility in direct contact with its unassimilable opposite amounts to a shock of consciousness:

> Where icy and bright dungeons lift
> Of swimmers their lost morning eyes,
> And ocean rivers, churning, shift
> Green borders under stranger skies . . .

The "I" of the poem describes himself (two stanzas further on) as a "derelict and blinded guest" staving off the death of the self by contriving unstable visions. His sense of unity with the crushing powers of magnetic, indifferent reality is a contrivance of wishful reluctance—reluctance to surrender past hopes and faiths:

> Waiting, afire, what name, unspoke,
> I cannot claim . . .

At this point "Voyages" swings into its Miltonic conclusion, discussed earlier on. The ending is doubtless overly rhetorical and pressing, given the fundamental subjectivity of the sequence in its final aspects. Clearly, Crane felt the need for a peroration that would recall and countersink the prayer at the end of "Voyages" II and other kindred moments earlier in the sequence. It is interesting to see how he revised the original poem, "Belle Isle," on which "Voyages" VI is based. That poem betrays a certain resemblance to "Repose of Rivers," as a poem of rediscovered memory, the recovery of a lost past. It remembers back to a time of shared discovery of transforming love and "absolute avowal," since betrayed by the lover. But it insists on the permanence of the transformation despite the betrayal.

> There was the river;—now there is
> Only this lake's reluctant face
> For me to search back to an island
> Sown by the river at its base.
>
> And remembering that stream of pain
> I press my eyes against the prow,
> Waiting . . . And you who with me also
> Traced that flood,—where are you now?
>
> An absolute avowal, yet termed
> By hours that carry and divide,
> You whispered as we passed Belle Isle
> And entered this too placid tide.
>
> That sharp joy, brighter than the deck,
> That instant white death of all pain,—
> How could we keep that emanation
> Constant and whole within the brain!
>
> Yet, clearer than surmise,—a place
> The water lifts to gather and unfold,
> Seen always—is Belle Isle the grace
> Shed from the wave's refluent gold.
>
> It is the after-word that holds
> Hushed willows anchored in its glow:
> It is the unbetrayable reply
> Whose accent no farewell can know. . . .[2]

Unpublished during Crane's lifetime, "Belle Isle" contains some very weak lines (for instance, 3–4, 7–10, and 17) but has the makings of a serious, concentrated love poem. It has tonal affinities, elusive but real, with such other poems as D. H. Lawrence's "Hymn to Priapus," which antedates it by about a dozen years, and Robert Lowell's "Water," written many years later. These are two poems of difficult love-consciousness, the former coping with the knowledge of having "betrayed" a loved woman who had died, the latter with a relationship that simply could not hold. In recasting "Belle Isle" for "Voyages," Crane kept its vision

2. Brom Weber, *Hart Crane* (New York: The Bodley Press, 1948), p. 391.

of a sacred place where love prevailed but not its sense of love's physicality. In this sense he purified its sense of exaltation, of having reached a sort of Empyrean:

> Still fervid covenant, Belle Isle,
> —Unfolded floating dais before
> Which rainbows twine continual hair—
> Belle Isle, white echo of the oar!

Also, he changed the final stanza in an important respect, by altering the first line from "It is the after-word that holds" to "The imaged Word, it is, that holds." The change is from an impression of intimate communication after lovemaking to one of inspired creation—from gratified mutual reassurance to a Platonically cherished ideal.

This shift from pure subjectivity to a Platonic abstraction overweights the ending of "Voyages." Had the epic dimensions of *The Bridge* been involved, the proportions would have been different—just as, without the pure inwardness of "Voyages" to keep it in order, *The Bridge* oversimplifies and falters in crucial ways. Nevertheless, "Voyages" does on the whole find its natural equilibrium. It all but succeeds in becoming that rarest of works, a purely lyrical sequence that engages passionately with matters of utmost human significance.

2. Olson's *The Maximus Poems*

Hart Crane could hardly have assimilated the discoveries of his elders, Eliot and Pound, when he began writing his sequences in 1923. Although in certain respects he was indeed what he has often been called, an American Rimbaud, this was basically because of his gifts—evident in his best work—for rapid associative movement by what he called "the logic of metaphor" and for bold and drastic emotional energy. He tried to harness these gifts, in his sequences, to the typical curve of a Romantic ode—moving into a "positive" positioning at the end with all the celebratory fireworks he could muster. The effort ran counter to his genius, and it is quite possible that, had he lived beyond his early thirties, he would have abandoned it and followed through to complex balancings of affective streams rather than going for the assertion of visionary transcendence.

But if Crane was still haunted by an outmoded structural ideal, the work of Charles Olson, coming to fruition in *The Maximus Poems* a long generation later, moved in quite another direction. Olson turned toward pure faith in the magnetic, implicitly form-discovering autonomy of the associative process—not limited to metaphoric association, however, but rather extended to the mind's moving every which way, intellectually and didactically as well as affectively. He chose a region in his native Massachusetts, Gloucester; placed himself at the center of it (as Williams, less assertively, had done with Paterson); and, as it were, proliferated from there. Where was the sequence going? Nowhere definable, except to build into something like an organic body through the proliferation just noted, one as it were of "cells" made up of intimate communications. "Letters," Olson frequently calls them: a suggestion of a method less formally conceived and more improvisatory than words like "cantos," "books," and "hymns." And the look of the volume is far more informal even than that of *Paterson.* Poems of all shapes and sizes sprawl across the unpaginated large pages; inconsistencies in titles and poem divisions abound; and the typography in general reinforces the notion of "open form."

Before turning to questions of style and a discussion of this tumultuous poetry, we should specify that we shall be dealing with only the first of the three "Maximus" volumes, supplemented by one poem from the second volume, "Maximus to Gloucester, Letter 27 [withheld]."[3] *The Maximus Poems* (1960) and *Maximus Poems IV, V, VI* (1968) were published before Olson's death in 1970. *The Maximus Poems: Volume Three* (1975) was edited "at Charles Olson's request from among his papers" by Charles Boer and George F. Butterick. The 1960 volume has its own definite if erratic unity. The poems in the later volumes contain much of interest as a sprawling aftermath of the initial impulse. A number resemble poems in the first book, a number turn to more open engagement with myth and fantasy, and a number are more frankly sexual than the 1960 group. But that first Maximus book, with a restored "Letter 27," is a closer model of one kind of modern poetic sequence than the less controlled later books, although they provide material aplenty for a Poundian open sequence in the making. Thus, when we refer to *The Maximus Poems,* we are speaking only of the thirty-seven poems of the 1960 volume plus "Letter 27." Also, although many of the poems are simply given titles without their letter numbers being designated, and although the last numbered letter is "Letter 23," for simplicity of discussion each poem is assumed to be a letter in a sequence of 38 letters. "Letter 27" then falls between "So Sassafras" and "History is the Memory of Time." Of course, calling the poems "Letters" when they are not so designated imposes form where it is definitely not wel-

3. Charles Olson, *The Maximus Poems* (New York: Jargon/Corinth Books, 1960); *Maximus Poems IV, V, VI* (London: Cape Goliard Press; and New York: Grossman Publishers, 1968); and *The Maximus Poems: Volume Three* (New York: Grossman, 1975).

come; indeed, the term "open form" seems particularly apt for Olson's work. Although it is obviously applicable as well to the sequences of Yeats, Pound, Eliot, and Williams we have discussed, all of these seem to be moving toward some encompassment, some closure, however tentative, that will hold against the chaos of things and of the mind's own turmoil. But Olson has little of their sense of art as an assimilation of chaotic being yet a hold against it as well.

Olson's own critical observations and, almost uniformly, Olson criticism in general throw little light on the actual character and quality of his poetry but are much more concerned with theoretical observations couched in a curiously ambiguous, half-intellectual, half-colloquial terminology. Reference to his well-known essay "Projective Verse" (1950) will sufficiently suggest his amalgam of ordinary free-verse theory (breath sweeps, free association, and organicism) and abstract pronouncements (especially the famous cliché presented as the invention of Robert Creeley but actually, as we have pointed out, as hoary as Romantic theory itself: "FORM IS NEVER MORE THAN AN EXTENSION OF CONTENT").[4] Other comments, for instance the description in *Mayan Letters* (1953) of "the methodology of the Cantos" as "a space-field where, by inversion, though the material is all time material, [Pound] . . . has turned time into what we must now have, space & its live air,"[5] have an interesting sweep of energetic thought without adding to what was already known. In both his prose and his poetry, what Olson contributes to modern poetics is, precisely, the idiom of his own energetic and rather heavily intellectual sensibility—a hurried channeling of various confused bits of twentieth-century poetic assumptions and attitudinizing, together with a Poundian tendency to harangue and, at worst, a deadly didacticism in which even the melody of passionate intellectual interest is lacking. This didactic blight is clearly seen in a passage such as the one closing "Letter 23":

> What we have in this field in these scraps among these fishermen,
> and the Plymouth men, is more than the fight of one colony with
> another, it is the whole engagement against (1) mercantilism
> (cf. the Westcountry men and Sir Edward Coke against the Crown,
> in Commons, these same years—against Gorges); and (2) against
> nascent capitalism except as it stays the individual adventurer
> and the worker on share—against all sliding statism, ownership

4. Charles Olson, *Selected Writings* (New York: New Directions, 1966), p. 16.
5. Ibid., p. 82. (*Mayan Letters* was first published by Divers Press, Mallorca.)

getting in to, the community as, Chambers of Commerce, or theocracy;
or City Manager

This is like Eliot at his most slackly discursive, but without even Eliot's minimal music. If this passage were representative of the whole of *The Maximus Poems,* there would be little point in quoting it and even less in devoting any time to the sequence. Much—most—of the poetry bears a much closer resemblance to verse of intrinsic interest, and so it is probably unfair to make this our first quotation. Yet the tone of the passage is one of the dominant ones in the sequence. It is that of an insistent, totally immersed "village explainer" (Gertrude Stein's phrase for Pound). This explainer presses local facts upon us we would certainly never have inquired about. And then he goes further—he presses interpretation on us we would never have asked for either. And indeed, it is one important role of the sequence to play such a part, assuming the tone of a fanatical localism that reads (correctly, no doubt, despite the parochial twist) the workings of historical forces of universal import into specific Gloucester events.

Our quotation exactly follows the line-divisions of the printed text, for no clue is given to suggest the passage is intended as prose in the usual sense of the word. The spacing here is the same used earlier in the poem between lines that are clearly presented as verse; nor does Olson use smaller type or quotation marks to set the passage off, as he does elsewhere to show a deliberate prose formation or a quotation of someone else's prose. At the same time, certain tricks of punctuation suggest a controlled rhythmic pacing: for instance, the unorthodox use of the first three commas and the omission of a period at the end. So also do the many sound-echoes, especially at line-ends, such as the repeated "er"-sound in "fishermen," "mercantilism," "adventurer," "ownership," and "Manager." Finally, the passage has a Poundian allusiveness, and indeed the reference to Coke is a kind of high music to one sort of Poundian ear. It may well be that to a pure mind all this will seem utter nonsense, even as an argument that Olson felt this passage to be nine lines of verse whose "form" is only an "extension of content." But the considerations we have mentioned, and the deliberate clumsiness of the style, do fit in with the work's implied poetics. Three poems further on, in the "letter" called "So Sassafras," the opening lines are hardly less prosaic although unmistakably laid out as verse. We quote only the very start of this eighteen-line passage:

Europe just then was being drained swept by the pox so sassafras
was what Ralegh Gosnold Pring only they found fish not cure
for fish so thick the waters you could put no prow'd

> go through mines John Smith called these are her silver
> streames Spanish north Latitudes to avoid Sable course
> W 2 minutes north fetch Cape an or Isle Shoals . . .

Olson draws on an Elizabethan ship's log and other documentary sources, mixing their phrasing with his own and evoking an atmosphere of practical day-to-day thinking and concerns and logbook style. Through these lines a reader of vivid imagination may visualize a world of actively adventurous men in the process of making an important discovery for English commerce. The language is replete with energetic verbs and syntactic shifts suggestive of altered awareness of an extroverted kind. Despite the rhythmic jerkiness, too, the emphatic alliteration, internal rhymes, and half-rhymes provide an interruptedly musical movement reinforced by such phrases as "for fish so thick the waters," "silver / streames," "Sable course," and "fetch Cape an [Ann]." Yet the passage, certainly by the time we reach line 18, forces our patience. It quickly becomes clear that the main effort is less the recovery of a quality of experience than a rhetorical assertion of an idea: namely, that the English explorers and settlers were prototypes of modern capitalism. The closing lines of this same passage read:

> . . . the Puritan coast was was fur and fish frontier cow-towns
> GLOUCESTER Queen of the fishtowns Monhegan Damariscove & 1622:
> Cap an all after her Weston Thompson Pilgrimes grabbing
> & Richard BUSHROD beating em to the westerly side of sd Harbour
> George derby agent and JOHN WHITE old minister John poking out the
> green stuff or whatever money looked like don't mistake there wasn't
> money Wow sd Pilgrimes ONE HALF MILLION BUCKS in 5 years from
> FURS at the same time FISH pulling all of Spain's bullion out of

The rhetorical dimension of *The Maximus Poems* is its intractable aspect. Its deadweight side is avoided, but *threatens* to dominate, at the very start, in the title and opening of "I, Maximus of Gloucester, to You." "Maximus" puts us on guard at once when he describes his tutelary role before the poem proper gets under way:

> Off-shore, by islands hidden in the blood
> jewels & miracles, I, Maximus
> a metal hot from boiling water, tell you
> what is a lance, who obeys the figures of
> the present dance

There is an implied promise here, by the way, of introspective depths to come in the sequence. That is, we shall have an exploration of the "islands hidden in the blood" as well as of whatever is meant by the ambiguous clauses "what is a lance, who obeys the figures of / the present dance." The deliberately awkward syntax often employed in these poems, to suggest the mind's inner gropings and adjustments of formulation as it discovers its own processes (and also, possibly, to suggest a careless swagger of personality as if to disdain all petty care for precision or lyrical consistency), has also been introduced in these few lines.

Theoretically, it should be possible to write a sequence of place without either intruding the author's personality on the progression of the work or substituting sheer quantity of information or tendentious wisdom for affective repossession and equilibrium. In practice, it seems a little difficult to keep the enterprise in the clear. Immersion in a body of knowledge that has called for research (however unsystematic the method and unscholarly the purpose) perhaps must perforce lead to an innocent pedantry and egotism, the twin sins of any expert who cannot carry knowledge lightly. At any rate, most of the vices and virtues of *The Maximus Poems* can be found in the opening proclamation just quoted and the succeeding six numbered parts of "I, Maximus of Gloucester, to You." All six parts center on aspects of Gloucester—its houses, its harbor, its past—while weaving a complex texture of contexts of feeling and thought around this center. Thus, we find very simple local impressions and notations all along the way:

> the roofs, the old ones, the gentle steep ones
> on whose ridge-poles the gulls sit, from which they depart,
>
> And the flake-racks
>
> of my city!

or:

> when bells came like boats
> over the oil-slicks, milkweed
> hulls
>
> And a man slumped,
> attentionless,
> against pink shingles
>
> o sea city)

But at the same time the succeeding sections concern themselves with a wide variety of perspectives—the first with the pursuit of blessing and reassurance; the second with organic form in nature ("love is form"), as exemplified by a bird's accumulation of all sorts of substances in building a nest, a symbolic motif (applicable to the poem's making as well) picked up again later in the fourth and sixth sections; the third and fifth with the displacement of "that which matters" in a people's regional memory and life by a cheapened national culture whose emblems are billboards, spray guns, juke-boxes, and other tokens of "pejorocracy." Thus the first poem of the book is an overture relating its tonal motifs of affectionate cherishing of the real Gloucester and its affinities with everything organic and culturally intrinsic in its history (such as a "carefully carved" figure under the bowsprit of an old boat) to opposite ones of dismay and indignation. The second and third Letters develop this latter tone, at the same time strengthening Olson's (i. e., Maximus's) identification with the region, partly through his sympathetic closeups of local persons. Letter 4, "The Songs of Maximus," provides lyrical echoes of these expanded developments of the poem's basic sets of concern and feeling. Here we have Olson close to his best, especially in the third and fourth of these "Songs." The former of these celebrates something like a bohemian New England version of voluntary poverty, as a form of resistance to commercialism and the corruptions of affluence. Cantankerousness and sweetnaturedness coincide in this song, and the life of deprivation is somehow felt to be a mode of being true to the inner meaning of Gloucester's past as a place of hard work, thrift, and the self-regard of craftsmen:

> This morning of the small snow
> I count the blessings, the leak in the faucet
> which makes of the sink time, the drop
> of the water on water as sweet
> as the Seth Thomas
> in the old kitchen
> my father stood in his drawers to wind (always
> he forgot the 30th day, as I don't want to remember
> the rent

The closing section of this song takes on a more revolutionary tone, and what we have called "sweetnaturedness" goes out the window, as the sequence moves into a mood of reckless hatred. This section (the last eighteen lines) begins:

> "In the midst of plenty, walk
> as close to
> bare
> In the face of sweetness,
> piss

In the time of goodness,
go side, go
smashing, beat them, go as
(as near as you can

tear

—as if to say: The only way to make a fresh start and get out from under the whole betrayal of hard-earned self-regard and deep-rooted local ways is to repudiate all facile affirmation within the present system of things. And that must be done with determined nastiness, an offensive "grossing out" (in the American slang phrase for shocking the complacent and unsophisticated would-be genteel sensibility) of the bourgeois.

Song 4 restores the human balance by providing a lover's image of a woman whose near-poverty and Yankee self-reliance are evoked by the description of her dress, and whose romantic aura and desirability are suggested by the very same lines and by the dreamy echo of Yeats that introduces the song. A hostile reader might wish to turn Song 3's language of deliberate ugliness of spirit against the contrary tonality here; but the feeling induced by the shift between the two songs is one of earned delight as a result of the repudiation of any false sense of ease. The happier playfulness of Song 4's opening is marked by the surprising allusion to Yeats's "The Lake Isle of Innisfree." (American regionalist rancor has truly let down its guard here!) Almost at once the mood blends with a tone of ardent discovery that leads, by the song's end, to the sequence's largest, most encompassing vision.

I know a house made of mud & wattles,
I know a dress just sewed
(saw the wind

blow its cotton
against her body
from the ankle
so!
it was Nike

And her feet: such bones
I could have had the tears
that lovely pedant had
who couldn't unwrap it himself, had to ask them to, on the schooner's deck

and he looked,
the first human eyes to look again
at the start of human motion (just last week
300,000,000 years ago

She

was going fast
across the square, the water
this time of year, that
scarce

And the fish

To recapitulate the progress of this little poem: It begins with a whimsical echo of Yeats's "I will arise and go now, and go to Innisfree, / And a small cabin build there, of clay and wattles made." Yeats's poem, incidentally, has despite its very ninetyish coloration strong affinities with Thoreau's *Walden*, as the poet notes in his *Autobiography:* "I had still the ambition, formed in Sligo in my teens, of living in imitation of Thoreau on Innisfree, a little island in Lough Gill."[6] The connection with Yeats is not merely whimsical; it involves a kindred sense of a need to live in a stripped-down way and reorient oneself by confronting essences of existence: the position developed in contemporary American local terms in Songs 3 and 4. The simple Yeatsian echo at the start of the latter song is transferred to the memory of the woman in her homemade dress crossing the town square in Gloucester, a vision comparable in its stunning effect on the speaker to that of the "woman in our town" in Williams's *Paterson,* V,ii. In fact a considerable complex of associations is set up here. The woman was "going fast / across the square," much like the Muse figure in *Paterson.* She is "Nike," the winged goddess of victory whose ample beauty still guards the great staircase in the Louvre and the thought of whom takes us back to the era of earlier civilizations antedating modern conceptions of property and the modern state. Hence the association of all the emotions evoked by the sight of a lovely statuesque woman walking against the wind with the archaeologist's thrilled recovery of a relic from the ancient world (and also with female figures carved on the prows of old schooners—a motif introduced in the earlier Letters of the sequence). It is a hard time of year for fishermen, the local resources are scarce, but the woman crossing the square in her own vivid reality summons up all the sense of human possibility, past and present, that is the total bearing of cultural memory.

We have been trying to suggest some key characteristics of the texture of Olson's sequence: both the obstacles it presents and its strictly lyrical aspects, and also the range of the consciousness at work in these poems. To speak of its structure as a whole, however, is to face the issue of its double nature as part lyrical sequence and part didactic poem. The

6. *The Autobiography of William Butler Yeats* (New York: Doubleday Books, 1958), p. 103. See also p. 47.

latter aspect is a matter of the large amount of space given to information about the founding and the early history of Gloucester, much punctuated by statistical and documentary materials, and to political, social, and moral argument of the kinds we have mentioned. Nevertheless, Olson (fortunately) never succeeded in sloughing off the work's dependence on emotional and subjective intensities for its major effects poetically, despite the presence of many passages like the one closing "Letter 23," and despite his ambition—proclaimed in the same poem—to be a special kind of poetic historian:

> I would be an historian as Herodotus was, looking
>
> for oneself for the evidence of
>
> what is said . . .

Yet even in "Letter 23" poetry gets in the way of thought-centered exploration, even if only minimally as in the closing section beginning "What we have in this field in these scraps among these fishermen, / and the Plymouth men, is more than the fight of one colony with / another . . ." There is, incidentally, a curious echo here of Eliot's more discursive rhythmic style in the flattest passages of *Four Quartets*—an example of his pervasive influence in even the most reluctant quarters. We have already of course noted—strange endeavor!—other "clues" that the passage is conceived as poetry despite its anti-poetic bearing. It is, as it were, a prosaic climax in a poem that has struggled to become prose (kneaded, however, into different shapes of versiform dough) from its very start:

> The facts are:
>
> 1st season   1623/4   one ship, the *Fellowship*   35 tons
>
> with Edward Cribbe as master (?—cf.
>
> below, 3rd season)
>
> left 14 men Cape Ann . . .

Still, the most moving and memorable parts of the sequence shake clear of this kind of not unpleasant but rather anesthetized formation, whose chief affect is of a scholar's half-stunned blinking after reading too many nautical logs, business contracts and charters, legal documents, and newspapers. Creating rhythmic balances and typewriter-designs out of such materials is a poetic fringe-art, like any other technique, unless undeniably assimilated into affective structure. In *The Maximus Poems* the assimilation is only partial although the poet tries to justify his method in the first poem by comparing it to a bird's foraging for nest-building materials ("feather to feather added . . ."). And just as a bird will occasionally discard some choice item, only to retrieve it later, so Olson apparently changed his mind about using "Letter 27." By carefully numbering it and indicating that it had been withheld, Olson presumably was restoring it to its rightful position between "So Sassafras" and "History is the Memory of Time."

The reason for the original withholding seems clear enough: the poem presents Olson's parents in a demeaning light. And yet, in part for this very reason, "Maximus to Gloucester, Letter 27" can well be considered a key poem in the sequence. Its concentration is unusual, it presents a series of affective facets that keep its rhetorical insistence within a context of feeling in a genuinely revelatory way, and it contends with an inherited humiliation that needs to be righted and that helps justify the sequence's repossession of Gloucester's history from the viewpoint of its long-exploited working folk. It seems best at this point to quote the whole poem as a central illumination of the sequence from which it was originally excluded:

> I come back to the geography of it,
> the land falling off to the left
> where my father shot his scabby golf
> and the rest of us played baseball
> into the summer darkness until no flies
> could be seen and we came home
> to our various piazzas where the women
> buzzed
>
> To the left the land fell to the city,
> to the right, it fell to the sea
>
> I was so young my first memory
> is of a tent spread to feed lobsters
> to Rexall conventioneers, and my father,
> a man for kicks, came out of the tent roaring
> with a bread-knife in his teeth to take care of
> a druggist they'd told him had made a pass at
> my mother, she laughing, so sure, as round
> as her face, Hines pink and apple,
> under one of those frame hats women then
>
> This, is no bare incoming
> of novel abstract form, this
>
> is no welter or the forms
> of those events, this,

Greeks, is the stopping
of the battle

It is the imposing
of all those antecedent predecessions, the precessions

of me, the generation of those facts
which are my words, it is coming

from all that I no longer am, yet am,
the slow westward motion of

more than I am

There is no strict personal order

for my inheritance.

No Greek will be able

to discriminate my body.

An American

is a complex of occasions,

themselves a geometry

of spatial nature.

I have this sense,

that I am one

with my skin

Plus this—plus this:

that forever the geography

which leans in

on me I compell

backwards I compell Gloucester

to yield, to

change

Polis

is this

Despite a fair amount of derivativeness—the first three verse-paragraphs reminiscent of Lowell's *Life Studies*, the opening of the last unit reminiscent of Eliot's "Journey of the Magi" ("set down / This set down / This"), and touches of Williams and even Donne—this poem brings out uniquely a sense of total immersion in whatever America and Gloucester are. "The geography of it" is the place as experienced by a child and by his mother and father under idiosyncratic local conditions circumscribed topographically (as in the couplet that makes up the second verse-unit), but also circumscribed by the conditions of life and the behavior presented so vividly. Olson addresses the "Greeks" as a modern Herodotus, wrenching a sense of identity out of the welter of reality's "battle": the identity of a self but also of a place, a "polis" or social organism. Moreover, he responds to a pressure to redefine that identity ("forever the geography / which leans in / on me I compell / backwards I compell Gloucester / to yield, to / change / Polis / is this") with a counter-pressure of his own that would reorient the past toward a less humiliated condition. The poem begins with a thick piling-on of emotional colors and sharp memories. Then, with a certain elegance of braking thought, it shifts into a subtler self-defining process. Finally, it insists on the necessity for muscular effort to change the current of national and regional self-awareness. What that current is has been bitterly defined and repudiated in "So Sassafras" (Letter 26), which focuses on the corruption, greed, personal licentiousness, racial feeling, and exploitation that are Gloucester's historical heritage. "History is the Memory of Time" (Letter 28) picks up this same stream of feeling and rhetoric, ending with the lines:

> They should raise a monument
> to a fisherman crouched down
> behind a hogshead, protecting
> his dried fish

Placed between these two poems, the excluded Letter 27 draws the poison of history into the life-sense of the poet and his family and community, making it personal and at the same time trying to counteract it by a purity of vision and sympathy. It had been prepared for by the third of "The Songs of Maximus" ("This morning of the small snow"), which, however, took on a nearly *lumpen* cast in its efforts to repudiate the prevailing culture. It was also anticipated by "Letter 5," a somewhat mean-spirited assault on Vincent Ferrini, a local poet and magazine editor whose work "Letter 5" demolishes with half-friendly contempt as far from the real thing and naturally alien to the true spirit of Gloucester. A number of scattered passages foreshadow "Letter 27" as well. But six poems do so most pertinently: "Maximus, to Himself" (Letter 12); the four poems (Letters 17–20) called "On first Looking out through Juan de la Cosa's Eyes," "The Twist," "Maximus, to Gloucester, Letter 19 (A Pastoral Letter," and "LETTER 20: not a pastoral letter"; and "Maximus, to Gloucester" (Letter 25). All these poems reinforce "Letter 27" without quite providing the central focus of its subtle linkage of private shame and memory with the largest perspectives of the sequence.

"Maximus, to Himself" (Letter 12) is one of the three or four truly introspective poems in the volume. It takes us at once into a sense of psychological and cultural inadequacy and displacement, reflected in its rhythmic and syntactic imbalances that project an awkwardly lurching, graceless existence:

> I have had to learn the simplest things
> last. Which made for difficulties.
> Even at sea I was slow, to get the hand out, or to cross
> a wet deck.
> The sea was not, finally, my trade.
> But even my trade, at it, I stood estranged
> from that which was most familiar. . . .

The poem then proceeds to find its balance, *nevertheless*, just within the existential apperception of these conditions. A sense of cordial resistance and self-reliance within uncertainties of every kind emerges, marked especially in the lines

> But sitting here
> I look out as a wind
> and water man, testing
> And missing
> some proof

and in the short second section that concludes the poem:

> It is undone business
> I speak of, this morning,
> with the sea
> stretching out
> from my feet

There is no period at the end of the poem, which is left open and "stretching" outward with its "undone business" to be gauged in a very personal way that has to do with the sense in which one is at the center, as a recording instrument and as one embodiment of the culture, of whatever there is to observe and to do. The quiet affirmation of a private and artistic purview and task here is balanced further on by "Maximus, to Gloucester, Letter 19 (A Pastoral Letter"—a vehement poem, presented as a spiritual gun-duel on the streets of Gloucester between the poet and a clergyman out to capture him for his church. The clergyman is presented as a vile embodiment of the pervasive commercialism that makes salesmen, with the surface cordiality and smoothness of modern businessmen, and institutional representatives out of people whose mission should lie elsewhere. Olson's comic presentation of himself in Letter 12 as totally inadequate to the world's practical demands and yet his own man is paralled here by the parody of a duel in the sun in a Western film. The hero will not back down to another kind of worldly demand, for conformity, of the kind that cows tradesmen and, too often, the young.

"LETTER 20: not a pastoral letter" is a companion piece to "Letter 19," but far more elusive in presentation and bearing. It is indirect where "Letter 19" is completely direct, a simple parable of the moral life—the imperative to resist at the price of losing social approval and perhaps affluence. Yet it too speaks in parables drawn from literal experience, and with a certain confessional ingredient. The poet has cheated a good and honorable man, so generous "he was embarassed / to ask for the rent"; and has been on good terms with a thieving fellow-crewman aboard ship. That is, he shares in our general complicity although he knows the truer way of outmoded heroes is still the necessary way. One can choose, or create, a preferable reality:

> The world of Hannas
> (the world of Earp)
> went with the blueberries,
> the chestnuts
>
> with the openness the exploiters
> had not beat out, was still walking, was going places
> in street-cars

A wider net of choice is cast in another companion-pair, the Juan de la Cosa poem and "The Twist" (Letters 17 and 18). The former, based on the work of a fifteenth-century mapmaker, is in part a delightful play of imagination over the images in de la Cosa's "mappemunde" that suggest the active presence of mythical imagination in his time; and in part a grim recalling of the murderously dangerous character of old-time transatlantic sailing. The latter poem has to do with the internal map of Olson's own world, retained from childhood but altered as Olson grew older. The world of his memory and emotional life, in this poem as elsewhere in the sequence, seems profoundly affected by a difficult marital relationship. The poems, and the choices, and the sense of hopelessness riding the sequence despite all its braveries, are colored by an erotic and romantic longing and love, and a tenderness and concern for children, that make the sequence in some degree a poem of a normal, hard-pressed adult sensibility continually trying to find its way amid all the obstacles. One of the obstacles, doubtless, is the ingrained resistance to social assumptions bred in part by a sense of inferiority to the "better," wealthier citizenry of Gloucester. Hence the parochially self-conscious stance about the town's favored families in "Maximus, to Gloucester" (Letter 25), asserting the priority of the poet's vision and sense of Gloucester to theirs—with the implication that he, not they, is the real heir to the place. All these shifts of perspective and tonal variations have been preparing us for "Letter 27," the poem of confession and reconciliation, bringing all the streams together, that was originally omitted from the book.

If "Letter 27" is the climactic culmination of the sequence, the alternative climax—in a distinctly minor key—comes in "Maximus to Gloucester, Sunday, July 19" (Letter 37). The tone here is quietistic and elegiac. The poem brings to the surface, for a more sustained moment than before, the essentially depressive mood that has dragged against the work's energies since the first Letter noted the difficulty of coping with "pejorocracy" and a juke-box civilization. Two other examples of this mood, from earlier Letters, are especially relevant. In "Maximus, to Gloucester" (Letter 14), the murderous Elizabethan slave-trader and privateer John Hawkins is contemplated, and then the poem moves by natural association into thoughts of death and of the persistence of Hawkins's kind of mercilessness in our modern world:

With the gums gone, the teeth
are large. And though the nose is then nothing,
the eye-sockets

And now the shadow
of the radiator on the floor
is wolf-tits, the even row of it
fit to raise
feral children.

You will count them all in,
you will stay in the midst of them,

you will know no law, you will hear them
in the narrow seas.

And the poem just before this one, "The Song and Dance of," contrasts the exoticism, romanticism, and voluptuousness of Cyprian and Cyrenaic civilization with the hard-driving practicality and profiteering that underlie Gloucester's values:

Venus
does not arise from
these waters. Fish
do.
And from these streams,
fur.

It was the hat-makers of La Rochelle, the fish-eaters of Bristol
who were the conquistadors of my country, the dreamless
present

Both these passages are charged with the sadness of irrecoverable loss, in the one instance of a culture's most generous possibilities (introduced by the image of a death's-head); in the other, of a life of spontaneous pleasure without guilt or self-consciousness. Combined with the poems we have noted of personal frustration, humiliation, and suffering, such passages carry forward throughout the sequence that depressive melancholy that constitutes its deepest emotional pressure. It seems very possible, by the way, that the special rhythm of the final unit in the passage just quoted ("It was the hat-makers . . . the dreamless present") provided Geoffrey Hill with a model for his extended lines that mix contemporary speech and allusions with traditional and learned phrasing and feed wry and comic tones into essentially appalled notations.

To return now to Letter 37, the penultimate poem in the sequence. The occasion around which the poem revolves is the annual memorial service in Gloucester for "all fishermen / who have been lost at sea." This year the ceremony is held despite the fact that "for the first time not one life was lost." The latter fact distances or muffles what might otherwise be a more emphatically tragic gravamen in the poem, and allows it to maintain a philosophical calm within which a certain irritated acknowledgment of the ceremony's validity is balanced against the shallowness and bad taste involved. The poem begins:

> and they stopped before that bad sculpture of a fisherman
>
> —"as if one were to talk to a man's house,
> knowing not what gods or heroes are"—
>
> not knowing what a fisherman is
> instead of going straight to the Bridge
> and doing no more than—saying no more than . . .

In fact, bad temper is very much part of the affect in the poem, giving it a dimension of personal feeling that resists the pompous or merely sentimental temptations of the occasion and endows its recognition with a cross-grained and bitter hectoring note:

> Let them be told who stopped first
> by a bronze idol
>
> A fisherman is not a successful man
> he is not a famous man he is not a man
> of power, these are the damned by God

Still more bitter language and imagery follow, and then a touch of elegiac philosophizing that might have sprung fully formed from the *Pisan Cantos:*

> When a man's coffin is the sea
> the whole of creation shall come to his funeral. . . .

An unwonted gentleness soon enters the poem, and then a tribute to fishermen that is quietly subtle and understanding:

> this afternoon: there are eyes
> in this water
>
> the flowers
> from the shore,
>
> awakened
> the sea
>
> Men are so sure they know very many things,
> they don't even know night and day are one
>
> A fisherman works without reference to
> that difference. It is possible he also
>
> by lying there when he does lie, jowl
> to the sea, has another advantage: it is said,
>
> 'You rectify what can be rectified,' and when a man's heart
> cannot see this, the door of his divine intelligence is shut

Thus, just before the sequence-end, a universally tragic affect is introduced and at the same time softened and given the idiosyncratic coloration for which, in a sense, the work has striven mightily all along. "April Today Main Street" (Letter 38), which concludes the sequence, is a firm poem of some substance, a reprise (with new details) of certain historic perspectives and contemporary closeups of Gloucester, and yet merely an afterbeat that helps contain the distanced melancholy of Letter 37. It ends with a question centered on an old sloop that

> to this hour sitting
> as the mainland hinge
>
> of the 128 bridge
> now brings in
>
> what,
> to Main Street?

The elegiac mode pervades the sequence and, by the end, gives it more shape than seems possible when one is bushwhacking through it for the first time. In *The Maximus Poems* the sense of history itself, as it impinges on and converges with individual destinies, is shot through with elegiac tones—not only simple laments and an occasional wash of nostalgia but also fiercely fatalistic outbursts such as we noted at the end of Letter 14. So, too, there is an almost querulously questioning coloration, with a subtly elegiac dimension, that appears at many strategic points—for example, in the passage just quoted with which the sequence ends. It projects a sort of fearful nostalgia for the passing moment, or else a saddened insecurity about the poem's strongest moods of assertiveness, even the triumph of poet over clergyman in the "Pastoral Letter" (Letter 19) and the defense of one man's lonely integrity against the conforming political crowd in "John Burke" (Letter 35).

Part Five

Poetry of Psychological Pressure

Chapter Thirteen

The Meditative Mode: Wallace Stevens and W. H. Auden

1. "Psychological Pressure"

Here, in the three chapters at the close of our book, we shall attempt to sketch a variety of sequence-modes that reflect, to a great degree, the stresses of the past few decades. In a book devoted to the genre as a whole we cannot, obviously, hope to be exhaustive—so great has been the proliferation of sequences in recent years. We have chosen sequences at a particular level of intensity that we feel represent three major modes: the meditative, the confessional, and the "main stream" (that is, work whose structural character parallels the sequences of Yeats, Pound, Eliot, and Williams). As a further center of emphasis, we shall touch on certain strong current tendencies with markedly political or ideological dimensions; and on other tendencies toward a narrowly self-limiting poetry centered on private experience and feeling projected with minimum resonance.

In discussing so wide a range of work under the rubric "psychological pressure," we are calling attention to an aspect of the most powerful lyrical poetry, past and present, which we feel has been too little attended to—especially since so much specifically modern work vitally depends on the presence and workings of this pressure. Very early on, we had occasion to note how often, in a sequence of power, we are confronted by a speaker literally or metaphorically *in extremis*. We noted, for instance, the mortal agony of Ramon Guthrie's "Today Is Friday" (in *Maximum Security Ward*—to be discussed more fully in the last chapter), and the burden of acutely tragic consciousness in Yeats's civil-war sequences and a number of other works. The forerunners and early exemplars of the modern sequence had already revealed the drastic engagement of the genre: the epic proportions of its struggle with depressive awareness. Those proportions were clearly anticipated in the tragic volatility of the progression of *Maud* and the terrific battle against remorse in the dynamics of *Poems of 1912–13.* They were assumed unquestioningly in the poetry of painful sense-responsiveness and overwhelming human empathy in *Song of Myself* and in the confrontation of psychic catastrophe in the Dickinson fascicles.

Throughout Part Five we shall find war-consciousness a tormenting pressure in the poetry of the most diverse writers. The unleashing of war at a previously unimaginable pitch of destructiveness is after all one of the key stresses, if not *the* key stress, of our age. From Yeats's work on, our sequences embody the responses of individual sensibility to war over more than fifty years: from World War I and the Irish "troubles" through Vietnam and into the present. In some sense all the sequences we shall consider henceforth are products of the war age, reflecting horror at war itself or at the depths of human bestiality revealed in our wars or indirectly related to the effects on modern sensibility of this unremitting presence in our world. The most obvious reflex lies in the variety of pressures to reshape our society, which has so signally failed to remove its own worst scourge. Ordering pressures in the sequences may involve a hatred of advanced technology and an intense desire to live a simpler life, closer to nature and to more primitive (idealized) human values; or of course to bring about revolutionary change, whether social or economic or sexual or all these and others besides.

Psychological self-awareness is of necessity a given of our time. The modern exacerbated and self-analytical sensibility, the need to cope with a sense of anomie that floats freely in the cultural atmosphere, the enormous disillusionments and alienation (which now pervade even revolutionary politics because of the ever-widening circles of terrorism that have virtually broken down ideological distinctions), and the vast tide of moral relativism unaccompanied by humane reorientation are facts of our psychological ambience. Moments of sheer delight or exuberant sense-response and consciousness are felt not only in their own right but also as signs of a lost innocence of feeling and of spontaneous participation—of our compromised humanity. Thus, the enchanting "Brag, sweet tenor bull" passage opening *Briggflatts,* the ecstatic song "The boat responded / Gaily" near the end of *The Waste Land,* the charmed closeups of natural things in *Pisan Cantos,* the precious memories of childhood in *The Maximus Poems*—all these passages and many kindred ones in our sequences have both the charm of their intrinsic purity and the pang of lost worlds they evoke.

The workings of psychological pressure in meditative sequences are more elusively operative than elsewhere, since this mode is the least presentative in its overt character. By definition the meditative sequence has a great deal of speculative, intellectualizing language—enough, usually, to diminish considerably the energy of its dynamics. Even the finest of Stevens's sequences, "The Auroras of Autumn,"[1] calls its designation as a sequence into doubt because so often it is not presentative but ratiocinative in character. Here, for example, is its eighth section:

> There may be always a time of innocence.
> There is never a place. Or if there is no time,
> If it is not a thing of time, nor of place,
>
> Existing in the idea of it, alone,
> In the sense against calamity, it is not
> Less real. For the oldest and coldest philosopher,
>
> There is or may be a time of innocence
> As pure principle. Its nature is its end,
> That it should be, and yet not be, a thing
>
> That pinches the pity of the pitiful man,
> Like a book at evening beautiful but untrue,
> Like a book on rising beautiful and true.
>
> It is like a thing of ether that exists
> Almost as predicate. But it exists,
> It exists, it is visible, it is, it is.
>
> So, then, these lights are not a spell of light,
> A saying out of a cloud, but innocence.
> An innocence of the earth and no false sign
>
> Or symbol of malice. That we partake thereof,
> Lie down like children in this holiness,
> As if, awake, we lay in the quiet of sleep,
>
> As if the innocent mother sang in the dark
> Of the room and on an accordion, half-heard,
> Created the time and place in which we breathed . . .

Now, all this is hardly Yeats crying out against "dragon-ridden" times when "a drunken soldiery / Can leave the mother, murdered at her door, / To crawl in her own blood, and go scot-free." It is not Sylvia

1. Wallace Stevens, "The Auroras of Autumn," in *The Palm at the End of the Mind: Selected Poems and a Play,* edited by Holly Stevens (New York: Vintage/Random House, 1972), pp. 307–16. The text is more accurate than that of *The Collected Poems.*

Plath calling herself "Lady Lazarus" after her third unsuccessful attempt at suicide, or John Berryman probing the undiscovered source of his barbed sense of guilt. There is nothing sensational here, nothing even like the anguish of Milton's dream of his "late espoused Saint" or of Donne's self-diminution in "A Nocturnall upon St. Lucie's Day." Moreover, the eight unrhymed loosened iambic pentameter tercets constitute a formal pattern repeated in all ten sections of "The Auroras of Autumn." Visually and aurally, this form never varies to match the shiftings of tonality either within sections or from one section to another—shiftings that are unobtrusive in any case.

Nevertheless, Part VIII works with considerable poignancy as it moves from its initial abstractions toward a would-be memory, presented in an extended simile, of a "holy" moment of family communion and of childhood innocence. The need to feel and assert the reality of states of innocence is a keen pressure in the poem. It is related to the need to see a reassuring symbolism in the "lights" recurred to in the sixth stanza. These are the northern lights or aurora borealis, the "auroras of autumn" of the title, which were introduced in Part I of the sequence and have been pondered ever since.

The poignancy, and its causes, are especially embedded in certain lines—for instance, in the parallel couplet

> Like a book at evening beautiful but untrue,
> Like a book on rising beautiful and true.

The sardonically wistful character of these lines is the reflex of a shared psychic and poetic search based on the pressure, in an age of skepticism and disillusionment, of a primitive need for the consolations of certainty. We speak of a "shared" search because the complex of feeling and the sense of hopeless urgency here are paralleled in many other poems of our age, notably those of T. S. Eliot, Hart Crane, and Kenneth Fearing. This sort of parallelism is not a matter of derivativeness but of convergence. (Similarly, the manipulation of abstractions in the opening lines—"There may be always a time of innocence. / There is never a place"—shares its baffled effort toward a desired, elusive certainty with various passages in *Four Quartets* and other modern works.) And Stevens's closing stanza is charged with the same primitive need we have mentioned, this time presenting vividly as an experienced reality what the subjunctive grammatical formation emphatically denies:

> As if the innocent mother sang in the dark
> Of the room and on an accordion, half-heard,
> Created the time and place in which we breathed . . .

This closing tercet is, exquisitely, at once irresistibly evocative and self-ironic. For generations born before the mid-1930s the accordion was, *par excellence,* the popular instrument of the most sentimental childhood memories of street-music. Here the accordion-image suggests an illusion, created in retrospective imagination, of the "innocent mother" providing for her children an enfolding music of innocence in a significant, enduring moment of time. In a leap of association belied by the poem's calm surface, this complex of feeling is connected with the natural human desire to find reassurance in the appearance of the northern lights. The desire is seen not as a call for supernatural revelation ("a spell of light, / A saying out of a cloud") but as a willed sense of "innocence" in the impersonal universe. This sense is reinforced by the opening tercet of the ninth poem, which actually completes the sentence:

> And of each other thought—in the idiom
> Of the work, in the idiom of an innocent earth,
> Not of the enigma of the guilty dream.

As we shall see in considering this sequence more fully, later in this chapter, other sections bring out these values more richly and intensely. Our point here is simply that a powerful psychological pressure is subtly at work, finally bursting through openly, in the movement of this passage that at first seems rather distantly speculative.

The insistence on "innocence" and repudiation of "malice" as intrinsic to whatever the universe and its processes may be (as embodied in "these lights") comes under the same aegis of the optative mood created by the poem's assertion by metaphor and by its closing grammatical legerdemain. If we ask, why the need for such insistences? what malaise prevents more direct affirmations?—then the rest of the sequence provides some of the answers in the recurrently hinted suggestions of unresolved guilt and nostalgia based on Freudian family tensions and, in odd touches here and there, on a historical sense touched by the Marxian emphasis on exploitative ruthlessness and brutality. The language of guilt and of the oppression of irreversible historical wrongs is handled so tactfully and tangentially by Stevens that one is surprised again and again, in his best work, by the strength of its emergence even when masked by his aestheticism. Perhaps, though, we would do well to turn to the work of another poet, W. H. Auden, for further development of this motif. Auden shares with Stevens a sometimes maddening verbosity, despite his great skills, that goes against the grain of his greatest possibilities as an artist. Those possibilities are usually revealed in short poems, or in sustained segments of larger works, in which a purity and immediacy of style combine to forestall the temptation to pontificate. It would have been splendid had Auden developed at least one sequence rigorously disciplined in the way his poem "On This Island" is disciplined, by a clean precision of tones uncluttered by pontificating. But there is indeed almost no other single poem of his that quite so decisively shows him heir to the lyric tradition of Campion and Shelley, and thus capable of conveying the pressure on modern poets to be transmitters rather than explainers—to attune the whole burden of awareness to the task of being an instrument of sensibility. The first stanza gives the program, which includes refusal to let any moment slip away:

> Look, stranger, on this island now
> The leaping light for your delight discovers,
> Stand stable here
> And silent be,
> That through the channels of the ear
> May wander like a river
> The swaying sound of the sea.[2]

We shall return to this poem further on in the context of the volume in which it first appeared (considered as a possible sequence). In its twenty-two lines, "On This Island" concentrates remarkably a sense of place that is as cherished as the presence of a beloved person, and elicits the same desire to keep the experience completely alive in the aroused memory:

> And the full view
> Indeed may enter
> And move in memory as now these clouds do,
> That pass the harbour mirror
> And all the summer through the water saunter.

One might argue, with only a modicum of exaggeration, that the poem condenses brilliantly the matter of entire neo-regionalist sequences.

But Auden is at his next-best, and sounds depths and preoccupations not found in "On This Island," in a poem like "The Shield of Achilles," which we shall be discussing in relation to his sequence "Horae Canonicae." This poem projects the major elements of the burden of awareness we have mentioned, elements familiar to humanitarian consciousness and humanistic dismay but so presented as to have the stir of immediacy and the shock of disillusionment. Without these qualities "The

2. Quotations by Auden are from W. H. Auden, *Collected Poems,* ed. Edward Mendelson (New York: Random House, 1976), except when otherwise indicated. Texts and titles are those given in this volume wherever possible.

Shield of Achilles" would be merely a harangue. The purview it provides explicitly (which includes "the full view" mentioned but not spelled out at the end of "On This Island") is the dismal prospect of modern civilization. The poem conveys—that is, breathes and sweats—the rise of totalitarian militarism that governs the lives of the people even in the world's freer enclaves; the torturing and repressions; and the cultural regression and prevalence of ingrained brutal attitudes—in other words, the defeat of all idealistic hopes for universal peace, democratic freedoms, and the progress of enlightened attitudes and loving concern among individuals and among the peoples. The power of Auden's poem resides partly in its simple distancing from mere rhetoric by means of a Homeric frame that speaks for itself. We see Thetis revisiting Hephaestos to get a new shield for her son Achilles; but this time the shield he makes substitutes—for Homer's famous visions of human inventivensss and useful arts, of "ritual pieties," and of athletes and musicians and dancers—closeups of blankfaced armies marching to unexplained grief, of innocent martyrs, and of children inured against all moral sensitivity. The poem avoids sentimentality by its forthrightness and concreteness, its adult tone of compassionate and bitter understanding, and its rigorous but not showy formal channeling. Four stanzas, including the first and last, consist of predominantly three-stress lines rhyming *xaxaxbxb;* these are the stanzas involving Thetis and Hephaestos directly, and they combine a quick semi-narrative pace with highly lyrical, celebratory phrasing. The five other stanzas, longer-lined and carrying the main weight of the pain of our current realities, are in rime royal, recalling Chaucer's use of the form in his tale of *Troilus and Criseyde* and eminently suitable for a "tale" of human betrayal and noble purposes reduced to rubble:

> Barbed wire enclosed an arbitrary spot
>   Where bored officials lounged (one cracked a joke)
> And sentries sweated for the day was hot:
>   A crowd of ordinary decent folk
>   Watched from without and neither moved nor spoke
> As three pale figures were led forth and bound
> To three posts driven upright in the ground.
>
> The mass and majesty of this world, all
>   That carries weight and always weighs the same
> Lay in the hands of others; they were small
>   And could not hope for help and no help came:
>   What their foes liked to do was done, their shame
> Was all the worst could wish; they lost their pride
> And died as men before their bodies died.

Against the heavy depression of such stanzas, with their language of keen intelligence and noble sentiment borne down by implacable forces, and of wit used as a hold against total breakdown, we have the traditional buoyancy of outlook embodied in the stanzas presenting Thetis' expectations:

> She looked over his shoulder
> For vines and olive trees,
> Marble well-governed cities
> And ships upon untamed seas,
> But there on the shining metal
> His hands had put instead
> An artificial wilderness
> And a sky like lead.

Characteristically, as in this opening stanza, the delighted anticipation of Thetis is countered at the end by the existential scene revealed on the shield, a scene unfolded in detail by the longer-lined rime-royal stanzas that follow. In the two closing stanzas, though, the poem's morale has been overwhelmed by that repeated revelation and its post-Holocaust implications—so much so that even the Homeric past and all its glamour have been exposed as containing the seed of our modern horror:

> A ragged urchin, aimless and alone,
> Loitered about that vacancy; a bird
> Flew up to safety from his well-aimed stone:
> That girls are raped, that two boys knife a third,
> Were axioms to him, who'd never heard
> Of any world where promises were kept,
> Or one could weep because another wept.
>
> The thin-lipped armorer,
> Hephaestos, hobbled away,
> Thetis of the shining breasts
> Cried out in dismay
> At what the god had wrought
> To please her son, the strong
> Iron-hearted man-slaying Achilles
> Who would not live long.

Auden's virtuosity is brilliantly displayed here. The closing stanza, in the form that has carried the more exalted and celebratory lyricism of the poem, now uses that very mode to give one final torque to the movement—the wrenching insight into the whole of history and the murderousness implicit in our most revered myths and art. The pure lyricism of delight ("Thetis of the shining breasts") is still present but swamped by the revulsion that has overtaken it.

It is clear enough, in a poem like this, that the poetry of psychological pressure is not merely a matter of private turmoil or malaise, to be discerned only in confessional poetry and Romantic intensities of expression and introspection. It has every sort of political and cultural nerve-endings, as our own personal lives do; and as, indeed, confessional poetry reveals again and again by presenting the poet's literal self as an embodiment of historical tensions and forces. That our major sequences of whatever cast exist and take shape in relation to such pressure is a central fact of the art.

2. Stevens's "The Auroras of Autumn"

The meditative sequence, insofar as it can in fact be so named, is a relatively rare beast, and a suspect one when it does not approach the level of intensity of the other sequences we have taken up. As we have noted, the speculative, ratiocinative mode invites the danger of shifting away from presentation and sliding into sheer discourse. A corollary of this tendency is that it takes more naturally to the methods of the continuous long poem than to the more demanding dynamics of the sequence. A consistent voice, a rational argument, the logical exploration of related themes, can provide a surface that not only conceals a work's lyrical structure (the relation of its crucially decisive units of affect) but smothers it to death. Frequently, too, even with relatively independent poems or sections, the meditative sequence sustains a certain formal consistency that militates against the wide shifts of tone and texture natural to the sequence. These problems occur even in so fine a work as Stevens's "The Auroras of Autumn."

Hardly the best known of Stevens's longer works but certainly the finest, "The Auroras of Autumn" comes closest in its dynamics to a true sequence. Despite its repetitive form (ten parts each twenty-four lines long, divided into tercets and hewing closely to a five-stress norm) its tonal range and relatively unthematic structure make it Stevens's best model for would-be authors of meditative sequences. It is consistently interesting, and in being so contrasts sharply with the excruciating tedium of many of his long poems such as, say, "Notes Toward a Supreme Fiction." All too frequently, poetry of this latter sort suggests a mind so swaddled in its ponderings that the poetic instinct mummifies out of sheer empathy with its surroundings.

In pursuit of this contrast for a brief moment of clarification, we may compare the opening movements of the two works just mentioned. "Notes Toward a Supreme Fiction," if we ignore its eight lines of prefatory dedication to a friend, begins with a section called "It Must Be Abstract." We quote the first numbered part of this section:

> Begin, ephebe, by perceiving the idea
> Of this invention, this invented world,
> The inconceivable idea of the sun.
>
> You must become an ignorant man again
> And see the sun again with an ignorant eye
> And see it clearly in the idea of it.
>
> Never suppose an inventing mind as source
> Of this idea nor for that mind compose
> A voluminous master folded in his fire.
>
> How clean the sun when seen in its idea,
> Washed in the remotest cleanliness of a heaven
> That has expelled us and our images . . .
>
> The death of one god is the death of all.
> Let purple Phoebus lie in umber harvest,
> Let Phoebus slumber and die in autumn umber,
>
> Phoebus is dead, ephebe. But Phoebus was
> A name for something that never could be named.
> There was a project for the sun and is.
>
> There is a project for the sun. The sun
> Must bear no name, gold flourisher, but be
> In the difficulty of what it is to be.

Read sympathetically (the one demand any poem is absolutely entitled to), these lines have their force and quality: the tone of a philosophy instructor addressing a disciple ("ephebe") with kindly, condescending wit and a certain display of profundity, complexity of mind, elegance of phrasing, and ultimate somberness of perspective. The *lesson* has to do with the utter disparity between the human need to name and characterize impersonal nature and see in it, somehow, reflections of ourselves (meanings we can grasp and encompass) and the intractable independence of that nature. One must, it is suggested, slough off all supposed, inevitably misleading "knowledge" and cultivate a cold, rigorous clarity. Underlying this insistence is something more than the cultivation of scientifically detached objectivity. It is humanity's pressure to adjust to our enforced liberation from past assumptions—from "a heaven / That has expelled us and our images" and from the idea of God. "The death of one god is the death of all," and so we must try to avoid calling the sun

"Phoebus" and giving it humanizing or myth-making attributions and simply cope with "the difficulty of what it is to be" as best we can.

Now, certainly we have in these lines a remote echo of *Paradise Lost*, with the sense of alienation and vast tasks lying ahead given a peculiarly modern turn. The pressure of irrecoverable loss makes itself felt, but how faintly compared with the lines concluding Milton's epic and how far from the spirit of "Some natural tears they dropped" and the stark and nobly tragic simplicities of those lines generally! Still, the pressure of that great loss is nevertheless present in the Stevens passage, and with it a call to new kinds of heroism of attitude. Elsewhere in "Notes Toward a Supreme Fiction," usually at points far away from one another, we find occasional moments of piercing feeling that create more emphatic tonal centers. One example comes in a few lines of Part V of the opening section:

> You lie
>
> In silence upon your bed. You clutch the corner
> Of the pillow in your hand. You writhe and press
> A bitter utterance from your writhing, dumb,
>
> Yet voluble dumb violence.

Another, fuller example comes in the tenth poem of the final section ("It Must Give Pleasure"), which begins with the line "Fat girl, terrestrial, my summer, my night." In defiance of the lesson to the "ephebe" that begins "Notes Toward a Supreme Fiction," or at any rate as a qualification of its advice to abjure anthropomorphic sentiments, this is a love poem to things of earth after all. Moreover, it is followed by an epilogue that attempts to relate the concerns of the whole work to those of the war. (The poem appeared in 1942.) The poet's "war," we are told, never ends, whereas the soldier's must, and yet they "meet" and are interdependent:

> How simply the fictive hero becomes the real;
> How gladly with proper words the soldier dies,
> If he must, or lives on the bread of faithful speech.

Mawkish, self-conscious shades of what was once a noble breed of straightforwardly patriotic poetry! William Collins's "Ode Written in the Beginning of the Year 1746" ("How sleep the Brave, who sink to Rest") may sleep bravely on without fear of displacement by this weak, downright meretricious turn of Stevens's: an adaptation of his metaphysical gropings on the subject of language to a momentarily political end, and a useless effort at that. We have noted that he has concerned himself, in this work, with the conflict he feels between a need for ruthless intellectual self-discipline and a sense of intimate, even voluptuous connections between human sensibility and that infinitely desirable "fat girl," nature. Incredibly, and with unintended self-irony, he tried at the last moment to convert the unwieldy labor of projecting this conflict into morale-building war poetry. We may be sure this verbal charge of the U.S. Marines to the rescue neither influenced the thinking of G.I. Joe nor contributed much to the poem's effort to transform elementary philosophical discourse into audibly breathing poetry. Despite many admirable stylistic features, and despite the case that can be made for the affective dimensions of intellectualized poetry that at times does reveal a driven quality, such poetry retains a special, academic flavor for the most part and suffers real damage from its discursive orientation, at least insofar as its possibilities as a lyrical sequence are concerned.

"The Auroras of Autumn," on the other hand, crackles with unusual personal intensity and with brilliantly gripping imagery. These qualities work together to keep its discursive tendency—not really negligible here either—under control. Even the inevitable Stevensian ponderings carry a spark of special poetic energy:

> Shall we be found hanging in the trees next spring?
> Of what disaster is this the imminence:
> Bare limbs, bare trees and a wind as sharp as salt?
>
> (IX)

But to return to our comparison with the first section of "Notes Toward a Supreme Fiction," here is the corresponding section of "The Auroras of Autumn." In these lines the poem does not knit its didactic brows in a rapture of instructiveness and the intricate pursuit of a simple thought. Instead, it opens its gaze upon the night sky and the northern lights and with wondering, primitive, imagining eyes presents what it sees, so that the abstractions that are part of the marveling, questioning awareness are themselves sensuously and mythopoeically perceived. The perception and awareness here are nakedly direct and simple, however complex and alert to ambiguity the ultimate mind at work:

> This is where the serpent lives, the bodiless.
> His head is air. Beneath his tip at night
> Eyes open and fix on us in every sky.
>
> Or is this another wriggling out of the egg,
> Another image at the end of the cave,
> Another bodiless for the body's slough?

This is where the serpent lives. This is his nest,
These fields, these hills, these tinted distances,
And the pines above and along and beside the sea.

This is form gulping after formlessness,
Skin flashing to wished-for disappearances
And the serpent body flashing without the skin.

This is the height emerging and its base . . .
These lights may finally attain a pole
In the midmost midnight and find the serpent there,

In another nest, the master of the maze
Of body and air and forms and images,
Relentlessly in possession of happiness.

This is his poison: that we should disbelieve
Even that. His meditations in the ferns,
When he moved so slightly to make sure of sun,

Made us no less as sure. We saw in his head,
Black beaded on the rock, the flecked animal,
The moving grass, the Indian in his glade.

The streaming bands of light in the night sky seem movements of a giant serpent who is invisible except for the eyes in its ever-moving head, so that we seem to be watched by them from "every sky." At the same time, what is being presented is only the literal impression of the luminous streamers in the sky: a serpentlike form only, of a "bodiless" being whose "head is air." The cosmic view, too, is literal—the whole night sky—and the mythic imaginings of the second stanza, reminiscent at once of primitive cosmogonies and of Plato's cave, are psychologically natural associations. Then the little shock of recognition in the third stanza picks up from the phrasing in the poem's opening sentence but relocates—or rather widens—the place "where the serpent lives." At first it was *here*—in the sense that the whole cosmos, especially the visible night sky, is "this" place. But suddenly "this" includes the physical earth itself, which has become more than just the vantage point from which we look at the auroras but their source and dwelling place as it is ours.

So now, suddenly, the third stanza shifts our attention to the wonder of the way the furthest-reaching and wildest encompassments of imagination are so earth-bound (and therefore endemic to ordinary human reality), yet lend a certain coloration of strangeness to the familiar: "these fields, these hills, these tinted distances, / And the pines above and along and beside the sea." The poem keeps stretching toward realization of the dreams of an ultimately controlling pattern of form, meaning, and vital inspiration that gives organic shape to the "maze"—the dazzling vision of the fifth and sixth tercets, reaching toward a sense of a divinity tormentingly out of view yet "relentlessly in possession of happiness." At the same time, it insists on everything in empirical awareness that resists being seduced by such a vision; and beginning with the sixth stanza introduces other "serpent"—or snake—associations: that snakes are poisonous sometimes; that they depend on the sun for warmth; that both in our native American history and in myth and legendry they evoke primitive ways of life and thought unacceptable outside a lost world of the remote past. Stevens makes a decisive associative shift when he uses the word "poison" in line 19. Intellectually, the allusion is to the "poison" of skepticism and disillusionment, historic aftermaths of the failure of religion and philosophical idealism. Poetically, the shift is to affects of danger, fear, and superstitious narrowness in the closing tercets, affects that balance the dreaming, buoyant, glittering mysteries of the poem's beginning. The psychological pressure—the poisonous depressiveness of self-aware disillusionment and a dwindled view of cosmic possibilities—asserts itself with great force in this opening poem and prepares us for the powerful psychological sections of the sequence that follow.

Of these, Poems II and VI are the most consistently powerful, setting the basic emotional keys for the rest of the sequence. The combined exalted vision, metaphysical speculation, and wary distrustful unease of the first section give way, in the second, to a fierce scrutiny of a haunting, deserted beach cabin, where the "wind is blowing the sand across the floor." In its leisurely way, the poem focuses more and more closely on *whiteness*—the whiteness of cabin and of flowers ("a little dried"), also suggestive of "winter cloud / Or of winter sky." At last the focusing becomes obsessive, and the poem builds up the kind of horror at whiteness we find at the heart of Frost's "Design" or Melville's "The Maldive Shark" or even *Moby-Dick*. The four closing tercets move from macabre abstraction to pure desolation:

> Here, being visible is being white,
> Is being of the solid of white, the accomplishment
> Of an extremist in an exercise . . .
>
> The season changes. A cold wind chills the beach.
> The long lines of it grow longer, emptier,
> A darkness gathers though it does not fall
>
> And the whiteness grows less vivid on the wall.
> The man who is walking turns blankly on the sand.
> He observes how the north is always enlarging the change,
>
> With its frigid brilliances, its blue-red sweeps
> And gusts of great enkindlings, its polar green,
> The color of ice and fire and solitude.

At the base of the horror and desolation, as the third of these stanzas may suggest, is the pain of irretrievable personal loss, the inexorability of change. The first half of the poem presses this feeling, with a kind of wounded nostalgia that makes of the deserted cabin and its whiteness something tomblike, an insubstantial family memorial chapel, as it were, at least in what the sight or thought of it evokes. If we see the acutely private feeling of loss at the start of the following stanzas in relation to the cold impersonal beauty projected at the end of the passage just quoted, we shall also grasp more fully the shift in Poem I from grandeur of vision to protective wariness in the face of danger. Poem II begins:

> Farewell to an idea . . . A cabin stands,
> Deserted, on a beach. It is white,
> As by a custom or according to
>
> An ancestral theme or as a consequence
> Of an infinite course. The flowers against the wall
> Are white, a little dried, a kind of mark
>
> Reminding, trying to remind, of a white
> That was different, something else, last year
> Or before, not the white of an aging afternoon,
>
> Whether fresher or duller, whether of winter cloud
> Or of winter sky, from horizon to horizon.
> The wind is blowing the sand across the floor.

That initial note, "Farewell to an idea," suggests a reluctant yielding to reality. For a moment one might think it an extension of the mood at the end of Poem I, the reluctant repudiation of a humanized vision of cosmic glory—and so it is, but its bearing swiftly pivots toward everything associated with the cabin and all its reminders of a once flourishing life. (A family life, *one's* family life of the past, the poem gives us to suppose without ever specifying.) There was an "idea," perhaps, once, that nothing would ever change, that the family and its members (the mother and father and children who appear in later sections) would not fall prey to time. Meanwhile, though, the starkly focused deserted cabin is treated in such a way that the chilling imagery of whiteness hardens and grows ever more sinister until absorbed into the even more overwhelming imagery of "frigid brilliances" in the final stanza.

In the wake of these opening poems the next two each repeat the opening note of Poem II: "Farewell to an idea." Poem III is pure elegy: "Farewell to an idea . . . The mother's face, / The purpose of the poem, fills the room"—thus the beginning, almost sentimental, identifying the inmost direction of the poem with the pressure of reassuring memory darkened by loss. And later the darkness finds its image:

It is evening. The house is evening, half dissolved.
Only the half they can never possess remains,

Still-starred. It is the mother they possess,
Who gives transparence to their present peace.
She makes that gentler that can gentle be.

And yet she too is dissolved, she is destroyed.

And again:

And the house is of the mind and they and time,
Together, all together. Boreal night
Will look like frost as it approaches them

And to the mother as she falls asleep
And as they say good-night, good-night. Upstairs
The windows will be lighted, not the rooms.

The language and tones of this third poem approach the slightly ambiguous, pained and yet distanced quality that one finds occasionally in the sophisticated popular songs of Bob Dylan, the Beatles, and others. It makes for a far purer poetry, of course, and yet the combinations of simple feeling, elusively imagistic flashes, and subtler intellectual and psychological suggestions invite comparison. The fourth poem repeats the opening phrase of the two preceding poems: "Farewell to an idea . . ." But in shifting from the memory of the mother to that of the father, it leaves notes of tenderness and the pang of loss behind. The father is the Freudian father, Oedipally viewed as powerful and refusing access to his power, as the mother was (equally Oedipally) viewed as altogether giving and accessible—"the mother they possess," a phrase Freud would have leaped upon and torn bodily out of the poem. But the father's creative action, presented as divine, is seen in a context of abstract thought despite vivid images, and in Poem V, in contrast to the mother, he becomes an even more abstractly conceived, Prospero-like embodiment of an illusion of his power. He "fetches pageants out of air"; this and many other like images at once suggest that the northern lights are a cosmic performance arranged by an elusive divinity and that they are a forbiddingly impersonal phenomenon. But for the moment we have shifted from something intensely personal, the vision and reluctant farewell centered on the lost mother, to a half-satiric farewell to the old anthropomorphic notion of an omnipotent male god.

After this foray into the archetypal, the sequence returns to its deeper affective movement. The next section picks up the image of the theater of imagination introduced in the two preceding poems, but suffuses the presentation with humanized impressions of the skyscape. The result is a sense of an innocently reassuring magnificence, "lavishing" itself on change, with a Whitman-like colloquial edge:

> As light changes yellow into gold and gold
> To its opal elements and fire's delight,
>
> Splashed wide-wise because it likes magnificence
> And the solemn pleasures of magnificent space.

But in the second half of this poem, the "theatre" becomes less reassuring as images of active destruction and terror replace these agreeably entertaining ones:

> The theatre is filled with flying birds,
> Wild wedges, as of a volcano's smoke, palm-eyed
> And vanishing, a web in a corridor
>
> Or massive portico. A capitol,
> It may be, is emerging or has just
> Collapsed. The denouement has to be postponed . . .

Then, for a moment, there is relapse into didacticism ("This is nothing until in a single man contained") for the length of a single sentence. But then the poem recovers itself mightily. Once again we are in the situation of the very start of the sequence, when the poem opened its gaze on the amazing night sky. Only here the literal situation (of someone opening a door onto the night and seeing the sky on fire) is projected more intimately and terrifiedly. The paradoxically cold flames of the auroras, awesomely spectacular and symbol-fraught at the beginning, are now crushingly threatening:

> This is nothing until in a single man contained,
> Nothing until this named thing nameless is
> And is destroyed. He opens the door of his house
>
> On flames. The scholar of one candle sees
> An Arctic effulgence flaring on the frame
> Of everything he is. And he feels afraid.

The word "frame" provides an interesting associative conversion here. Its immediate association, in this passage, is most literally with the doorframe; and this association blends into a feeling of confinement within strict boundaries despite the cosmic scope of the scene—in part, as if one existed within a picture-frame pressing in on whatever world the painting one dwells in presents; and in part in recognition that the body frames and limits the self and all its awarenesses. In addition, there is a resonance with Poem III, which juxtaposes a sensibility lapped in peaceful cotton-wool (that magically serene childhood, remembered so nostalgically, that none of us ever had) with a coldly objective recognition of the illusoriness of such "memory": "The house will crumble and the books will burn." The closing image in Poem III is of the house, earlier identified as the house "of the mind," under attack by appallingly indifferent natural forces:

> Upstairs
> The windows will be lighted, not the rooms.
>
> A wind will spread its windy grandeurs round
> And knock like a rifle-butt against the door.
> The wind will command them with invincible sound.

As we have suggested, there is a resonance between this torque-effect and that in Poem VI. In the latter, the word "frame" serves as a vortex into which all the preceding imagery (house, shelter, enclosure) of the sequence pours: the cave, and even the serpent-skin, in the first poem; the cabin in the second; the rooms and house in the third; and the house-theater-mansion complex of images in the fourth and fifth. But now the lonely body is suddenly in a space far more terrifying than that of Poem IV, where the strong father sat "in space, wherever he sits, of bleak regard."

Poems VII, VIII, and IX confront the tension between the frail human shelter, of bodily self and of imagination, and the uncontrollable power of inhuman nature. We have seen, earlier on, how Poem VIII presses this tension into something like a near-apology for the "innocence" of the impersonal universe. Poem IX presses even further in this willed conversion of disastrous reality into innocence. Here the Freudian family—"sticky with sleep" and unconscious of the "guilty dream" and of its coming betrayal by adolescent attachments to new loves outside the family—is identified with the ruthless innocence of universal process. Then the closing poem distances itself from mortal terror, in the first seven stanzas, by ringing witty changes on the various limits of human possibility, as they might be seen by an experimentally minded divinity. Lines 1–12 in particular are surprising in their projection of the sets of human circumstance tested, as it were, by a scientifically speculative god. Their whimsical buoyancy exactly balances a completely opposite quality of awareness: the bitterness (echoing tones of loss and fear and disastrous bewilderment in earlier parts of the sequence) kept in sight through words and phrases such as "unhappy," "misery," "the finding fang," and "solemnize the secretive syllables." The distancing in these twelve lines, then, is not a denial of the terror with which the sequence contends. Rather, it is but a momentary shift of emotional perspective toward that terror:

> An unhappy people in a happy world—
> Read, rabbi, the phases of this difference.
> An unhappy people in an unhappy world—
>
> Here are too many mirrors for misery.
> A happy people in an unhappy world—
> It cannot be. There's nothing there to roll
>
> On the expressive tongue, the finding fang.
> A happy people in a happy world—
> Buffo! A ball, an opera, a bar.
>
> Turn back to where we were when we began:
> An unhappy people in a happy world.
> Now, solemnize the secretive syllables.

At the end, just a few lines further along, the sequence reasserts its deeper and darker intensity as it strives to contain its mutually opposing energies. Its god, or archetypal imagination, meditates

> The full of fortune and the full of fate,
> As if he lived all lives, that he might know,
>
> In hall harridan, not hushful paradise,
> To a haggling of wind and weather, by these lights
> Like a blaze of summer straw, in winter's nick.

This brilliant closing coalesces summer and its dried flowers with boreal frost, and the "loud, disordered mooch" of Part V with the "naked wind" of Part IV, and idle reverie with appalled insight. Indeed, it brings together all the disparate streams of the whole work, in a manner more tightly controlled than in most of the sequences we have been considering—a quality signaled by the formal regularity of all ten parts.

But though it lacks the full freedom the genre invites and calls for, it is on the whole clear that "The Auroras of Autumn" comes close to being a true sequence, unique in Stevens's writing. Its structure is essentially dynamic; for it moves by way of successive reorientations of awareness and modulations of feeling, far more than by way of logical development or thematic accretion. The shiftings are at once subtle and vivid, and are richly enabled by the work's confessional brooding, with its dense overlays of emotional coloration: grief, nostalgia, tenderness, guilt. This complexly layered brooding follows hard upon the dazzling display, in

Part I, of the poem's apparent initial pressures of mingled exhilaration and a darker sense of deceptive cosmic promise, both "poisoned" by existential fatalism. The repeated transition to the realm of private memory—"Farewell to an idea"—in three succeeding sections gets to the "ghostly origins" (to use a phrase from another Stevens poem, "The Idea of Order at Key West") of these pressures. Thus the sequence is launched into its fiercely personal introspective voyage into repossession of primary, passionate states of childhood and pubescent consciousness. The final set of the work, its sense of the precise, precarious balance of the immediate moment down to the most minute "haggling of wind and weather," combines all the preceding tonal motifs in an affective equilibrium comparable, despite every obvious difference, to that of Shakespeare's *The Tempest.*

3. Auden's *Look, Stranger!*

Nothing in Auden's work is exactly comparable to "The Auroras of Autumn." His interest in longer structures was, from the viewpoint of the sequence in our modern sense, disproportionately discursive and expressed itself, basically, in the long, continuous poem. Nor did he reveal any interest, such as Yeats had, in shaping lyrically ordered volumes. So little do his volumes hang together, in fact, that he felt free to reshuffle them according to such arbitrary principles as alphabetization by first lines (the 1945 *Collected Poetry*) or rough chronology (the later collected volumes). Nevertheless, certain volumes and certain modulations toward the sequence bear a relation to the problems and general character of the genre. The volume *Look, Stranger!* (1936)[3] may well be considered an unrealized sequence, somewhat as *A Shropshire Lad* is, and with similar inhibiting factors at work. The "Sonnets from China" (a later, shortened version of the sonnet sequence in the 1939 *Journey to a War*) might have been a sequence with a trifle more depth and intensity. The closest Auden came to such a structure, probably, was in "Horae Canonicae," the final section of *The Shield of Achilles* (1958); and yet this work is less compelling as a whole than the two earlier ones because so dominated, except in the sheerly beautiful opening poem and in a few other touching or especially singing passages, by an almost arch cleverness combined with a certain doctrinally determined predictability.

It may perhaps bespeak a certain bankruptcy of British poetry in the thirties (despite a small number of superb lyric poems) that the best

3. Published in the United States as *On This Island* (New York: Random House, 1937).

examples of modulations toward the sequence the period affords are probably the spotty *Look, Stranger!* and the sonnet sequence in *Journey to a War*. Auden himself dismantled *Look, Stranger!* when he arranged the 1945 *Collected Poetry* so eccentrically. Nor did he reassemble it as a self-contained unit in the *Collected Shorter Poems 1927–1957*, in which 23 of the original 31 poems are reprinted, together with 18 other poems and the "Sonnets from China" sequence of 21 poems, under the simple heading "PART TWO (1932–1938)." It is true that twelve of the first thirteen poems here are from *Look, Stranger!*, but they are thoroughly rearranged and not set off in any way. (In contrast, the "Sonnets from China" are all numbered, under the single title.)

Yet despite the later dismantling, the 1936 volume has its shaky coherence and special qualities. For one thing, it contains several of Auden's finer poems, especially "On This Island" but also such pieces as "O What Is That Sound," "Through the Looking-Glass," "Autumn Song," "His Excellency," "May," "A Bride in the 30's," and "Prologue" (reprinted as "Perhaps" in the 1945 *Collected Poetry* but not later). For another, its incompletely solved problems as a sequence are interesting ones. It is shot through with tonalities that sometimes create a vivid interaction. One double tonal stream dominates the first seven poems in this way. It involves, on the one hand, a pure celebration of England: her traditions and memories, her landscape (gloriously, in the fifth poem, "On This Island"), even her place-names. "Prologue" speaks for instance of "This fortress perched on the edge of the Atlantic scarp," of "wind-loved *Rowley*," and, in a less Shakespearian and only half-humorous vein, of Newton "in his garden watching / The apple falling towards *England*," and then, classically, of "*Merlin*, tamer of horses."

Combined with this strong neo-regionalist sense of England's historical and mythical past—deeply patriotic, really, despite the younger Auden's revolutionary attitudes—is a grimmer corollary: the despair of a bankrupt present charged with violence and fear. The well-known poem "O What Is That Sound" was introduced fairly early in the sequence—just, in fact, after "On This Island," to counterbalance its open joy and happy anticipation. Its essential character as a ballad, despite its modifications of the ballad stanza (namely, each of the stanzas rhymes *abab* and is made up of three four-stress lines and a final two-stress line), links it with the reverent love of tradition in "Prologue." At the same time, its ambiguous central situation, around which its accumulated suspense and catastrophic awareness develops in a series of frightened questions and hardly reassuring replies, embodies the shocked tension of the crisis of the 1930s. In one sense, the poem symbolizes the loss of faith in prevailing political and economic systems, panic at the successes of fascism and threat of war, and horror at the future—a horror comparable to that in Yeats's "The Second Coming." In this context Auden's foreboding, like Yeats's, is perforce European and, beyond that, fully international. The worldwide economic and political crisis informs without literally entering the poem, and with it the prophetic sense of mass cruelty of a new order—the smell of the war and the Holocaust, not yet fully in view but already smashing down the past:

> O it's broken the lock and splintered the door,
> O it's the gate where they're turning, turning;
> Their boots are heavy on the floor
> And their eyes are burning.

A fairly weak sestina, "Paysage Moralisé," rounds off this opening group of seven poems. It echoes—negatively—the thrilled love of England, "the mole between all Europe and the exile-crowded sea," in "Prologue."[4] The opening line betrays the poem's dispirited mood: "Hearing of harvests rotting in the valleys"—a typical Depression-motif, of course. There is nothing in "Paysage Moralisé" to match the subtly heroic vision in "Prologue" of "some possible dream"—even "at this very moment of our hopeless sigh" (both poems share this sense of present crisis)—of sternly disciplined action toward a new state of things:

> And out of the future into actual history,
> As when *Merlin*, tamer of horses, and his lords to whom
> *Stonehenge* was still a thought, the *Pillars* passed
>
> And into the undared ocean swung north their prow . . .

Thus "Prologue," which began with the striking lines of prayer:

> O Love, the interest itself in thoughtless Heaven,
> Make simpler daily the beating of man's heart . . .

Instead, "Paysage Moralisé" is saturated with loss, as if it were literally an aftermath of the realization projected in "O What Is That Sound." It is a poem of stylized disillusionment, barely saved from mawkishness by the discipline of its form and by its care to sustain a tone of elegy for lost hopes and of lingering readiness to revive them. Really more private in its sorrowing confusion than the surface language would at first suggest, it seems to reach toward some unuttered confession.

We have lingered for a moment over this poem because it makes,

4. Because it provides the latest version by Auden of this poem, we are following the text (renamed "Perhaps") given in *The Collected Poetry of W. H. Auden* (New York: Random House, 1945), pp. 89–90.

with the poem that follows it, "Now the leaves are falling fast" (later titled "Autumn Song"), a double bridge to the highly personal group of five poems—"Through the Looking-Glass" "The Watchers," "A Misunderstanding," "His Excellency," and "Who's Who"—that introduces an entirely new center of feeling in the volume, suggesting (without forcing the matter) parallels of an intricate sort between the inner psychic life, with its balked needs and capacity for love and joy, and society, considered as an organism. The elusive strain in the first three of these five poems, with their combined ardor, furtiveness, and unconventionality, is not merely comparable with the tone of the neo-regional and socially engaged poems; it seems an alternative expression of the same complex of feeling. One thinks, inevitably, of the similar barriers and associations in Housman, and of the similar inability to carry a sequence to something like completion. Auden's gift for poignant first lines seems very close to Housman's both in tone and in musicality: "Out on the lawn I lie in bed," "Now the leaves are falling fast," "Earth has turned over; our side feels the cold," "Now from my window-sill I watch the night"—these and many other beginnings have the authentic Housman-tug. An insistent poetic intellectuality and love of complex formations, together with the somewhat more advanced moral tolerance of his day, make Auden, finally, a very different kettle of verse from Housman. Yet the sense of shameful perversity or implicit disgrace in a number of his early love poems is something shared with, and perhaps in part learned from, his predecessor, and so is the masked character of its presentation:

> You are a valley or a river-bend,
> The one an aunt refers to as a friend,
> The tree from which the weasel racing starts.
>
> Behind me roars that other world it matches,
> Love's daytime kingdom which I say you rule,
> His total state where all must wear your badges,
> Keep order perfect as a naval school.
> Noble emotions, organized and massed,
> Line the straight flood-lit tracks of memory
> To cheer your image as it flashes by,
> All lust at once informed on and suppressed.
>
> ("Through the Looking-Glass")

The group of five poems (IX–XIII in the original ordering) is not solely, or mainly, devoted to this "challenge to the shifts of love," as another line in "Through the Looking-Glass" has it. Rather, it embodies a turning to inward feelings generally: a teacher's tender concern for the boys under his care ("The Watchers"), clearly the reflection of Auden's own experience; the realization of dreams of love that always, in the event, seems a misgauging of the situation ("A Misunderstanding"); the bitter ambiguities of success ("His Excellency" and "Who's Who"). Yet a certain reverberation has been set going that colors all these tonal motifs and is picked up again in such later poems of *Look, Stranger!* as poems XIX–XXIII. The first of these, not reprinted later,[5] is a sonnet that asks:

> Yes, we are out of sight and earshot here.
> Are you aware what weapon you are loading,
> To what that teasing talk is quietly leading?
> Our pulses count but do not judge the hour.
> Who are you with, from whom you turn away,
> At whom you dare not look? Do you know why?

The second, another sonnet ("Two Climbs"), speaks again of a kind of fulfillment of a disreputable desire that nevertheless remains incomplete:

> Fleeing from short-haired mad executives,
> The sad and useless faces round my home,
> Upon the mountains of my fear I climb

and:

> it was eyes we looked at, not the view,
> Saw nothing but ourselves, left-handed, lost,
> Returned to shore, the rich interior still
> Unknown: love gave the power, but took the will.

And still further on, in poem XXVI (not reprinted), we have one of the two poems in the book in which lovers' joy has been fully achieved, although here the imagery is of forbidden love and of warfare and spying—a compromised ecstasy, as it were, like that of victory in revolution or war:

> That night when joy began
> Our narrowest veins to flush
> We waited for the flash
> Of morning's levelled gun.
>
> But morning let us pass
> And day by day relief

5. For this and other poems not reprinted later, see *The English Auden*, ed. Edward Mendelson (New York: Random House, 1977). The sonnet is on p. 149 ("To lie flat on the back with the knees flexed").

Outgrew his nervous laugh;
Grows credulous of peace

As mile by mile is seen
No trespasser's reproach
And love's best glasses reach
No fields but are his own.

No special resolution, musically or emotionally speaking, is reached in *Look, Stranger!* that could be said to hold its varied strains of feeling and awareness in some sort of acutely sustained equilibrium. The work is not really organically interactive as a whole, and after the thirteenth poem ("Who's Who") it meanders among its various preoccupations, with an occasional splendid poem, an occasional overly discursive one, an occasional touching one that goes on and on forever, however. But the mixture and the felt reciprocities make the work deserve to be kept intact with whatever revision, and so do the scattered poems of lyric power or superb grace that punctuate the book. In one instance, at least—poem XXVII: "Fish in the unruffled lakes"—Auden's virtuosity brings him to a successful adaptation of Yeats's style and an unqualified avowal of delight in love. The poem ends:

Sighs for folly done and said
Twist our narrow days,
But I must bless, I must praise
That you, my swan, who have
All gifts that to the swan
Impulsive Nature gave,
The majesty and pride,
Last night should add
Your voluntary love.

But perhaps this adaptation helps reveal the deepest affective stream in the book: its passionate need to align its sensibility and range of style and general scope with the whole poetic tradition. This is obviously not just a matter of modeling oneself on Yeats but of saturation in the tradition, so that at one moment one catches the flavor of an Elizabethan air, at another a turn like something of Jonson's or Marvell's or Dryden's, at another an exalted Homeric or Sophoclean phrasing. Another way of putting this is to say that Auden, in writing a poetry of the troubled inner self or of social disturbance and hope for revolutionary change, was never content to be merely anecdotally self-analytical or rhetorical, but wrote with the sound of the masters (that is, of the whole of our poetry) in his ears and spirit. The book is a struggle to win through to a perfection of sensibility and perspective in these terms.

4. Auden's "Sonnets from China" and "Horae Canonicae"

As we have noted, Auden did keep his "In Time of War" sonnet sequence intact, although he cut seven, picked up the Forster sonnet to close the sequence, rearranged the others slightly, rewrote them fairly heavily, and changed the title to "Sonnets from China" as an indication of the "war" of primary reference. (In 1938 the Sino-Japanese War was in progress and the long internal revolutionary struggles of modern China were in mid-career. Auden and Christopher Isherwood spent several months in China in 1938 not as participants in these struggles but as writers preparing a travel book.) Both titles were slightly misleading, since most of the sonnets have to do neither with war nor with China; and even when they do so, they are not war-poems or poems of description or reportage or "Chinese" poems in any qualitative sense, but essentially contemplative poems that could as well have been written at home under stimulus of the daily newspapers and one's own personal preoccupations and sympathies. One of the poems dealing more explicitly than usual with war, Sonnet XII in the *Collected Poems,* may serve as an example:

> Here war is harmless like a monument:
> A telephone is talking to a man;
> Flags on a map declare that troops were sent;
> A boy brings milk in bowls. There is a plan
>
> For living men in terror of their lives,
> Who thirst at nine who were to thirst at noon,
> Who can be lost and are, who miss their wives
> And, unlike an idea, can die too soon.
>
> Yet ideas can be true, although men die:
> For we have seen a myriad faces
> Ecstatic from one lie,
>
> And maps can really point to places
> Where life is evil now.
> Nanking. Dachau.

The poem has its force: ironic at first, then compassionate, and finally ironic again, but desperately and angrily so. When the second, harsher irony takes over the poem, the sonnet-form itself is twisted and compressed, as though (although sustaining the Shakespearean rhyme-scheme) it had become impatient with its own limits. There is a kind of shock, too, in the words "true" and "really"—a sudden realization that ideas have physical impact in their own right: can kill, can (in the image of that concrete abstraction, a map) focus precisely on a pinpointed, spotlighted world of pain: the bombing of Nanking by the Japanese, the concentration camp at Dachau (the next two sonnets zero in on sacrificed and wounded soldiers). The poem tries to compel the apparently impersonal systems and apparatus of war to reveal the hideous personal realities on which they are based. Sonnet XII, then, is a fairly active, formally improvisational, meditative poem, with an obvious rhetorical purpose, as well, of shaking its readers into seeing these realities and responding to them somehow. In a way the rhetoric is like that of "On This Island," but directed toward a political end rather than toward awakening the senses to themselves and their objects.

The whole sonnet-sequence is meditative and rhetorical in this manner. The first ten provide a pocket-history of social evolution—or rather, a fanciful, symbolic jumble to account for our present predicament and prospects. This history-jumble begins with the Creation and the expulsion from Eden; moves on to the invention of language (our great weapon and self-destroyer), the development of peasant societies, the rise of warrior-heroes who become manipulative rulers, the emergence and demystification of poetry, and the successive discoveries and degradations of Renaissance, Enlightenment, and democratic values; and sums up the results as successful resistance to God's purposes and subjugation by invisible psychological demons who are really the old "vanquished powers" (dragons and kobolds and such) in their modern form. Auden's discursive bent here is obvious, however leavened by wit, charm of phrasing, and sudden intensities.

Perhaps the most winning sonnet is the first. It contrasts man to all the other creatures, who came into being completely adapted ("Bee took the politics that suit a hive, / Trout finned as trout, peach moulded into peach") and stayed "content with their precocious knowledge." But as for our species—

> . . . finally, there came a childish creature
> On whom the years could model any feature,
> Fake, as chance fell, a leopard or a dove,
>
> Who by the gentlest wind was rudely shaken,
> Who looked for truth but always was mistaken,
> And envied his few friends, and chose his love.

This changeable, possibly vicious, possibly saintly, vulnerable, erring thinker and lover—*Homo sapiens*—is the ambiguous, treacherous creature who in later sonnets tries to objectify and rationalize murder and oppression so as to conceal the human suffering involved. Man is by definition self-defeating, and Poem IX gives us a parable of how even

God is unable to get through to him, and how man uses God's own messenger (presumably the organized Church), whom he has corrupted into a teacher of "so many ways of killing." Poem X then shows man laid open to psychic devastation by his conquest over the supernatural. What gives all these sonnets a lyrical cast is their overall wryness, in varying degrees of intensity and coloration (pity, irritation, tragic feeling, or whatever). Sonnet X has this quality in abundance, from its ruefulness at the loss of a magic universe to its air of hilarious yet grim logic at the very end: "The vanquished powers were glad"—

> To be invisible and free; without remorse
> Struck down the silly sons who strayed into their course,
> And ravished the daughters, and drove the fathers mad.

The next six sonnets (XI–XVI), sometimes centering on Chinese circumstances or on aspects of war, are essentially poems of existential despair paradoxically and hopelessly struggling for its opposite state. Looking at China—but in the most general way—Sonnet XI laments: "History opposes its grief to our buoyant song." That word "buoyant" must be restricted to occasional and natural sources, not to human beings in the mass:

> Certainly praise: let song mount again and again
> For life as it blossoms out in a jar or a face,
> For vegetal patience, for animal courage and grace:
> Some have been happy; some, even, were great men.
>
> But hear the morning's injured weeping and know why . . .

Neither the "quick new West" nor the deeply traditional East ("the flower-like Hundred Families . . . in the Eighteen Provinces") has learned to counteract "the will of the unjust" and the State's "fairly noble unifying lie." This opening salvo, or preachment, is followed by the bitter Sonnet XII—a horrified recoil from the whole military-political embroilment of the moment—and then by the savagely ironic Sonnet XIII, which fuses the fatuous indifference of upper-class war-mongers to individual suffering with a celebration of the anonymous Chinese soldier whose innocent courage must, despite betrayal and his own ignorance, have made its contribution to a better future. As an elegy for an unknown soldier in the confused modern wars and revolutions, this sonnet is at least in touch with the noble tradition of Collins's "How Sleep the Brave":

Far from a cultural centre he was used:
Abandoned by his general and his lice,
Under a padded quilt he turned to ice
And vanished. He will never be perused

When this campaign is tidied into books:
No vital knowledge perished in that skull;
His jokes were stale; like wartime, he was dull;
His name is lost for ever like his looks.

Though runeless, to instruction from headquarters
He added meaning like a comma when
He joined the dust of China, that our daughters

Might keep their upright carriage, not again
Be shamed before the dogs, that, where are waters,
Mountains and houses, may be also men.

The sardonic fifteenth sonnet elaborates on this despicable perspective that sees a gap between dull, dispensable peasants and true "men," with its picture of a world in which governments and armies exist to find the opportunity for slaughter. In the sixteenth, Auden-like modern sensibility is shown as forever frustrated:

Loss is their shadow-wife, Anxiety
Receives them like a grand hotel, but where
They may regret they must: their doom to bear

Love for some far forbidden country, see
A native disapprove them with a stare
And Freedom's back in every door and tree.

The final group of sonnets (XVII–XXI) consists of exercises in degrees and modes of affirmation, given the essential placement of morale in the lines just quoted from Sonnet XVI. The group begins with a torque-like shift of attention to the nature of popular songs. The wryness here lies in the perception that, "simple like all dream-wishes" and employing the "elementary rhythms of the heart," these songs accompany us even in the most tragic aspects of our lives: "The dying and the lovers bound to part // Hear them and have to whistle"—

Think in this year what pleased the dancers best,
When Austria died, when China was forsaken,
Shanghai in flames and Teruel re-taken.

France put her case before the world: *Partout*
Il y a de la joie. America addressed
Mankind: *Do you love me as I love you?*

After this fairly lightweight piece of heavy irony, which nevertheless, in the nature of such things, provides its little wrenching moment, Sonnet XVIII moves into a more intimately serious (and Keatsian) mood: our longing for the mythic past ("an ancient South, / A warm nude age of instinctive poise, / A taste of joy in an innocent mouth") and for our dreamt-of Future (where "each ritual maze / Has a musical plan"). Both are mere fictions, yet somehow involved in our condition as human beings who "live in freedom by necessity." While this sonnet loses edge and energy in a tiny welter of wise sayings at the end, it does prepare us somewhat for a rather strong affective moment in Sonnet XIX. Here, in the midst of confusion and disaster, the poet seems to speak out personally from the page—not as W. H. Auden, bright young pontificator, but as the underlying sensibility of all this weighing and brooding. Here an ideal mode of behavior is set up, in the image of Rilke's steadfast endurance during the period before he found refuge, in 1921, in the Château de Muzot in Switzerland and was able to complete the *Duino Elegies* and *Sonnets to Orpheus:*

To-night in China let me think of one

Who for ten years of drought and silence waited,
Until in Muzot all his being spoke,
And everything was given once for all.

Awed, grateful, tired, content to die, completed,
He went out in the winter night to stroke
That tower as one pets an animal.

Rilke's "drought and silence," his waiting—as he himself put it—to ripen like a fig, was hardly Auden's way. But the point here, of course, is the artist's need to endure and, finally, to create what he or she must create. Oddly enough, this brief passage seems to touch greater depths of feeling than all the notes concerning war and terror elsewhere in the sequence. It is clearly the real "conclusion," as it were, although two sonnets follow it. Sonnet XX is in a sense a statelier, more generalized reprise of Sonnet XIII ("Far from a cultural centre he was used"), contrasting anonymous heroes such as the "runeless" dead Chinese soldier in that poem with the "men of a sorry kennel, racked by guilt" who need to leave impressive monuments behind because they themselves are unloved and rapacious. This political piety is fortunately not the closing note, which comes in Sonnet XXI, dedicated to E. M. Forster and energetically peopled with characters from Forster's novels who embody both our compromised condition and our finer possibilities:

> Yes, we are Lucy, Turton, Philip: we
> Wish international evil, are delighted
> To join the jolly ranks of the benighted
>
> Where reason is denied and love ignored,
> But, as we swear our lie, Miss Avery
> Comes out into the garden with a sword.

With this witty turn of semi-*Angst* the sequence ends, its curiously didactic enterprise having reminded us of the world's ills and the necessity for and dangers of militancy. Throughout its slightly forced progression, streamlets of wry self-knowledge, helpless dismay and anger, and humorous or simply compassionate awareness provide pleasant or moving moments, although there are no adventures into the private abyss and only a few of the sonnets approach the breadth of vision and imagination of the best work in *Look, Stranger!* Another way of looking at the sequence is to think of it as acknowledging and attempting to define the pressures on the modern psyche without coping with them in any sustained artistic fashion. It is certainly not a developed poetic sequence except in the old sense of "sonnet-sequence," but its variations from conventional sonnet forms, together with its balancings of disillusion and hope, irony and directness, are a modulation toward the genre. The short units avoid redundancy through the variations—though not entirely—and also permit the *impression* of a new start with every poem. (The same sort of prolonged discourse without such frequent breaks would be intolerable.)

In fact, to be fair, the task Auden set himself and somewhat funked in these sonnets was a very difficult one, considering the sheer weight of centuries of moralizing behind their meditations. Every step of the way was over a bog of clichés; much of the poet's effort was to use his virtuosity and readiness of phrase to keep from sinking irrevocably. (He did sink, often enough, but revocably.) It took him some twenty years to get to the point where he could deal with almost complete effectiveness with this problem in a single masterpiece, "The Shield of Achilles." As we suggested earlier, his use of the Homeric frame there, and the interaction between the condensed lyrical stanzas and the more colloquially grainy expanded ones (themselves disciplined, however, by a superbly handled rime-royal pattern), enabled him to use his talents fully and to project emotional tensions while sparing himself confessional revelations—never his forte. "The Shield of Achilles" is unique in the Auden canon. It stands out in the book whose title-poem it is as the one poem of power, so that we accept the rest of the volume, almost unconsciously, as ancillary to it and justified by it: a matrix of other poems, more or less interesting and relevant. The success of this one poem, however—Auden's coming-through in this particular mode—may have freed him for the unusual internality of the sequence "Horae Canonicae" that closes the book. If Rilke is the hero of "Sonnets from China," Auden comes closer here than anywhere else to the *Duino Elegies*. The kinship lies in the dreaming associativeness, with a supposed rational thread as well, of these poems built around the varying states of human sensibility associated with different times of the day and framed by an awareness of Christian liturgy and preoccupations.

"Horae Canonicae" is given a slightly obtrusive surface structure because the seven poems are ordered according to the seven canonical hours starting with Prime—sunrise, the "vaunt of the dawn." The succeeding poems work through the day to sunset—Terce, Sext, Nones, Vespers—and then to the night office of Compline and to cockcrow of the next day, Lauds. So divided, the day centers both on the shifting states of sensibility we have mentioned and on an execution, with strongly Christian overtones. But within the constraints of the time-ordering and the religious framework the poems offer considerable variety and freewheeling association. There is formal variety as well. The first, fourth, and sixth poems employ the same stanza: 16 alternating nine-syllable and seven-syllable lines, with irregular rhyming and echoing patterns. "Terce" is also syllabic verse, with alternating eight- and eleven-syllable lines. The three-part "Sext" is in irregular rhymed couplets; "Vespers" is in fairly rhythmical prose broken into short units; and the delightful "Lauds," with its refrain, repeated lines, and off-rhymes, is based, as Monroe K. Spears points out, on the medieval Spanish *cossante*.[6] All of which is to say that Auden's virtuosity is given full play in these intricately managed poems and that their baroque character is an important quality of their artistry: an architectural dimension, visual as well as aural, that allows for originality of design in a way the "Sonnets from China" could not.

"Prime" sets off the sequence with a richly swinging welter of sound that is a lyrical evocation of the state between dreaming and waking. It catches the disorientation of waking into full daylight life, a moment in which innocence is miraculously recaptured even by the most worldly minds. The meter itself contributes to this sense of disorientation: a "correct" syllable count here depends on elision, so that in the opening

6. *The Poetry of W. H. Auden: The Disenchanted Island* (New York: Oxford University Press, 1963; 1968), p. 320.

stanza's last line, for instance, "body and" counts as two syllables ("bod-yand") although the words are pronounced normally: "Between my body and the day." Actually, this practice leads to an irregularity in the number of syllables. It is in any case hard to hear lines of seven or nine syllables as metrical, and so the perceived rhythm depends on other elements: the way the lines swirl around a four-stress norm, the numerous sound-echoes, and the semantic and syntactic parallels. Auden's reliance on the comma complicates the impact on first reading, so that the syntax turns out to be effective mimicry of the happy confusion between self and world as one opens one's eyes:

> Simultaneously, as soundlessly,
>   Spontaneously, suddenly
> As, at the vaunt of the dawn, the kind
>   Gates of the body fly open
> To its world beyond, the gates of the mind,
>   The horn gate and the ivory gate
> Swing to, swing shut, instantaneously
>   Quell the nocturnal rummage
> Of its rebellious fronde, ill-favored,
>   Ill-natured and second-rate,
> Disenfranchised, widowed and orphaned
>   By an historical mistake:
> Recalled from the shades to be a seeing being,
>   From absence to be on display,
> Without a name or history I wake
>   Between my body and the day.

The pure lyricism of the opening lines, evoking a realm of mysterious spiritual space, is restrained only by one quiet word, "kind," that here suggests a different world of sophisticated human awareness. We are thus gently prepared for the typically Audenesque battery of phrases that swarms in with "nocturnal rummage" in line eight. This is self-depreciatively ironic, politically conditioned, existentially melancholy language of the self-conscious worldly self. The two states, of disembodied transport among the unbounded images of the mind and the universe, and of the knowledgeable, limited empirical self, are in this stanza perfectly juxtaposed, while the sensibility at work finds itself just between them, not yet having resumed a daily identity. And although the whole stanza is a perfectly clearly developed, though extended, sentence, its length and involutions of thought and its mixture of Shelleyan expansion and Eliot-like grumpy pondering project a mostly happy confusion. To be "without a name or history" is to be full of possibility, like a newborn child, rather than full of misery and anomie.

The sense of being "without a name or history" is developed in the second stanza. It involves absolute innocence, "Adam sinless in our beginning, / Adam still previous to any act." If the first stanza carried a companionable reminiscence of Eliot, however, the second is a bit too much in the grip of the older Anglican master. We can hear "Ash Wednesday" and *Four Quartets* in

> And I know that I am, here, not alone
>   But with a world and rejoice
> Unvexed . . .

But Auden mostly rights the balance in his idiosyncratically immediate and witty way. These lines come in a passage they somehow mar but cannot overwhelm:

> . . . in complete obedience
> To the light's laconic outcry, next
>   As a sheet, near as a wall,
> Out there as a mountain's poise of stone,
>   The world is present, about,
> And I know that I am, here, not alone
>   But with a world and rejoice
> Unvexed, for the will has still to claim
>   This adjacent arm as my own . . .

The problem of echoes is a special one in Auden, for his virtuosity is of a kind that draws on every resource so knowingly that it is a sharing of methods in the interest of the poem. Thus, in the passage just quoted, we also find a clear echo of Hopkins's "The Wreck of the *Deutschland*" in lines three and four.[7] And in the third and last stanza, in which the ego reasserts itself and becomes again aware of identity, mortality, and guilt, the heaviness of all this is leavened by a vivid little passage much indebted to Marianne Moore's precise poetic observations:

> The eager ridge, the steady sea,
> The flat roofs of the fishing village
>   Still asleep in its bunny,
> Though as fresh and sunny still, are not friends
>   But things to hand . . .

7. We refer especially to images in Hopkins's fourth stanza ("at the wall / Fast, but mined with a motion, a drift" and "I steady as a water in a well, to a poise, to a pane") but also to the precise idiom of paradoxical tension shared by the poems in both passages.

But it was, in the first instance, Auden's own image "near as a wall" that brought into view the Hopkins association of wall and poise and stone (all part of his complex of "horror of heights," which may be subliminally present in the condition Auden is describing); and in the second instance, it was the deceptiveness of the surface impression, its very concreteness and specificity of detail, that linked Auden's sense that the "things to hand" are not really "friends" despite their pleasing conformation to Moore's comparable feeling (in "A Grave," for instance, in which the surface charm and variety of a cliffside ocean-view is contrasted with its hidden yawning murderousness). These considerations are matters of proportion, and the obvious borrowing of a tone is always troublesome, especially in a poet whose strength is not that of a powerful and original imagination and style but of extraordinary skill and a finely engaged humanity. But Auden's genius was an assimilative one, and each example needs to be weighed both sympathetically and on its own merits.

Putting it sympathetically and on its own merits, then, the sequence remains relatively lively throughout because of its sharply concrete observation, its flexible craftsmanship, its pleasure in out-of-the-way words such as "bunny" for small ravine (and suggesting the language of intimate and comforting affection as well) and "fronde" for discontent and petty malice, its spirited intelligence, and its witty notes on human behavior. At the same time, most of the poems build to a rather ponderous ending. (The exception is the song of returning dawn, "Lauds," which is purely joyous.) In "Prime" it is "dying / Which the coming day will ask"; in "Terce," the "victim . . . knows that by sundown / We shall have had a good Friday"; in the three parts of "Sext," each one ends with a reference to "this" death and dying; in "Nones," we see "all the creatures / Now watching this spot" of ritual death; and in "Vespers," the violence and injustice at the dead heart of society are summed up in prose: "For without a cement of blood (it must be human, it must be innocent) no secular wall will safely stand." "Compline" avoids reference to death at the end; its weightiness lies more in its imagery of devotion and in the one word that dominates its conclusion, where we have the kind of wish that caps "Little Gidding." The imagery of this passage is somewhat homelier than Eliot's—a "picnic" rather than the allegorical fire and rose—but the theological term "perichoresis" refers not so much to its root meaning of "going around" as to a kind of mutual indwelling of God and communicant. Typically, though, Auden reinforces the root sense at the same time, a bit playfully:

> That we, too, may come to the picnic
> With nothing to hide, join the dance

As it moves in perichoresis,
Turns about the abiding tree.

In the sketchy narrative of the day that the sequence takes us through (along sparsely dotted lines), "Prime" touches the moment of arising. "Terce" rather sardonically suggests the trivial personal concerns and habits of public and private figures getting their day started—for instance,

After shaking paws with his dog
(Whose bark would tell the world that he is always kind),
The hangman sets off briskly over the heath . . .

And:

Gently closing the door of his wife's bedroom
(Today she has one of her headaches),
With a sigh the judge descends his marble stair . . .

and we see the poet, who's "taking a breather / Round his garden before starting his eclogue," and each of the rest of us, too, hoping that we can "get through this coming day." "Sext" then turns to a triptych showing the day in full swing: the ordinary person lost in benign vocations; the powerful figures who make history and decide matters of life and death for the masses; and the crowd, the many against the one, who shed individuality and thus absolve themselves of social guilt:

Whatever god a person believes in,
in whatever way he believes

(no two are exactly alike)
as one of the crowd he believes

and only believes in that
in which there is only one way of believing.

This tone of mingled disdain and hopeless, possibly fear-laden recognition is familiar in Auden and ultimately, together with sheer disillusionment with the working of all kinds of power, is at the heart of his most painful affects. We find it again, more violent in its intensity, in "Nones," the central poem of the sequence and the one of most passionate force:

The faceless many who always
Collect when any world is to be wrecked,

> Blown up, burnt down, cracked open,
> Felled, sawn in two, hacked through, torn apart.

"Nones" rigorously explores the many facets of human complicity and feeling involved in the race's murderousness, as symbolized by the execution that is so important in the poem's day. In so doing, the poem concentrates the various tones of guilt with which the sequence is so often preoccupied (as we saw in the first stanza of "Prime"). But it also gets through to something else, an equally savage but subtler feeling, that underlies this sequence and much of Auden's best other work. This is, to put it crudely, an overwhelmed apperception of the irreversible and impersonal working of fatality, for which even the most brilliantly individual personality and the most intimately personal feeling are momentary vehicles, to be nurtured, used, and discarded. We have recurred to Auden's ready internalization, in contexts of his own making, of other poets' tonalities. In these poems, he has internalized a whole major organic set of modern urban sensibility and its art. "Prime" introduces the issue:

> I draw breath; that is of course to wish
> No matter what, to be wise,
> To be different, to die and the cost,
> No matter how, is Paradise
> Lost of course and myself owing a death . . .

"Terce" advances the feeling through one hellish turn of thought:

> At this hour we all might be anyone:
> It is only our victim who is without a wish,
> Who knows already (that is what
> We can never forgive. If he knows the answers,
> Then why are we here, why is there even dust?) . . .

"Sext" fixes the position by a simple imagery of the terrifying certainty of things: the expressions of people "forgetting themselves in a function."

> You need not see what someone is doing
> to know if it is his vocation . . .

But it is "Nones" that fully develops the dread aspect of the present moment seen as foreclosed, an aftermath of the irrevocable past. It begins:

What we know to be not possible,
  Though time after time foretold
By wild hermits, by shaman and sybil
  Gibbering in their trances,
Or revealed to a child in some chance rhyme
  Like *will* and *kill*, comes to pass
Before we realize it. We are surprised
  At the ease and speed of our deed
And uneasy: It is barely three,
  Mid-afternoon, yet the blood
Of our sacrifice is already
  Dry on the grass . . . .

"Nones" does not "resolve" the predicament, but simply absorbs and contains it in its inescapable reality. The poem puts the matter more drastically, more in terms of the outflanked will of humanity, than does Stevens's "Sunday Morning" with its longing for immortality and its "proof" that "Death is the mother of beauty." But the ending of "Nones"—the one accommodation possible for those who would deal unflinchingly with fatality and still maintain morale—is strongly reminiscent of the one in "Sunday Morning":

Not knowing quite what has happened, but awed
  By death like all the creatures
Now watching this spot, like the hawk looking down
  Without blinking, the smug hens
Passing close by in their pecking order,
  The bug whose view is balked by grass,
Or the deer who shyly from afar
  Peer through chinks in the forest.

These lines have a slightly self-depreciative edge, and they add wryness by making the "creatures" co-watchers of "this spot" of horror instead of solely the aesthetic objects Stevens makes of them, and there is a saving humor at two or three points. The passage does not tell us to make the most of what we have, with a tragically conditioned delight, in Stevens's manner. But it affirms, by *its* manner, the universal kinship of the creatures and gives a loving turn to the tone at the very end, restoring something like the innocence of the start of "Prime."

This is the true ending of the sequence, but the work perseveres through three more poems. "Vespers" breaks down the brooding sensibility into two alternative selves, the aesthetic-centered and the pragmatically revolution-centered. They are at odds with one another yet reciprocal, and each is a crucifier and victimizer in spite of its dreams and ideals. This is the prose-poem of the sequence, brittle in its witty balances and to some degree trapped by them: "In my Eden a person who dislikes Bellini has the good manners not to get born: In his New Jerusalem a person who dislikes work will be very sorry he was born," and yet both are "accomplices" who "in spite of themselves"

> to remind the other (do both, at bottom, desire truth?) of that half of their secret which he would most like to forget,
>
> forcing us both, for a fraction of a second, to remember our victim (but for him I could forget the blood, but for me he could forget the innocence) . . .

"Vespers" is a semi-comic interlude in the poetic progression of "Horae Canonicae" that nevertheless returns with something like a shock to the sense of a guilty, compromised, and unredeemable state. "Compline," at the moment of bringing the sequence full circle out of the full waking state back into dream, drifts off from the painful clarities of the day into confusion about what it all really amounted to and into the hope, as consciousness slips away, of being spared after all and joining that eternal "perichoresis" beyond our bondage to the aftermath of original sin. "Lauds" is a sweet afterbeat of the whole embroilment, catching at the renewed sweetness of things offered by each dawn. With this poem all brooding disappears and the sequence sings joyously in relief, celebrating what is celebratable and letting the rest go for the moment, and overcoming the tension between individual self and mass-impersonality by gentle, unquestioning juxtaposition:

> Among the leaves the small birds sing;
> The crow of the cock commands awaking:
> *In solitude, for company.*

Etc., and lovely. And the unresolved reconciliations of "Nones" have been even more distanced by this point. But their resonances continue, a private counterpoint to the objectified vision of disaster in a cruel human universe and its fiercely sad music in "The Shield of Achilles."

Chapter Fourteen

The Confessional Mode: Robert Lowell and Others

1. Lowell's *Life Studies*

Since confessional poetry provides the most obvious instances of the working of psychological pressure, perhaps we should have taken up confessional sequences at once in the preceding chapter. We have particularly wished, however, to head off any implication that the two things—one of them a shaping agent in all lyrical structure, the other a special modern tendency—are identical. Psychological pressure, in its poetic meaning, has a double function. It is a constant force in a poem's dynamics, and it is also the active element, directly revealed within the work, that generates the dynamic process to begin with. We were therefore at pains to discuss meditative sequences first, for they are far too often treated as though they were pure discourse of a certain subtlety that, for some odd reason, had been set to verse.

In Stevens's and Auden's work, we had occasion to note, there is a certain perception, or perspective, as elusive as it is pervasive in both poets—an insight linking them with the confessionals. It has to do with "innocence," a term both Stevens and Auden engage with to mediate between the terror of an indifferent universe and the sense of unconscious guilt that accompanies our helpless compromise with the paces life puts us through. The universe is "innocently" ineluctable in its processes; we are "innocent" in being what we have to be—in our complicity with the process that tears us out of the childhood cocoon, for instance, or with the whole character of our society with its hideous wars and injustices. This is paradoxical only from the viewpoint of our inescapably anthropomorphic gigantism, no doubt, but what can we do? We are human, and these realities leave a taste of blood in our mouths. The sense of self-betrayal, through the very processes of enlightenment and liberation that were supposed to bring a new joy to humanity (and in important ways certainly have done so, though neither unqualifiedly nor to all alike) was summed up in Auden's tragicomic poem just before World War II: the tenth of the "Sonnets from China":

So an age ended, and its last deliverer died
In bed, grown idle and unhappy; they were safe:
The sudden shadow of a giant's enormous calf
Would fall no more at dusk across their lawns outside.

They slept in peace: in marshes here and there no doubt
A sterile dragon lingered to a natural death,
But in a year the slot had vanished from the heath;
A kobold's knocking in the mountain petered out.

Only the sculptors and the poets were half-sad,
And the pert retinue from the magician's house
Grumbled and went elsewhere. The vanquished powers were glad

To be invisible and free; without remorse
Struck down the silly sons who strayed into their course,
And ravished the daughters, and drove the fathers mad.

The shift to confessional poetry from writing like this is a simple one: to the attempt to present the poet's own naked self and unrationalized, uncensored actual feelings and behavior. It is another turn of modern sensibility, foreshadowed early in the last century by *The Prelude* and other works. These, however, could cherish a sense of an "immortal spirit" that seems to make a purposed whole of "the terrors, pains, and early miseries, / Regrets, vexations, lassitudes interfused / Within my mind," as Wordsworth had it. Doubtless the chilled feeling that the "immortal spirit" either does not care what happens in one's mind or does not exist has lurked in the shadows of human thought from the start, and has created its own varieties of psychological pressure, but this feeling is central in the confessional poetry of the last generation. The chaos of the psychic situation becomes the ground of a reoriented art in which the beset self is the testing-ground and embodiment of all human possibilities. The terrors, pains, early miseries, regrets, vexations, and lassitudes that Wordsworth suffered are now the main proofs of one's existence; the degree of their intensity confirms their (and our) reality. The artistic problem is to make a genuine poetry out of the language of untrammeled self-awareness.

We may take Robert Lowell's *Life Studies* (1959) as the *locus classicus* of confessional poetry in the last generation. In Lowell's work he uses himself, in his day-by-day existence, as his central symbol, on a stage swept clear of Wordsworth's universal moral supervision of all nature, and even of Stevens's wistfully hypothetical Prospero-props and Auden's dissolving absolutes. The demons of the mind have prevailed, but they are no longer demons but the givens of a neurotic and oversensitized personality, self-deprecating and harshly candid as well about his parents, the women he has loved and married and divorced, his children, a dying uncle, and anyone else who comes under scrutiny. The harsh candor is often mixed with affection; and love and contempt—whether toward oneself or others—often seem alternative forms of one another. By the same token, the dramatic force and lyric beauty of Lowell's best writing are impossible to sort out from its lacerating ironies. This is not a matter of malicious glee at the misfortunes of others, nor yet of morbid delight in one's own humiliation. Yet there was something akin to these, a kind of diabolism in a minor key, in the way Lowell was stung into almost joyful alertness by his own pain and that of others. There was something life-giving for him in the keen disturbance he evoked: a nervous awakening, at once paradoxical and all too natural, like that which Keats experienced when seized by the thought of death.

The passion for this sort of disturbed awareness is characteristic of modern sensibility in many ways—reticence and social "decency" sacrificed to acutely observant frankness. Moreover, Lowell carried to an extreme our sense that the private life embodies the national life. For him this premise was an expanded dimension of his not always welcome empathy, which had a cannibalistic edge and imposed the feeling that everyone else, and the country, and history as a whole were somehow identical with Lowell and explainable in his private terms.

We are talking about poetry, of course, not gossiping about Lowell. The problem we have been discussing generates a certain resistance to confessional poetry, which cannot succeed unless it gets beyond merely talking about oneself. If the details of one's private life become something more, poetically, the reason will be the affective language resonating through the poem. In the process "oneself" (Robert Lowell, the Irish poet Austin Clarke, Allen Ginsberg, W. D. Snodgrass, Sylvia Plath, John Berryman, Anne Sexton) operates like any other dominating symbol in a work. That is, "I" becomes something other through the medium that absorbs its associations; it becomes a magnetic cohesive center for all the emotional and subjective currents running through the work.

Lowell outdid other confessional poets in this colonizing of possible identities. He assumed that everything about himself, such as the manic bouts leading to frequent hospitalization, expressed modern America and its crises. Often the poems associate his private malaise with his social class's loss of leadership to sturdier if cruder newcomers. Two passages in "Waking in the Blue," about his confinement in a posh mental institution, will illustrate:

> hours and hours go by under the crew haircuts
> and slightly too little nonsensical bachelor twinkle
> of the Roman Catholic attendants.
> (There are no Mayflower
> screwballs in the Catholic Church.)

and

> . . . the pinched, indigenous faces
> of these thoroughbred mental cases,
> twice my age and half my weight.
> We are all old-timers,
> each of us holds a locked razor.[1]

Life Studies (1959), in which this poem appeared, remains Lowell's most striking book and most successful sequence, the one in which his own idiomatic bent took full control. Its impact outdid that of Allen Ginsberg's *Howl* (1956) and *Kaddish* (1961). Ginsberg's work expressed humiliation, grief, and hysterical revolt leavened by a wild humor, and it opened the way into the new libertarianism of the ensuing two decades. He and the other Beats made *Life Studies* appear less shocking in its directness and self-loathing than it would otherwise have done. But that is to oversimplify the issue somewhat. Lowell's superior poetic artistry and range of awareness bit deeper into the psyches of his fellow poets and his readers generally than did Ginsberg's prophetic outcries and maledictions.

In addition, *Life Studies* crackled with points of contact with the living American past. It bore many traces of vital older traditions (literary and puritanical, but patriotic too) to which Lowell, despite his nonconforming warp, was still loyal. He was still the poet who had spoken unselfconsciously, in "The Quaker Graveyard at Nantucket," of "our North Atlantic Fleet." It was not *their* fleet, not the fleet of Moloch or of monstrous capitalist interests as Ginsberg might have had it, but *ours*. However self-abasing he often is, Lowell does not relinquish the distinction

1. Unless otherwise indicated, all quotations by Lowell are from *Selected Poems* (New York: Farrar, Straus and Giroux, 1976). The original divisions referred to, including the prose section called "91 Revere Street" (not reprinted in *Selected Poems*), are those of *Life Studies* (New York: Farrar, Straus and Cudahy, 1959).

of being heir-presumptive to the whole tradition. The confident assumption of elite spokesmanship is a rather important tonal positioning that runs through *Life Studies*. Usually it is masked in wryness or self-irony or sardonic hostility, and so we have a mixed tonal stream; but still this proprietorship colors the things being poetically presented. Thus, it is clear in the passages just quoted from "Waking in the Blue" that "we" are Mayflower descendants, "thoroughbreds," and "old-timers"—"mental cases" though "we" may also be—as opposed to the "Roman Catholic attendants," doubtless sons of immigrants, with their undistinguished appearance and manners. "We"—*ergo* the nation—are degenerate.

This particular tonal aspect accounts for a good deal of the structure and bearing of *Life Studies*.[2] The whole volume—that is to say, the whole of the sequence—is divided into four major parts. Part One contains four poems and begins with "Beyond the Alps," which presents a swirl of associations during a journey, Lowell's descriptive note tells us, "On the train from Rome to Paris. 1950, the year Pius XII defined the dogma of Mary's bodily assumption." The first stanza is indicative:

> Reading how even the Swiss had thrown the sponge
> in once again and Everest was still
> unscaled, I watched our Paris Pullman lunge
> mooning across the fallow Alpine snow.
> *O bella Roma!* I saw our stewards go
> forward on tiptoe banging on their gongs.
> Life changed to landscape. Much against my will
> I left the City of God where it belongs.
> There the skirt-mad Mussolini unfurled
> the eagle of Caesar. He was one of us
> only, pure prose. I envy the conspicuous
> waste of our grandparents on their grand tours—
> long-haired Victorian sages bought the universe,
> while breezing on their trust funds through the world.

The circumstances of travel here are fortunate ones, and underlined as such by the slightly ironic mock-complaint that "our grandparents" were even better off than we, so much so that they were of the class of eminent Victorians "breezing on their trust funds through the world." The stanza is redolent of the kind of privilege taken for granted throughout the sequence. It is also redolent of something else: a sense of involvement in all manifestations of power or madness. Thus, "the skirt-mad Mussolini," with his satyriasis and violence, is cannibalized as "one of

2. On the structure of *Life Studies*, see also M. L. Rosenthal, *The New Poets* (New York and London: Oxford University Press, 1967), pp. 28–66.

us." Not that Lowell limits his identifications to those we have mentioned, but in his work one sinful or humiliation-laden empathy leads to another!

The structure of *Life Studies* embodies the process. The first three poems of Part One all have to do with aspects of power and privilege—their ruthless, corrupt, or stupid misuse, certainly, but in a context of familiar and shared concerns. Part Two, a prose autobiography named "91 Revere Street," recounts young Lowell's miseries as a member of an elite New England family in which the father had failed in his naval career and in his professional and business undertakings, and in which every kind of psychological problem prevailed. Witty, touching, and self-absorbed, the prose section provides the background for the dense malaise that is the atmosphere of the sequence as a whole—especially the sense of real but decayed connections with whatever might constitute power and prestige among the "best" and richest American families, and the poet's vision of himself as a ruined being nurtured in the wreckage of a destroyed world. In Part Three a series of four poems links the sensibility pervading the sequence to four outstanding modern writers out of phase with their age. Part Four, the title-section (in two numbered sections), returns to the passions and humiliations introduced in Part Two. It moves from childhood days through the deaths of Lowell's parents; then, in the last eight of its fifteen poems, it brings us more and more frankly into his adult pathology and struggles. The poems are not linked as continuous narrative, nor in any specified thematic relationship; and in fact about a third of the book (not counting "91 Revere Street") is not in itself confessional but takes on the confessional coloration through juxtaposition with the main body of poems. But there is a clear emotional progression through the four sections: from an atmosphere of disintegration, violence, and madness in Part One, to one of absurd disorder and early sorrow in the prose of Part Two, to one of neglected possibilities leading to tragic degeneration or coolly self-protective detachment on the part of some of our finest spirits in Part Three, and at last to an accumulation of private anguish and demoralization and guilt in Part Four—in which, also, a tiny hint of a way to health is inconclusively introduced.

As we have several times remarked in less extreme terms, the counterpart of the work's unconscious aura of inherited advantage of every sort is its display of disgrace, morbidity, and humiliation, within which lurks something like a pride of rampant sexuality, whether in the satyrisiac notes we have mentioned or in the fear of sexual failure. This very deep-going affective motif asserts itself emphatically at the end of each of the book's four parts, even the prose section "91 Revere Street." Thus, the fourth poem of Part One is sharply different from the three preceding it, which have to do, in their varied ways, with the sense of postwar Europe in full decline, with a comparable state in Renaissance Europe, and with the state of the Republic during the Cold War and at the moment of Eisenhower's Inauguration in 1953 ("and the Republic summons Ike, / the mausoleum in her heart"). The section concludes with "A Mad Negro Soldier Confined at Munich," in which the supposed speaker presents himself in terms that parallel Lowell's self-portraits in poems further along, in a state of frenzy or gross self-revulsion—in or out of the hospital. At the same time, the monologue contains deliberately planted notes suggesting that international madness and confusion are the order of the day and that the soldier's state and—by associative extension in the context of the whole sequence—Lowell's are an expression of that condition:

> "Cathouses talk cold turkey to my guards;
> I found my *Fräulein* stitching outing shirts
> in the black forest of the colored wards—
> lieutenants squawked like chickens in her skirts.
>
> Her German language made my arteries harden—
> I've no annuity from the pay we blew.
> I chartered an aluminum canoe,
> I had her six times in the English Garden.
>
> O mama, mama, like a trolley-pole
> sparking at contact, her electric shock—
> the power-house! . . . The doctor calls our roll—
> no knives, no forks. We file before the clock,
>
> and fancy minnows, slaves of habit, shoot
> like starlight through their air-conditioned bowl.
> It's time for feeding. Each subnormal boot-
> black heart is pulsing to its ant-egg dole."

The frenzy, raunchy exhibitionism, and ironic self-abasement in this passage from the poem make the "mad Negro soldier" the first of the three alternative personae Lowell employs in the sequence at strategic points. It is unusual for a poet to put a dramatic monologue in quotation marks, but Lowell does so here and in the other two instances—"Words for Hart Crane," at the end of Part Three; and " 'To Speak of Woe That Is in Marriage,' " the book's penultimate poem. (The final poem, "Skunk Hour," is reserved for the poet's literal self-abasement and descent into voyeurism and absolute demoralization.) The object of the quotation marks, apparently, is to make sure we know that the supposed speaker is not Lowell himself. This is particularly necessary since the language is not consistently that of the supposed speaker, as can be seen in line five and the final stanza of the passage just cited. In general, the passage is an adaptation of an educated, highly tensed literary vocabulary to the randy colloquialisms of college youths and to incompletely assimilated black street-talk. The inconsistency would be a flaw if the poem were clearly attempting to present itself as a realistic rendition of how the soldier would normally talk. We could not, in any case, make a valid judgment; conceivably the "mad Negro soldier" could be a Harvard Ph.D. But consistency is not the issue. All that counts is the dynamic shifting of tone that brings the poem, at last, to a sophisticated sense of automatized reduction of personality—expressed with such elegant articulateness in the teeming imagery of the closing stanza. (A poem does not have a voice; it uses voices when relevant.) In a sense the quotation marks indicate detachment from the affect of the poem. In another sense, they (and the title) are but the flimsiest gesture toward self-disguise, like a mask consisting of a pair of earmuffs.

"Words for Hart Crane," also set in quotation-marks, serves a similar function. A little presumptuously, Lowell gives Crane the words to describe his desperate and driven state and his relation to American poetry. The projection involved is perfectly plain, but the earmuffs do remind us that Lowell is not literally calling *himself* an American Catullus or Shelley, or the twentieth-century Whitman; nor is he describing himself as an aggressive homosexual. And yet, as projected here, the ideal of the alienated genius whose sexual proclivities are somehow part of his romantically bohemian magnetism is clearly an attractive mirror-image:

> "When the Pulitzers showered on some dope
> or screw who flushed our dry mouths out with soap,
> few people would consider why I took
> to stalking sailors, and scattered Uncle Sam's
> phony gold-plated laurels to the birds.
> Because I knew my Whitman like a book,
> stranger in America, tell my country: I,
> *Catullus redivivus,* once the rage
> of the Village and Paris, used to play my role
> of homosexual, wolfing the stray lambs
> who hungered by the Place de la Concorde.
> My profit was a pocket with a hole.
> Who asks for me, the Shelley of my age,
> must lay his heart out for my bed and board."

The third poem in quotation marks, " 'To Speak of Woe That Is in Marriage,' " is another sonnet-variation, this time generally in five-stress rhyming couplets. Here the speaker is a woman whose husband's pathology seems fairly close to that of the "mad Negro soldier." Again (in line 6, where the tone is that of a third person commenting on the situation rather than of the woman as she presents herself elsewhere in the poem) we have a certain inconsistency of language, though not so drastic as that in "A Mad Negro Soldier Confined at Munich." The effort to see the husband's behavior through the wife's eyes provides another confessional mirror-image through the kind of unwelcome empathy we have mentioned. Her suffering is presented with such relish, such muscular energy in describing *his* actions, that the mask of speaking woman is really a window on the man. (We are not really very far from Edgar Lee Masters's method here.)

> "The hot night makes us keep our bedroom windows open.
> Our magnolia blossoms. Life begins to happen.
> My hopped up husband drops his home disputes,
> and hits the streets to cruise for prostitutes,
> free-lancing out along the razor's edge.
> This screwball might kill his wife, then take the pledge.
> Oh the monotonous meanness of his lust. . . .
> It's the injustice . . . he is so unjust—
> whiskey-blind, swaggering home at five.
> My only thought is how to keep alive.
> What makes him tick? Each night now I tie
> ten dollars and his car key to my thigh. . . .
> Gored by the climacteric of his want,
> he stalls above me like an elephant."

There is an element of caricature in each of these poems that gives them a bold, cartoonist's clarity and grotesqueness, a pointing of comic possibility in their "hopped up" energy, their harsh incidental remarks and sardonic mood, their puns ("must lay his heart out"), and their scattered other wordplay and wisecracks ("Cathouses talk cold turkey," "I knew my Whitman like a book," "This screwball might kill his wife, then take the pledge"). Yet they are far from being comic poems. Their concern is with hideous distortions and displacements of all that might make for being happy and good. Lowell's closing poem, "Skunk Hour," with its picture of a local society gone to decay and neglect and of the poet's own squalid behavior "watching for love-cars," has a similar combination of witty turns and deep, horrified seriousness:

> One dark night,
> my Tudor Ford climbed the hill's skull;
> I watched for love-cars. Lights turned down,
> they lay together, hull to hull,

> where the graveyard shelves on the town. . . .
> My mind's not right.
>
> A car radio bleats,
> "Love, O careless Love. . . ." I hear
> my ill-spirit sob in each blood cell,
> as if my hand were at its throat. . . .
> I myself am hell;
> nobody's here—
>
> only skunks, that search
> in the moonlight for a bite to eat.
> They march on their soles up Main Street:
> white stripes, moonstruck eyes' red fire
> under the chalk-dry and spar spire
> of the Trinitarian Church.

A strong urban spirit tends to mark confessional poetry, and this perhaps accounts for its psychological self-awareness, its volatile sense of the ludicrous, its skeptical knowledgeableness, and its feeling of political embattlement. Our modern sense of enormous, uncontrollable issues about which we can never do enough—and anyway we're damned if we act and damned if we don't—pervades these poems. Lowell, something of an activist in his own quite individual way, describes his experience as a conscientious objector in World War II humorously but powerfully in "Memories of West Street and Lepke," a strategic poem in *Life Studies*. He calls himself, in retrospect, "manic," but rejoices in having been held alongside his lowly and criminal and eccentric prison mates. Still, although he reports comic and brutal details with a certain joy, his poem gravitates toward a tragic vision of America gone murderously bad and out of touch with her own meanings. The vision is concentrated in the figure of one inmate, "Czar" Lepke of the infamous Murder Incorporated, on whom the lens zooms in at the very end:

> the T-shirted back
> of *Murder Incorporated's* Czar Lepke,
> there piling towels on a rack,
> or dawdling off to his little segregated cell full
> of things forbidden the common man:
> a portable radio, a dresser, two toy American
> flags tied together with a ribbon of Easter palm.
> Flabby, bald, lobotomized,
> he drifted in a sheepish calm,
> where no agonizing reappraisal
> jarred his concentration on the electric chair—

hanging like an oasis in his air
of lost connections. . . .

These lines seem at first so casual and relaxed one hardly realizes their serious direction or even their rigorous technical control. Some of the ironic allusions are dated now; younger readers are unlikely to spot the echoes of Henry Wallace's claim that ours is "the century of the common man" or of John Foster Dulles's Cold War call for an "agonizing reappraisal" of the international situation. No matter. It is clear the language hardens into a Dantean picture, at once witheringly sardonic and not without pity, of the stupefied murderer in his little hell. Lepke, with his patriotic and religious symbols, his superior access to information, and his other possessions and privileges, becomes an emblem of one aspect of modern America: her "lost connections" with her own past despite her power. Earlier in the poem, Lowell had spoken of *himself* as having been "out of things"—a foreshadowing of the Lepke passage, and one refusal among many in the poem of any role of clear-eyed saint as peculiar to the poet. The subtle association of himself with Lepke, both sharers in our depravity and forgotten greatness, is one of Lowell's poetic triumphs.

This sort of building up a poem to a pitch of intensity, after a calmly goodnatured start, is characteristic of Lowell at his best. Generally the poems begin in an atmosphere of nostalgic reminiscence. What is being recalled can be rather unpleasant, actually, but one has the impression at first of memories that have mellowed through the passage of time; often, the beginnings have a muted dimension of edginess held in check. After the whole poem has struck home, though, one realizes it was all done in the key of hysteria. The brilliant "My Last Afternoon with Uncle Devereux Winslow," for instance, begins with a funny vignette of five-year-old Robert being very difficult and spoiling his parents' holiday plans. It ends with a terrifying, macabre description of Uncle Devereux's appearance when "dying of the incurable Hodgkin's disease." The description preserves intact the irrevocably violent impact of this death-apparition—literally, for all the imagery suggests the bone-whiteness of a skeleton—on the child's mind.

Almost inevitably, the psychological pressure of the hysteria generating and controlling this sequence exhausted itself *and* the poet's artistic control over longer structures. Lowell never again did a whole book as successful as *Life Studies,* his great demonstration of what the confessional mode can do—its power to cram the world's riches into the tiny room of the poet's vulnerable self. The book gave him a claim to be what Pound called an "inventor": that is, a writer who has "found a new process, or whose extant work gives us the first known example of a process." Lowell had extended the frontier of poetic possibility through his method, and in so doing had found it necessary to produce our first confessional sequence.

2. Lowell's *Day by Day*

Life Studies succeeded for the obvious reasons that the sequence form allows for fragmented presentation and that, primarily lyrical in its overall movement, it consists of a series of independent centers and allows for an enormous range of emotion and connotation. These characteristics met Lowell's need to free his style by coming clean about his childhood and family drama, his destructive and degrading problems, his sense of what other writers showed him most about himself. His drive toward self-clarification was also a drive toward self-acceptance, symbolized in the bold, rank image in "Skunk Hour" of the unscarable mother skunk, strong with the life-force, in clear contrast to all the human degeneration and loss of nerve around her. In creating his sequence, Lowell had magnificent models, but *Life Studies* then became a barrier to his future development. He could hardly repeat the identical curve of feeling, with its tentative balances and its excitement of achievement at great personal risk, that was the book's special triumph.

He himself knew all this quite well. He tried by main force—that is, by writing masses of irregular, unrhymed "sonnets"—to storm the barrier to doing a new sequence. These poems constitute an endlessly proliferating poetic journal and probably helped him to cope with the increasing disorder of his private life at the start of his final decade. After the incoherent welter of the various groupings of sonnets he published in four volumes between 1969 and 1973, he prepared one final sequence, *Day by Day*, that was published in 1977, the year of his death. Even more painfully personal than the sonnets often are, it is in its way a recovery from their gasping diffuseness. It is comparable, in its three-part structure, to *Life Studies* but charged with a very different tone of deep sadness we might call self-elegiac. In the new sequence we have the aftermath of a great change: divorce, a new marriage, and removal from New York to England. Anguish now hovers about his choice and new circumstances; suffering has not, after all, disappeared with the great change, and Lowell's opening use of Ulysses, Penelope, and Circe as symbolic personae is telling. Nevertheless the work as a whole is more narrowly personal in its reference and allusions than *Life Studies*—that is, it lacks the earlier book's range and the self-transcendence that the distancing through memory can achieve. Humor and energy are still present, but the sense of a series of premonitory summings-up of a bad state of affairs is overriding, as is the work's gentle melancholy. The poem "Suicide" is particularly pathetic: not quite an apology for not having followed the examples of friends—fellow poets who have committed suicide (other poems in the volume *are* such apologies)—but a driven poem of death-obsession and fear. Placed between two pieces about separation and about remorse for the failure of love, it presents—to echo the point we have already made about the book's dominant tone—a terrified self-elegy that is the key to the whole sequence. Its emphasis is on predicament and regret rather than on acute life-excitement, or the origins of one's guilt, or the need to assimilate others' identities and backgrounds to oneself. One gets the sense of an aghast, rueful spirit, exhausted by all its beating of wings against its own limits, in the process of surrender. And yet the net result is a sequence far more purely lyrical than *Life Studies* in its movement although far too reliant, also, on our information about, and interest in, Lowell's marriages and friends and specific circumstances of every kind.

Behind *Day by Day*, doubtless, lie the retrospectively melancholy sequences of other poets written at relatively advanced ages, especially Hardy's *Poems of 1912–13*, Pound's *Pisan Cantos*, and Book V of Williams's *Paterson*. Perhaps, too, the attachment of Lowell's vision of himself (or his anti-self, to pick up Yeats's mystical jargon) to heroic images from the past owes its half proud, half self-dismissive (by contrast, say, to Ulysses) tonal complex to Yeats's *The Tower*. But of course Lowell's way is to put himself in front of his symbols, so to speak, so that it could almost be said that he is a symbol of *them* rather than the other way around. The first part of the sequence presents the whole complex of his marital situation: the self-absorbed poet crushed between two powerful women is seen in the circumstance—i.e., the predicament—of Ulysses returning to Penelope from Circe. The poems in Part I begin by presenting Ulysses-Lowell almost simultaneously in the six-section "Ulysses and Circe," whose reversible symbolism makes an equation between the fall of Troy and "my" emotionally disastrous divorce and marriage. The blending of the Homeric hero and the modern anti-hero is handled at the start by a sort of sleight-of-hand mixing of pronouns—that is, of "he" and "I" in shifting contexts. They are perfectly clear grammatically, once one sorts them out, but it is just the need to sort them out that creates the blending:

> Ten years before Troy, ten years before Circe—
> things changed to the names he gave them,
> then lost their names:
> *Myrmidons, Spartans, soldier of dire Ulysses . . .*
> Why should I renew his infamous sorrow?

He had his part, he thought of building
the wooden horse as big as a house
and ended the ten years' war.
"By force of fraud," he says, "I did
what neither Diomedes, nor Achilles son of Thetis,
nor the Greeks with their thousand ships . . .
I destroyed Troy."[3]

The passage is heroically stark in its promise, but thereafter pathos takes over in the book. A process of increasing self-debasement—sexual despair and jealousy, helpless remorse, a feeling of being despised—accumulates through the five poems of Part I. It is a new view of "Ulysses" to think that neither Circe nor Penelope finds him very interesting any longer, for the one has her own divine and slovenly independence and is of a generation that speaks another language than his, and the other has accepted a life devoted to other concerns than his pleasures and pains. In the second poem, "Homecoming," a tone like that of "Mauberley (1920)" enters the sequence—that is, of bitter, irrecoverable loss. But of course it comes through more personally:

the lash across my face
that night we adored . . .
soon every night and all,
when your sweet, amorous
repetition changed.

Or:

Sometimes
I catch my mind
circling for you with glazed eye—
my lost love hunting
your lost face.

It is only two poems further on (with the Homeric imagery dropped for good after the long opening piece) that "Suicide" enters the sequence. Ambiguously addressed to a "you" that is both death and a lost beloved friend, it strongly echoes "Death & Co." and other poems by Sylvia Plath. The echo is unfortunate, because the poem seems to refer to Plath literally and to take on her misery abjectly, without the curious angry pride that gave a certain glory to her poems. Yet the depressive

3. This and the following quotations are from *Day by Day* (New York: Farrar, Straus and Giroux, 1977).

cast is all the greater in "Suicide" because of its abject character; it should have been named "Fear of Suicide"—

> I go to the window,
> and even open it wide—
> five floors down, the trees are bushes and weeds,
> too contemptible and small
> to delay a sparrow's fall.

So the first movement of the sequence dwindles in morale until it closes with "Departure," a poem whose pathos is that of the voyager lacking all perspective or mission:

> "Caught in the augmenting storm,
> choice itself is wrong,
> nothing said or not said tells—
> a shapeless splatter of grounded rain . . .
> Why, Love, why, are a few tears
> Scattered on my cheeks?"

Thus "Departure" ends. (The whole poem is enclosed in quotation marks as a form of slight distancing from the self-pity marking its movement from beginning to end.) The melody here, as in so much of this book, is bittersweet: whenever this particular kind of forlorn music enters a poem, the phrasing tends to be more disconnected than usual, a way of sustaining a pitch of mood without full coherence, as we saw in the first quotation from "Homecoming" as well. "Departure" prepares us for the scattered points of reference in Part II of the sequence, which resembles the succession of sonnets in Lowell's first collection of them, *Notebook 1967–68* (1969), in being what we have called a "welter" of wry remembering and of bemoaning both the passage of time and one's own aging. These are poems of reminiscence, addresses to old friends some of whom have died, and one-way communications with past loves and wives. The mood parallels Pound's summoning up of old days in Provence and elsewhere in Canto 74: "we will see those old roads again, question / possibly / but nothing appears much less likely." And Lowell, in the concluding poem of Part II, "Endings":

> When I close my eyes, the image is too real,
> the solid colors and perspective of life . . .
> the tree night-silvered above a bay becomes
> the great globe itself, an eye deadened to royal blue
> and buried in a jacket of oak leaves.
>
> Why plan; when we stop?

The poignancy of this sprawling second movement is undeniable, and sometimes exquisite in its power to compel sympathy despite the constant allusion to important personages by name, and the inevitable assumption that gossip about oneself is by definition immortal gossip fit to wring tears from the gods. However, the wider range of allusion in Part II does "objectify" the sequence by providing at least surface displacements of attention from the poet's own suffering and complications of love-relationship to the lives and deaths of other persons. ("This year killed / Pound, Wilson, Auden"; "Robert Penn Warren talked three hours / on Machiavelli"; "To my surprise, John [Berryman], / I pray *to* not for you.") Then, in the long title-section, Part III, itself broken into three subdivisions, we return to the full development of the tonal motifs introduced in Part I.

The general movement of the first section of Part III is from an encompassing fatalism to a facing-up to the dead-end situation of the new marriage. The second section then moves, as before, into an uncomforting resort to memory but then returns, full circle, to the same catastrophic dead end. And the third section presents a state of thorough emotional breakdown and desolation, attempted return to the old New York world, and an effort at self-placement through the method of *Life Studies.* That is, the unwanted lover-husband traces the state of being unwanted back to his mother's behavior and essential rejection of him and also tries to achieve a state of depressive transcendence in the closing poems. So the poems of Part III are themselves a bit of a welter and out of control, and place too much demand on our interest in Lowell as a psychiatric problem and patient and a monstrously loving and lovable human disaster. And yet we find a genuine, sheerly aesthetic resolution in some of the simpler, purer lyric poems, somewhat reminiscent perhaps of the later work of Randall Jarrell but more in touch with the clearest streams of our traditional poetry of loss and mutability. These poems would include "The Day," "So We Took Our Paradise," and "This Golden Summer." The heroic dimensions implied at the book's start are never realized by the later development; there is nothing even approaching the deep recovery of a people's cultural past as in *Paterson* or *Mercian Hymns;* instead the work is lost in a Freudian maze of analytical tracings of a quite private nature. But in the few poems of the sort we have just mentioned, an unwonted delicacy of perception and association achieves, paradoxically, a tensile strength that holds and that helps them survive the morass of egotistical expansion surrounding them. The following two stanzas from "This Golden Summer," for instance, give us all we need to know about the pity of a shaken and explosive love-relationship. In another, more sensitively receptive age they might be all we would need to reveal a world of affective nuances and human terrors and delights.

Is our little season of being together
so unprecarious, I must imagine
the shadow around the corner . . .
downstairs . . . behind the door?

I see even in golden summer
the wilted blowbell spiders
ruffling up impossible angers,
as they shake threads to the light.

3. Austin Clarke's *Mnemosyne Lay in Dust* and W. D. Snodgrass's "Heart's Needle"

We have mentioned certain other authors of confessional sequences besides Lowell—namely, Austin Clarke, W. D. Snodgrass, John Berryman, Allen Ginsberg, Anne Sexton, and Sylvia Plath. Obviously, it is impossible to take up their work in as much detail as we have given to major sequences, but a few general observations, together with a few notes at least on their individual differences and accomplishment, may be useful. And the first thing to be said is that these sequences constitute a poetry of struggle, primarily at the level of holding on to identity, self-regard, and normal sanity. The struggle is against a hypersensitivity that not only enhances awareness—which is a gain unless life is made thereby unbearable—but also presses physically on the nervous system. One's own inner intensities and conflicts, as well as the world's crushing injustices and uncontrollable laws (social and natural), are converted into the language of neurosis, hysteria, and breakdown. This is counterpointed by a capacity for deep joy and peace, presented mainly as nostalgia or vision. It is also, however, implicit in the strictly formal dimensions of poems—their sound and rhythm—and associative reaching and leaps that can be profoundly satisfying even while some subjective hell is being opened up for us. It was Lowell who pointed the way for others to bring into the open their private experiences and memories, often enough exposing themselves rudely or embarrassingly, yet releasing themselves for the discovery of driving passions and inner stances that could be mobilized and converted into serious poetic dynamics.

The danger, as we have seen with Lowell, lies in the subordinations of poetic art to something like psychological exhibitionism: when the poet's self blots out the quality of the poem or makes it secondary to the sacred drama of his or her psychological case history, complete with therapy. One sees traces of this danger in *The Waste Land* and *Mauberley,* but it has been circumvented in both by the associative method adopted, the avoidance of autobiographical accounts, and the use of external sources of evocation (the "objective correlative"). The poets we have named as "other" confessional poets can claim only partial successes in mastering the difficult problem. The Irish poet Austin Clarke's *Mnemosyne Lay in Dust* (1966) is essentially an account of a bout of his hysterical amnesia and eventual recovery, not without lasting effects. He handles the poetic conversion in a number of unpretentious ways—first of all by simply using the third person. Also, he cultivates a grammatical compression that helps him project a state of neurotic awareness directly through his style, and he varies verse-forms skillfully in accordance with shifting perspectives of the sequence. (His use of *rime riche* and idiosyncrasies of idiom and syntax, too, give a curious distinctness to his writing that brings out its inner feeling without letting it slip into mere familiarity.)

At one point in Poem I, for instance, we have a direct description of the protagonist's symptoms, although the language is hardly clinical and the compression makes it even less so; at another, the presentation of the mental hospital to which he is being taken brings in a whole world of horror and madness quite independent of his condition, yet altogether relevant to him and suggesting a relation between his condition and that of Ireland and her history. These are the second and the final stanzas:

> For six weeks Maurice had not slept,
> Hours pillowed him from right to left side,
> Unconsciousness became the pit
> Of terror. Void would draw his spirit,
> Unself him. Sometimes he fancied that music,
> Soft lights in Surrey, Kent, could cure him,
> Hypnotic touch, until, one evening,
> The death-chill seemed to mount from feet
> To shin, to thigh. Life burning in groin
> And prostate ached for a distant joy.
> But nerves need solitary confinement.
>   Terror repeals the mind. . . .
>
> The eighteenth century hospital
> Established by the tears of Madam
> Steevens, who gave birth, people said, to
> A monster with a pig's snout, pot-head.
> The Ford turned right, slowed down. Gates opened,
> Closed with a clang; acetylene glow
> Of headlights. How could Maurice Devane
> Suspect from weeping-stone, porch, vane,
> The classical rustle of the harpies,

Hopping in filth among the trees,
The Mansion of Forgetfulness
Swift gave us for a jest?[4]

The allusion to Swift is a reference to Swift's founding of St. Patrick's Hospital in 1746, as the editor of Clarke's *Selected Poems* points out. He also quotes these lines from Swift's "Verses on the Death of Dr. Swift":

He gave what little Wealth he had,
To build a House for Fools and Mad:
And shew'd by one satyric Touch,
No Nation wanted it so much.

A knowledgeable Irish reader would be aware of the association set up by Clarke and would take it as a suggestion of the heritage that is an element of the pressures working within the poem, and also as inseparable from a sardonic harshness shared with Swift and with the nation as a whole because of its memories of violence and squalor. Clarke's use of *rime riche* (Devane-vane) and his unexpected, vibrant shifts of context draw attention from his suffering protagonist as such to the evocative life of his language. In this respect he certainly has the advantage of Lowell although lacking the latter's purest heights and his general poetic range. He also achieves a multiply suggestive density through his handling of stanzas of some length.

The second, sixth, and tenth sections of Clarke's sequence are perfect examples of this highly developed talent. Suddenly, in Poem II, we are plunged into a confused scene, half-hallucinatory, in which attendants straight-jacket Maurice after undressing him and plunging him into a steam-bath; and this scene fades into a cinema-like aftermath and into actual hallucinations that follow:

Straight-jacketing sprang to every lock
And bolt, shadowy figures shocked,
Wall, ceiling; hat, coat, trousers flung
From him, vest, woollens, Maurice was plunged
Into a steaming bath; half suffocated,
He sank, his assailants gesticulating.
A Keystone reel gone crazier;
The terror-peeling celluloid,

4. The text is that of Austin Clarke, *Collected Poems*, ed. Liam Miller (Dublin: Dolmen Press, 1974, in association with Oxford University Press; also reprinted in Austin Clarke, *Selected Poems*, ed. Thomas Kinsella (Dublin and Winston-Salem, N.C.: Dolmen Press and Wake Forest University Press, 1976), with silent corrections we have followed.

Whirling the figures into vapour,
  Dissolved them. All was void.

Drugged in the dark, delirious,
In vision Maurice saw, heard, struggle
Of men and women, shouting, groans.
In an accident at Westland Row,
Two locomotives with mangle of wheel-spokes,
Colliding: up-scatter of smoke, steel,
Above: the gong of ambulances.
Below, the quietly boiling hiss
Of steam, the winter-sleet of glances,
  The quiet boiling of pistons.

The crowds were noisy. Sudden cries
Of "Murder! Murder!" from a byway,
The shriek of women with upswollen
Bodies, held down in torment, rolling
And giving birth to foundlings . . .

This is the projection of torment, yet without an ounce of sentimentality or implied self-pity because the imagery takes over in its own right and the tumult of actual hallucination is brought into the poem's foreground. In the sixth poem the sense of emotional and moral pain is greatly intensified, and still the objectification is sustained as though the protagonist were detached from his suffering:

One night he heard heart-breaking sound.
It was a sigh unworlding its sorrow.
Another followed. Slowly he counted
Four different sighs, one after another.
"My mother," he anguished, "and my sisters
Have passed away. I am alone, now,
Lost in myself in a mysterious
Darkness, the victim in a story."
Far whistle of a train, the voice of steam.
Evil was peering through the peep-hole.

Not all the sections work as well as these we have mentioned. A few are not much beyond the vignettes offered in Henley's "In Hospital," and others focus on stages of Maurice's recovery a bit literally after all. But the effort is a pioneering one in modern Irish poetry. It was encouraged into being by the writing of Lowell and others, but resists the egotism—that is, the arrogation of universal spokesmanship—that so often marks American confessionalism. It also resists the superimposed cosmopolitanism of the Americans. For better or for worse, Clarke risks a poem that smells of local streets and suggests humble circumstances and concerns. Even his closing section (XVIII) reveals a constricted environment within which the return to relative health is less than exhilarating. Indeed, the sense of the local takes over so decisively that one realizes, in this somewhat unexpected context (though one should have known better) how deeply the sequence is saturated with an embittered neo-regionalism of the most fundamental kind, deliberately flat and matter of fact:

> Rememorised, Maurice Devane
> Went out, his future in every vein,
> The Gate had opened. Down Steeven's Lane
> The high wall of the Garden, to right
> Of him, the Fountain with a horse-trough,
> Illusions had become a story.
> There was the departmental storey
> Of Guinness's, God-given right
> Of goodness in every barrel, tun,
> They averaged. Upon that site
> Of shares and dividends in sight
> Of Watling Street and the Cornmarket,
> At Number One in Thomas Street
> Shone in the days of the ballad-sheet,
> The house in which his mother was born.

This sort of internal self-discipline (not aggrandizing the self, and maintaining a decent humility toward the workings of history and the realities of other people's lives) is highly un-American. As we have noted, history tends to be a symbol of Lowell. And, since they were, however individual, following similar poetic paths, it is a symbol as well of some other confessional poets. Thus, in W. D. Snodgrass's *Heart's Needle* (1959), whose title sequence was praised by Lowell as "beautifully perfect and a break-through for modern poets" in its "best parts" (jacket blurb), both the Cold War and the deaths of soldiers in winter fighting in Korea become symbols of the poet's marital problems—his divorce and relationship with his young daughter. The sequence begins:

> Child of my winter, born
> when the new fallen soldiers froze
> In Asia's steep ravines and fouled the snows,
> When I was torn
>
> By love I could not still,
> By fear that silenced my cramped mind

To that cold war where, lost, I could not find
my peace in my will . . .[5]

This winter-imagery, with the world situation being used to frame a private predicament, is repeated in most of the other nine poems of the sequence; the essential figure is pressed most emphatically in the ninth poem:

Our states have stood so long
at war, shaken with hate and dread,
they are paralyzed at bay;
Once we were out of reach, we said,
we would grow reasonable and strong.
Some other day.

Like the cold men of Rome,
we have won costly fields to sow
in salt, our only seed.
Nothing but injury will grow.
I write you only the bitter poems
that you can't read.

In these passages, not only history but Dante's famous *in his will is our peace* and Donne's trick of turning a light colloquial phrasing to emotional and lyric purposes—for instance, "This ecstasy doth unperplex / (We said) and tell us what we love"—are echoed and diminished by the reversal of reference. Their main function is simply to lend an impressive backdrop of one sort or another to enhance the poignancy of divorce, remarriage, and the difficult situation of divorced parents and their children. No political rage or anguish really marks these pages, nor is there a special vision kindled by Dante or Donne or anyone else. Rather, the only pressure that really counts poetically in this sequence is the private one, in brief but revelatory closeups of tension and grief, and in sharp glimpses into the significance that may be lurking in any apparently ordinary observation or experience. Poem 3 gives us a perfect example of the differences between the poetry that connects and the poetry that does not, both in Snodgrass and in many other poets.

The child between them on the street
Comes to a puddle, lifts his feet
And hangs on their hands. They start

5. Quotations are from W. D. Snodgrass, *Heart's Needle* (New York: Alfred A. Knopf, 1959).

At the live weight and lurch together,
Recoil to swing him through the weather,
  Stiffen and pull apart.

We read of cold war soldiers that
Never gained ground, gave none, but sat
  Tight in their chill trenches.
Pain seeps up from some cavity
Through the clenched teeth in sympathy;
  The whole jaw grinds and clenches

Till something somewhere has to give.
It's better the poor soldiers live
  In someone else's hands
Than drop where helpless powers fall
On crops and barns, on towns where all
  Will burn. And no man stands.

Clearly, the scene in the first stanza is well observed, a characteristic moment of apparent rapport among father, mother, and child. Yet the language catches the parents' estrangement precisely in its imagery of burden and tension. The verbs, especially, are both natural to the scene and psychologically suggestive of antagonism (beginning with the third line), a suggestion to which the phrases "hangs on their hands," "live weight," and "through the weather" contribute effectively. The next two stanzas bear some relationship to this opening, a feeling of suffering endured helplessly; but despite the seriousness of their subject-matter, they lack real immediacy of the kind that would make the juxtaposition of the complex little family vignette and the passage on the "poor soldiers" a vital one. More stanzas follow those we have quoted, reaching tortuously for a demonstrated connection between the two realms. Meanwhile, the first stanza's fine sense of lives under a strain, which needs no explanation and is quite self-sufficient, remains the only affect that counts in the poem.

Not to labor the point that emerges again and again in confessional poetry: the shift of emphasis that occurs in the structuring of sequences when major, insistent, continuing attention is given to private suffering—its literal character, its history, its psychological implications—creates a largely unsolved problem. If some intensely reverberating identity is felt between private matters and public concerns, it cannot simply be pointed out, however cleverly or ingeniously. It must be felt in the manner of, say, Yeats's civil-war sequences, where the terrors of bloody violence in the land and the poet's sense of his own inadequacy converge at crucial points—for instance, in "The Road at My Door," in which the line "Caught in the cold snows of a dream" perfectly fixes the relation of personal and national predicament. But in such work the pressure and feeling of "public" and other "larger" concerns (fatality, the loss of the past, dreams of ideal beauty or goodness, the sting of injustice, the depths of brutality, the natural world) tend to dominate the work and catch up the purely personal realizations in their orbit. When the emphasis and proportion shift, one almost always feels that the connections are forced and arbitrary, a kind of editorializing rather than poetic discovery. We may compare Snodgrass's lines on "cold war soldiers" with the shock of Yeats's "Caught in the cold snows of a dream." Only a very aroused confessional poetry, such as we find in Lowell's "My Last Afternoon with Uncle Devereux Winslow" or "Memories of West Street and Lepke," can overcome the sense that unnecessary reinforcements are being called up, and that all they do is to clutter the poem.

One is inclined to object that the only "break-through" (Lowell's word of praise for the "Heart's Needle" sequence) Snodgrass gives us is the report of having come through the crisis of divorce and separation from his child—and the guilt accompanying a new marriage that confirmed his abandonment of the old life together—with a certain strength and equilibrium. The closing poem shows father and daughter together at the zoo (that most dependable resort of divorced fathers and their children on days they spend together) once again. It ends:

> Well, once again this April, we've
> come around to the bears;
>
> punished and cared for, behind bars,
> the coons on bread and water
> stretch thin black fingers after ours.
> And you are still my daughter.

In a personal sense, probably more than in a poetic one, this is a touching passage, foreshadowed by others earlier in the sequence. Instances include the second poem, which begins "Late April and you are three" and broods over the child's ignorance that separation is already arranged for; and the fourth poem, a sensitively lyrical hovering over the vulnerability of children to even the slightest disappointments, with the mood genuinely reinforced by notations of the changing season and colder weather:

> Like nerves caught in a graph,
> the morning-glory vines
> frost has erased by half

still scrawl across their rigid twines.
Like broken lines

of verses I can't make.

So this sequence, in its "best parts" (to quote Lowell again), is in essence an autobiographical narrative subordinated to its lyrical realizations along the way. Its gentle poignancies give it a delicate, bittersweet texture of love, loss, and hope. Along with this genuine feeling, the sequence is padded with assertions of a fatalistic sense, though only thematically conveyed, of spiritual oppression that is somehow linked with the national and international political atmosphere. The genuine feeling does reveal itself despite this interference, but a bit thinly. It is best exemplified in a stanza like the one closing the fourth poem:

Night comes and the stiff dew.
I'm told a friend's child cried
because a cricket, who
had minstreled every night outside
her window, died.

Tenderness verging on sentimentality marks the work's basic music of feeling. The imposed thematic context, by which the world's woes, from war to weather, are meant to be felt as symbols of the poet's misfortunes, is an obnoxious intrusion. It is not as obnoxious here, however, as in other sequences that force the supposed connection with more punch but no more justifiably.

4. John Berryman's *77 Dream Songs* and Allen Ginsberg's *Kaddish*

To start with an obvious example, we have John Berryman's 1964 volume *77 Dream Songs* (the first three groups of poems in the expanded volume, *The Dream Songs,* which appeared in 1969). *77 Dream Songs* gives the impression of a semblance of structure, as opposed to the uncontrolled proliferation—385 poems—of the later book. Berryman alludes to the issue we have been discussing in his original closing poem, #77. He does not offer an apology or propose a problem here, but instead presents a self-confident view of what the sequence has been reaching toward:

Seedy Henry rose up shy in de world
& shaved & swung his barbells, duded Henry up
and p.a.'d poor thousands of persons on topics of grand

moment to Henry, ah to those less & none.
Wif a book of his in either hand
he is stript down to move on.

—Come away, Mr. Bones.

—Henry is tired of the winter,
& haircuts, & a squeamish comfy   ruin-prone proud national
    mind,   & Spring (in the city so called).
Henry likes Fall.
Hé would be prepared to líve in a world of Fáll
for ever, impenitent Henry.
But the snows and summers grieve & dream;

thése fierce & airy occupations, and love,
raved away so many of Henry's years
it is a wonder that, with in each hand
one of his own mad books and all,
ancient fires for eyes, his head full
& his heart full, he's making ready to move on.[6]

The liveliness and bounce here are so attractive they almost conceal the classic complacency of a talented, egocentric neurotic about his condition. In his prefatory note to the expanded volume, *The Dream Songs,* Berryman sought to distance himself from the "Henry" of the poem, who is, *of course,* not a literal representation of Berryman himself (but, *of course,* is also closer to Berryman's subjective self than any psychiatrist could ever have come):

> The poem then, whatever its wide cast of characters, is essentially about an imaginary character (not the poet, not me) named Henry, a white American in early middle age sometimes in blackface, who has suffered an irreversible loss and talks about himself sometimes in the first person, sometimes in the third, sometimes even in the second; he has a friend, never named, who addresses him as Mr. Bones and variants thereof.

Bearing in mind the fact that what counts in poetry is its dynamics, not its alleged subject-matter (except as this generates a context of feeling and awareness and constitutes a psychological pressure), we may safely revise this description somewhat. "Henry" is perforce Berryman's self-mirroring protagonist, seen in a glass that splits and distorts the reflection; thus, for instance, the "unnamed friend" is still another split-off fragment of the work's total self-awareness. From one angle, Henry

6. Texts are from John Berryman, *The Dream Songs* (New York: Farrar, Straus and Giroux, 1969).

certainly is a reflex of the literal poet—what in the world, otherwise, would we be able to make of the poem we have just quoted? From another, "he" is a minstrel-show figure—an obvious turn on the Romantic motif of the artist-as-clown, with a macabre dimension that the name "Mr. Bones" will help suggest. From yet another, the reflection is of an innocent primal self, uttering babytalk.

Berryman's disclaimer, then, is simply nothing more than a reminder that his art is not photographic reproduction or ordinary discourse. Also, Berryman's explanation is not at all about structure—the movement of the whole sequence, if it is felt to have a movement—but rather about the nature of "Henry." And indeed, the succession of "dream songs," a term suggesting a stream of reverie and fantasy, is primarily a succession of separate performances on the stage of the mind, some sheer buffoonery, some actings-out of deep emotion, and some declamations and political or philosophical musings and pronouncements. There is a faint suggestion in the seventy-seventh dream song of successfully completed therapy, comparable perhaps to the actual recovery of Austin Clarke's protagonist and the faint stirrings in that direction at the end of Lowell's "Skunk Hour." But we hardly have the "story" of a struggle for mental health at the heart of the dream songs—even less so than in *Life Studies,* in which that is not the point in any case. What Song #77 does suggest, however, is that the poems hitherto have presented a spirit of many moods, at once amoral ("impenitent") and too burdened by human feeling and great concerns; and at once self-ironic (he has "p.a.'d . . . on topics of grand moment to Henry") and bursting with self-regard ("these fierce and airy occupations, and love"; his "mad books"; his "ancient fires for eyes"). The "dream songs" constitute an enormously blown-up, carefully analyzed and displayed, proudly guilt-ridden, morbid, boisterous, and presumptuously representative self-image. Their serious buffoonery projects, as its best, real horror or joy; at its worst, it imposes on us the overblown annexation of other people's sufferings we have observed in Lowell: a kind of artistic demagoguery that confessional poets too easily find congenial. Songs #41 and #60 act out this annexation most crudely: the former on behalf of Jewish victims of the Nazis; the latter on behalf of black Americans, complete with vaudeville dialect:

> The cantor bubbled, rattled. The Temple burned.
> Lurch with me! phantoms of Varshava. Slop!
> When I used to be,
> who haunted, stumbling, sewers, my sacked shop,
> roofs, a dis-world *ai!* Death was a German
> home-country.
>
> (#41)

Afters eight years, be less dan eight percent,
distinguish' friend, of coloured wif de whites
in de School, in de Souf.
—Is coloured gobs, is coloured officers,
Mr Bones. Dat's nuffin?—Uncle Tom,
sweep shut yo mouf . . .
Bit by bit
our immemorial moans

brown down to all dere moans. I flees that, sah.
They brownin up to ourn. Who gonna win?
—I wouldn't *pre*dict.
But I do guess mos peoples gonna *lose*.
I never saw no pinkie wifout no hand.
O my, without no hand.

(#60)

We are not suggesting that dialect humor, even that of stage-dialect, is by definition censorably condescending; or even that comic vulgarity with a racial dimension is intrinsically unacceptable. Berryman's heart was certainly in the right place, and his vaudevillean and sentimental projections of would-be empathy are touching, and sometimes he is as funny as he wishes to be. But at the point at which the demands of art and those of encompassing human empathy converge, something else is called for than this kind of patter if the desired spokesmanship is to take place naturally and without philistinism. Berryman's method in these instances is too facile, the easy humanitarianism of the nightclub comic who depends on our understanding that he means no harm in playing for all the laughs he can get. And at the same time, there is the sentimentality, the playing for tears, of popular song. Berryman gets beyond these levels in his final turn in the second quotation, but when he does so it is too late for what he started out to do; he has moved into a note of horror suggested by what has gone before but not accounted for by it.

As in "Heart's Needle" (but more vibrantly and dramatically), what is most powerfully moving in the "dream songs" is the passionately evocative affective life achieved in some of them. Song #29 stands out sharply, among a few others not quite at its pitch of sustained feeling, as the purest poem in the volume—and of the whole 385 as well—in its concentrated control of a mixture of tones. Tragic in its bearing, it nevertheless makes use of buffoonery without letting it get out of hand. This buffoonery emerges in the occasional touch of babytalk, deployed to suggest Henry's innermost pathetic vulnerability. The mawkish babytalk element counterpoints the intellectual and psychological knowledgeableness of most of the poem's phrasing, connected perhaps in its guilt-ridden torment to Henry's identification with his father (whose suicide is described in Song #76, "Henry's Confession"). It is the "innocent" element in the poem's relentless pain:

> There sat down, once, a thing on Henry's heart
> só heavy, if he had a hundred years
> & more, & weeping, sleepless, in all them time
> Henry could not make good.
> Starts again always in Henry's ears
> the little cough somewhere, an odour, a chime.
>
> And there is another thing he has in mind
> like a grave Sienese face a thousand years
> would fail to blur the still profiled reproach of. Ghastly,
> with open eyes, he attends, blind.
> All the bells say: too late. This is not for tears;
> thinking.
>
> But never did Henry, as he thought he did,
> end anyone and hacks her body up
> and hide the pieces, where they may be found.
> He knows: he went over everyone, & nobody's missing.
> Often he reckons, in the dawn, them up.
> Nobody is ever missing.

We may contrast this extraordinary poem with a more typical *relatively* successful one, Song #53. If there were more of a sense of structure in the whole sequence, there might be powerful reciprocities between two such poems, sustained and carried along by strategically placed reinforcing poems all along. The wit and colloquialism would seem less arch in Song #53, and the post-Holocaust savagery of its final line less presumptuously facile. For the poems are connected in certain ways. The first stanza of #53 alludes to Henry's "unforgivable memory" and presents him as a heroically Blakean grotesque: "He lay in the middle of the world, and twitcht." He is seen for the moment as the wounded Achilles ("Pelides") requiring foot-medication. This hubristically imagined son of a man and a goddess is "human (half)"—and then we see him in his completely mundane, modern aspect, "down here as he is," subjected to everything mortally destructive in what the world has to offer us. Hence the shift, in the succeeding stanzas, to a series of quotations hostile to all human contact:

> He lay in the middle of the world, and twitcht.
> More Sparine for Pelides,
> human (half) & down here as he is,

with probably insulting mail to open
and certainly unworthy words to hear
and his unforgivable memory.

—I seldom *go* to *films*. They are too exciting,
said the Honourable Possum.
—It takes me so long to read the 'paper,
said to me one day a novelist hot as a firecracker,
because I have to identify myself with everyone in it,
including the corpses, pal.'

Kierkegaard wanted a society, to refuse to read 'papers,
and that was not, friends, his worst idea.
Tiny Hardy, toward the end, refused to say *anything*,
a programme adopted early on by long Housman,
and Gottfried Benn
said:—We are using our own skins for wallpaper and we cannot win.

In general, we cannot speak of *77 Dream Songs*—or, even more emphatically, of its fivefold longer successor—as a sequence of purest ray serene. It has something of the welter-character of Lowell's groupings of sonnets, which may indeed have been prompted by *The Dream Songs*. And it suffers, too, from the same problem that plagues any sonnet-sequence. That is, it repeats the same basic form over and over, and over and over, forgoing one of the great advantages of the true sequence: its playing off of significantly varied metrical, rhythmic, and even visual patterns against one another. Berryman uses one basic pattern, an eighteen-line poem divided into three stanzas of equal length—but with varied prosody and irregular rhyming—most flexibly. He makes it an excellent vehicle for Henry's changing moods, internal dialogues, anecdotes, and pronouncements. Or— as Song #71 puts it, a little wryly and a little proudly:

Spellbound held subtle Henry all his four
hearers in the racket of the market
with ancient signs, infamous characters,
new rhythms. On the steps he was beloved,
hours a day, by all his four, or more,
depending. And they paid him.

But the proliferative tendency, and the sense of a vaudeville performance punctuated by desperate sobbings, boisterous speeches, and cultural notations, militate against the formation of a finely interactive sequence.

Having said this, we should note the possible modulation toward structure in these seventy-seven poems. Song #1 (most are simply numbered, without titles) gets things going in the negative with its first two lines: "Huffy Henry hid the day, / unappeasable Henry sulked." And, at the very end, it finds a natural image and an image of human deprivation to give dignity to what starts out as adorable (from the poem's point of view, not the world's) petulance: "Hard on the land wears the strong sea / and empty grows every bed"—combining primitive and Wagnerian tones of rigorous endurance fairly authoritatively. Song #77 (following "Henry's Confession," which, we have suggested, perhaps is intended to give us a basic clue to the manic-depressive orientation and counterpointings of the poems as a whole) professes self-understanding and readiness to move on to new places: the great hope and purpose of the psychoanalyst's patient. The book as a whole is divided into three sections of 26, 25, and 26 poems respectively, and with some forcing (aided by the three fairly ambiguous epigraphs to the volume), we may characterize these groupings as (1) Henry's unhappy sense of himself and the world, including his identification with all the suffering and oppressed and sense of America's long betrayal of her ideals, (2) Henry's lamentations (really, more turns on the tones and motifs of the first group), and (3) mental hospitalization and thoughts of recovery (again, mixed with more of what the earlier poems have already offered). But there is little sense of large units of affective realization acting in relation to one another. We are dealing with interesting fragments and possibilities at best, and there is little profit in lecturing to these poems about what they might, or should, have become.

Allen Ginsberg's *Kaddish* ("For Naomi Ginsberg 1894–1956") makes no pretence at the kind of small-scale ingenuity practiced by Berryman. The simple, pious mourner's kaddish of Judaic tradition is, first, a chant of praise to God and then a prayer on behalf of the dead person and of the mourners. Ginsberg includes these elements in a five-part sequence that borders on a long poem. In it he leaves the restraint and unaffected humility of the traditional kaddish in another world entirely, although the sequence does give vent to the private hysteria, the rush of intimate memory, and the darker visions of existence that often are involved. In this release, barely contained within the limits of each part of the work (formal perfection is hardly one of its qualities), the sequence finds the affective elements that go into its ordering. The first part, "Proem: I," is an elegiac "psalm" addressed to Naomi, the poet's mother, that combines memories of her, evocations of the poet's present state of mind and circumstances, feelings of exaltation in sadness, and relief at a suffering woman's liberation from madness and physical suffering. The mélange of details and of styles of phrasing will be suggested by three lines far apart from one another in the "Proem": "Strange now to think of you, gone without corsets & eyes, while I walk on the sunny pavement of Greenwich Village"; "All the accumulations of life, that wear us out—clocks, bodies, consciousness, shoe, breasts—begotten sons—your Communism—'Paranoia' into hospitals"; and "Take this, this Psalm, from me, burst from my hand in a day, some of my Time, now given to Nothing—to praise Thee—But Death."[7] But all is dominated in this section by a rush of images invested by a sudden sense of strangeness:

> It leaps about me, as I go out and walk the street, look back over my shoulder, Seventh Avenue, the battlements of window office buildings shouldering each other high, under a cloud, tall as the sky an instant—and the sky above—an old blue place. . . .
> Toward the Key in the window—and the great Key lays its head of light on top of Manhattan, and over the floor, and lays down on the sidewalk—in a single vast beam, moving, as I walk down First toward the Yiddish Theater—and the place of poverty
> you knew, and I know, but without caring now . . .

At the end of "Proem" the rhythm tightens into a staccato intensity, each line broken by punctuation into several very short units of excited prayer, with one longer breath-sweep included to counteract any merely mechanical effect: "Thee, Heaven, after Death, only One blessed in Nothingness, not light or darkness, Dayless Eternity." In general Ginsberg's rhythms do not really approximate his desired models, which range from incantatory Biblical passages to the actual praying in synagogues to literary sources such as, most obviously, Blake and Whitman and, at times, traditions of revolutionary political oratory. His long lines and catalogues and parallel constructions do suggest these models, but the ear is not a finely attuned one—not even to his own improvised patterns. The main rhythmic and tonal emphasis is ordinarily established at the beginning of one of his lines, while the latter part tends to wander off on its own or drag along behind or else take a deep breath and provide a last burst of feeling at the very end.

Part II, called "narrative" in the table of contents, is humanly the most moving section of *Kaddish.* It is much the longest section, too, a scarified account, half prose-poetry and half a succession of informative notations, of Naomi's life—her girlhood as a Jewish immigrant from Russia, her Socialist and Communist background, but most of all her paranoid hysteria that wreaked havoc with her family and that must have had very much to do with her sensitive, loyal son's own psychological

7. Quotations are from Allen Ginsberg, *Kaddish and Other Poems 1958–1960* (San Francisco: City Lights Books, 1961).

problems and perhaps with the surge of homosexual awareness and desire he describes in the context of Naomi's story. Some of the details and anecdotes are achingly painful or hideous, some hilarious and infinitely touching, some brutally humiliating for Ginsberg himself and members of his family. The section performs the same function (feeding in information and a context of feeling the rest of the work draws on) as Lowell's "91 Revere Street" section in *Life Studies*, which by comparison is a model of genteel restraint. The historical roots affecting the innermost personality of Naomi, and therefore of Allen too as he is presented here, go back to pogroms, the passions of political factions stemming from the Russian Revolution and the Communist accusation that the Trotskyists were allied to the Nazis. The identifications of private personality and large social and political struggles are not arbitrary in this work, as they are to a great extent in Lowell and most of the other Confessionals. They are closer to the genuine, inescapable identifications of the Irish poets we have discussed; and as with the Irish poets, they are not so much tokens of the poet's great and noble and Christlike assimilation of the world's sufferings and injustices as an aspect of self-knowledge and of the poet's problems of personality.

For these reasons Ginsberg's spokesmanship for more people than himself seems more genuine and powerfully effective than that of the others. At the same time, his poetic power, both in the sense of the vast evocative power of a master of his craft and the sheer vitality of such a master's phrasing and emotional presence, just does not match the basic genuineness of the spokesmanship he assumes. You cannot make a silk purse out of lines like "O beautiful Garbo of my Karma—all photographs from 1920 in Camp Nicht-Gedeiget here unchanged—with all the teachers from Newark—Nor Elanor be gone, nor Max await his specter—nor Louis retire from this High School" or "Kicking the girls, Edie and Elanor—Woke Edie at midnite to tell her she was a spy and Elanor a rat. Edie worked all day and couldn't take it—She was organizing the union.—And Elanor began dying, upstairs in bed"—and worse. It is all a wrenching story of the underside of a great deal of our life, and there *are* scattered beautiful moments; but the wretched, exalted feeling of his extended monologue lies mainly in its compulsive outpouring of note upon note of event and of recollected ambience. The effect is of literal experience—a triumph of copiousness and overcoming of revulsion, but with no artistic control beyond that of varying the tonalities within this flood of notations and outcries. That variation is of importance, it does create a loose dynamics, but it is the dynamics of logorrhea on the analyst's couch.

Four shorter sections follow the "narrative." "Hymmnn" is a variation of the true kaddish: a hymn in praise of God, beginning conventionally but then introducing some very unorthodox dimensions while remaining true to the paradoxically devout tradition of accepting His will and therefore glorifying whatever has been ordained. Here the form is far more disciplined by a rigorous parallelism and by the example of the true kaddish than elsewhere. The first four lines will illustrate:

> In the world which He has created according to his will Blessed Praised
> Magnified Lauded Exalted the Name of the Holy One Blessed is He!
> In the house in Newark Blessed is He! In the madhouse Blessed is He!
> In the house of Death Blessed is He!
> Blessed be He in homosexuality! Blessed be He in Paranoia! Blessed
> be He in the city! Blessed be He in the Book!

The next section, numbered "III" and called "lament" in the table of contents (we note this confusing format without comment), returns briefly to the mood and line of "Proem," repeating its inconsistent parallelism and its tone of elegiac evocation ("only to have seen her weeping on grey tables in long wards of her universe," "only to have come to that dark night on iron bed by stroke when the sun gone down on Long Island"). Its movement is somewhat different, for it ends in a minimal vision of continuity combined with a stronger imagery of death's entropic finality and of the barrier to further imagination death interposes: "Creation glistening backwards to the same grave, size of universe, / size of the tick of the hospital's clock on the archway over the white door." Section IV (called "litany" in the table of contents) bursts into a hysterical incantation to the dead mother that breaks from the earlier patterns of long-lined successive waves of detail-studded and image-laden parallel utterances. The "litany" contains a fair number of relatively long lines (though not as long as in previous sections), but is dominated by the shorter ones. One is reminded of various antecedents, from Yeats's "The Statesman's Holiday" to Eliot's "Ash Wednesday" to Lorca's "Death of a Bullfighter," but again—as with the tradition of long-lined cadences—the whole sense of things and affect of the rhythm and its variations is quite different in *Kaddish*. In "litany" Ginsberg recapitulates the earlier points of emotive attention but with no suggestion of praising God for the horror of what his mother and he have endured. The mother is addressed (in a turned-about way) as a Christian poet might address the Virgin; but the tone of "litany" barely conceals a deliberately domestic and unpretentiously intimate address, and we soon move into a profane range of awareness:

> O mother
> what have I left out

O mother
what have I forgotten
O mother
farewell
with a long black shoe
farewell
with Communist Party and a broken stocking
farewell
with six dark hairs on the wen of your breast
farewell
with your old dress and a long black beard around the vagina
farewell
with your sagging belly
with your fear of Hitler
with your mouth of bad short stories
with your fingers of rotten mandolines

and so on for 52 lines, the last 29 of which all begin with the phrase "with your eyes"—all, that is, but line 52 itself, which brings the "litany" full stop at the thought of death, just as section III did:

with your eyes of stroke
with your eyes alone
with your eyes
with your eyes
with your Death full of Flowers

As "Ash Wednesday" is not in itself a sacred text, it is no blasphemy to note the very remote echo of the "Lady of silences" passage in Eliot at the start of the "litany." But very shortly the repeated initial "with," and the succession of unexpected items completing the parallel phrases, are much more reminiscent of the strange closing stanza in "The Statesman's Holiday." Of course, the series of items in Yeats's poem is whimsically and romantically raffish, as opposed to Ginsberg's deliberate grossness of detail—but the echo seems clear, and it is almost as though Ginsberg had set out to write something like the "old foul tune" Yeats's stanza mentions but does not provide:

With boys and girls about him,
With any sort of clothes,
With a hat out of fashion,
With old patched shoes,
With a ragged bandit cloak,
With an eye like a hawk,
With a stiff straight back,

With a strutting turkey walk,
With a bag full of pennies,
With a monkey on a chain,
With a great cock's feather,
With an old foul tune.
Tall dames go walking in grass-green Avalon.

Oh, what a difference between Yeats's beautifully controlled metrical pattern and catalogue, in his riddling song of joy in despair, and Ginsberg's lurching. For the "litany" might well have moved superbly, so that its echoing stark effects and its release into full sadness at the end could be fully alive and overwhelmingly so. That is, its absolute frankness of self-cleansing candor in the face of disgust, and the absolute tenderness that withstands all the disgust, might have come through far more beautifully. At any rate, the "litany" brings to a head the process of reconciling the need to celebrate a beloved, dead mother and the need to purge oneself of her deeply imprinted negative impression. This happens by way of a screaming incantation against the horror of things, a horror embodied in the mother as ultimate victim. Then the brief concluding Section V (called "fugue" in the table of contents) concentrates a final swell of love and compassion, but also a final bleak hardness and negativism translated into the repeated "caw caw caw" of crows "in the white sun over grave stones in Long Island." The dry call of the birds becomes interchangeable, by the poem's end, with our human cry to an unresponsive heaven: "Lord Lord Lord caw caw caw Lord Lord Lord caw caw caw Lord."

The formal variations from section to section, and the shifts of affective coloration and intensity they accompany, make it possible to think of *Kaddish* as a sequence. At the same time, it falls short of its possibilities, partly because of superficial surface continuities. The most obvious example, and the only one we shall make a point of here, is the long, catch-all second section—the "narrative" that is more like an ancient mariner's relentless monologue. The range among extremes of feeling in this section forcibly suggests the need for extensive revision to allow the isolation of certain affects in self-contained units. (In general, Ginsberg opts for mixed affects whenever he has the space for them. But even Whitman, a greater worker with complex tonalities by far, understood—at the very start of the rise of the modern sequence—the importance of isolating certain affects very purely, in sections of their own.) We shall not labor the matter, but will let one small instance suffice:

2 days after her death I got her letter—
　　Strange Prophecies anew! She wrote—"The key is in the win-

dow, the key is in the sunlight at the window—I have the key—Get married Allen don't take drugs—the key is in the bars, in the sunlight in the window.

Love,
your mother"

which is Naomi—

This passage comes at the very end of the "narrative" and almost lifts it out of its rambling desultoriness (inevitable because too much has been thrown together in a careless-sounding way, despite the special moments). But in itself the passage is truly transcendent, rooted in the realities that underlie the basic psychological pressure acting on the whole work and yet miraculously beyond them. This, and the other "special moments" we have alluded to, could have worked to enormous effect if isolated in a poetically more developed structure. The genuinely organic character of the essential elements in the sequence is not, however, in question. In this important respect, Ginsberg surpasses Berryman and even Lowell.

5. Sylvia Plath's "Final" Poems and Anne Sexton's "The Divorce Papers"

Two more confessional instances present interesting problems it seems useful to consider. They are Anne Sexton's "The Divorce Papers" (1976) and the group of twelve poems completed by Sylvia Plath during the period 28 January to 5 February 1963.[8] Both groups of poems were published posthumously, the one edited by Anne Sexton's daughter, Linda Gray Sexton, and the other by Sylvia Plath's husband, Ted Hughes. Neither group can be called a definitive text. "As her literary executor," Linda Gray Sexton writes in her "Editor's Note" to the volume *45 Mercy Street* (1976), in which "The Divorce Papers" first appeared, "I have altered the placement of a few poems. . . . Certain poems have been omitted, however, because of their intensely personal content, and the pain their publication would bring to individuals still living."[9] It seems likely that these observations would apply at least as much to the "intensely personal" and pain-giving "The Divorce Papers" as to anything else Anne Sexton wrote; and so we are dealing with an uncertain order-

8. Anne Sexton, *The Complete Poems* (Boston: Houghton Mifflin, 1981), pp. 509–35; Sylvia Plath, *The Collected Poems*, ed. Ted Hughes (New York: Harper & Row, 1981), pp. 262–73.
9. Reprinted in *The Complete Poems*, pp. 479–80.

ing and a possibly incomplete text, especially since Anne Sexton's death in 1974 left a manuscript she was still in process of revising.

The problem with the "final group" of Sylvia Plath's poems is comparable. We were formerly persuaded, on the whole, that the closing poems in her *Ariel* (1965) were a sequence in formation, and the idea was reinforced by Hughes's "Notes on the Chronological Order of Sylvia Plath's poems."[10] Hughes wrote that "the final group of poems dates from mid-January 1963." She would have composed them, then, within less than a month before her death on February 11. The final twelve poems, which seemed to cohere as a sequence, are (in the order printed in *Ariel*): "The Hanging Man," "Little Fugue," "Years," "The Munich Mannequins," "Totem," "Paralytic," "Balloons," "Poppies in July," "Kindness," "Contusion," "Edge," and "Words." Hughes's note, however, omits mention of the first three of these and the eighth, while he does name the others. In *Collected Poems* he places the four poems he did not name among pieces written earlier. The final group of poems, all dated 1963, now begins with "Sheep in Fog," first written (a note tells us) "on 2 December 1962," but "the last three lines . . . were replaced by the present three-line verse on 28 January 1963." The other poems of 1963, in the order printed in *Collected Poems*, and with their dates as noted by Plath, are "The Munich Mannequins," "Totem," and "Child" (all 28 January); "Paralytic" and "Gigolo" (both 29 January); "Mystic," "Kindness," and "Words" (all 1 February); "Contusion" and "Balloons" (both 4 February); and "Edge" (5 February).

At any rate, despite the substitution of four poems in the 1963 final group, and the reordering of the eight remaining poems as well, our original sense that the closing group is a sequence in formation[11] seems even more justified now. The final dozen do constitute a special group, completed within a period of nine days of intense concentration, and with many indications that the poet is readying herself for her coming suicide (only six days after the last of the poems, "Edge," was written). We do not of course know what she would have done with the poems in this group had she lived after all. She might well have augmented or diminished the number, and her ordering would very likely have been different. If we remember that with these poems, as with "The Divorce Papers," we are dealing with the work of a person on the verge of sui-

10. Ted Hughes, "Notes on the Chronological Order of Sylvia Plath's Poems," in Charles Newman, ed., *The Art of Sylvia Plath* (London: Faber & Faber, 1970). Hughes was the editor of *Ariel* (New York: Harper & Row and London: Faber and Faber, 1965) as well as of the later *Collected Poems*.

11. See M. L. Rosenthal and Sally M. Gall, " 'Pure? What Does It Mean?'—Notes on Sylvia Plath's Poetic Art," *The American Poetry Review*, 7 (May/June 1978), 37–40, written on the basis of the *Ariel* text.

cide, the whole question of the readiness of recently written work for publication becomes grotesquely magnified. We are in a very open field indeed. A truly remarkable aspect of these poems is the way in which, while their main pressure is so ineluctably deathwards, Plath was still able to objectify her several levels of feeling and self-awareness. She found room, for example, for the sweetly humorous "Balloons" and the determinedly ironic "Kindness," even as she was being pulled into what her poems call the "perfect" world of death—as hypostatized, especially, in "The Munich Mannequins," "Words," and "Edge."

As printed in *Ariel,* the final group began with three poems introducing the speaker's suicidal obsession in three different perspectives, marked by powerful images of horror and desolation. But the true pitch seemed to be set most decisively by the fourth poem, "The Munich Mannequins," which was the first poem in the *Ariel* grouping to be retained. Its bearing is immediate and inescapable, but so drastic from the start that "Sheep in Fog," the "new" opening poem, prepares us for it beautifully by its gentler dolor that intensifies slowly until, at the end, we realize we are in a realm of the profoundest unhappiness and lonely terror:

> The hills step off into whiteness.
> People or stars
> Regard me sadly, I disappoint them.
>
> The train leaves a line of breath.
> O slow
> Horse the color of rust,
>
> Hooves, dolorous bells——
> All morning the
> Morning has been blackening,
>
> A flower left out.
> My bones hold a stillness, the far
> Fields melt my heart.
>
> They threaten
> To let me through to a heaven
> Starless and fatherless, a dark water.

The combined impressionistic notations of a literal country scene (literal sheep, literal fog) and projection of overwhelming internal sadness without an object—but attaching itself to whatever touches the senses for the moment—gather arbitrary associations as the poem moves onward. The movement is leisurely in its grieving progress; but then, all at once, we are in the "heaven" of death—a heaven, presumably, because the object of all the fear-crammed yearning evoked in the first four stanzas. It is no state of paradisal ecstasy; the urge toward it presses the poem in the one direction only. The sequence has begun with a shivering premonition; the "dark water" image at the poem's close foreshadows its guilty echo at the end of "Child," three poems further on, and its doom-laden, decisively confirming echo in "Words," near the end of the sequence.

Then comes "The Munich Mannequins," uncompromisingly stark at once:

> Perfection is terrible, it cannot have children.
> Cold as snow breath, it tamps the womb
>
> Where the yew trees blow like hydras,
> The tree of life and the tree of life
>
> Unloosing their moons, month after month, to no purpose.

Only after this mordant overture, so bitterly melodic, are the "mannequins" of the title introduced:

> . . . in their sulfur loveliness, in their smiles
>
> These mannequins lean tonight
> In Munich, morgue between Paris and Rome,
>
> Naked and bald in their furs,
> Orange lollies on silver sticks,
>
> Intolerable, without mind.

As a result of this delayed introduction of the poem's central symbol, the opening lines have a prophetic mystery and power. They present "perfection" in terms of thwarted female sexuality, introduce life-images fraught with death association (yews and hydras), and suggest the futility of continuing life on any terms. As the poem proceeds, it confronts one essential issue of life-acceptance before allowing the mannequins to take the stage. This is the issue of love, dealt with obliquely and ambiguously:

> The blood flood is the flood of love,
>
> The absolute sacrifice.
> It means: no more idols but me,
>
> Me and you.
> So, in their sulfur loveliness, in their smiles
>
> These mannequins lean tonight

We have allowed the two foregoing quotations to overlap in order to stress the elusive connections involved. To go on living would be to allow the "blood flood" to continue—the literal pulsing of the heart's blood but also the menstrual cycles associated with the sexual life and now felt to be "to no purpose." They are "to no purpose" not only because no child will be born or even conceived, but also because the life cycle itself is seen as empty of meaning. To sacrifice it will be to fix one changeless image—that of oneself perfected in death—forever. "You" will have become dead and fixed for "me," at the same time, by the act that makes "me" the sole "idol" henceforth—sulfurously smiling in the hell of immobility. Granted, there is a certain ambiguity in these lines. It not only keeps open divergent meanings, but also conceals the childish suicidal motive of making those one loves pay exclusive attention to oneself at last. Then the mannequins, reminiscent of the horror-figures in a number of Plath's earlier poems from "The Disquieting Muses" on,[12] take over. They at once bring in associations endemic to *Ariel*, linking the speaker's psychic condition with whatever she has learned about war and Nazi terror ("Munich, morgue between Paris and Rome") and with the whole oppressive pattern of human unhappiness.

The poem finds sinister images for all its concerns, and uses them to rush us to the final negation telegraphed early on in "cold as snow breath." The concerns reveal themselves as sources of frustration and self-depreciation: the mechanics of power and subservience (symbolized in the obsessive Plath-symbol of shoe-blacking, which emerged with puzzling emphasis in "Daddy"); the assertiveness and common domesticity of German personality, rendered mordantly ironic by the grisly recent past; and the impersonality of the surfaces of existence, "glittering and digesting." ("Glittering" is a sinister, two-edged word in Plath's poems, invariably suggesting both seductiveness and destruction.) The concluding line, set off from the rest of the poem, is: "Voicelessness. The snow has no voice." The poem is a cold pastoral, no less than was Keats's Grecian urn. (Its vision too was of perfection, even of love, at the expense of life.) But the movement in Plath's poem is from dread through renunciation, and from revulsion to perfect stillness. The tones accumulate; the initial morbidity of imagination simply attaches itself to more and more contexts of deathliness.

The succeeding poems provide new centers of vision and feeling but never break out of this malign spell. The stress in "Totem" remains on the fascination and terror of death, again concentrated in an image of "glittering":

12. In *The Colossus* (London: William Heinemann, 1960; reprinted London: Faber and Faber, 1967). See *The Collected Poems*, pp. 74–76, in which "The Disquieting Muses" appears among poems written in 1957.

There is no mercy in the glitter of cleavers,
The butcher's guillotine that whispers, "How's this, how's this?"

Even in "Child," in which a mother addresses her child—"the one absolutely beautiful thing," whose "clear eye" deserves to be filled "with color and ducks" and images that are "grand and classical"—there is a surrender to the terror. For the mother knows what the true imprint of her suffering love-presence will be:

. . . this troublous
Wringing of hands, this dark
Ceiling without a star.

In the next poem, "Paralytic," the persona shifts—but the obsession does not. The paralytic of the title is a man, powerless to move or touch, for whom the world glides by in two dimensions. He wants to "relapse" into death. The poem thus projects a desired acceptance by the man of the death willed in "The Munich Mannequins" by the woman who speaks there. Earthly beauty is too much. (The affinities of intensity and visionary morbidity in Plath and Dickinson are striking—would there were world enough and time to go into them here.) That beauty—like "glittering" surfaces in other poems—"glides by like ticker tape," is hidden behind the "starched, inaccessible breast" of a nurse playing at being concerned for her patient, and is most predatory when most innocent:

I smile, a buddha, all
Wants, desire
Falling from me like rings
Hugging their lights.

The claw
Of the magnolia,
Drunk on its own scents,
Asks nothing of life.

"Paralytic" forms a peculiarly bitter little sub-group with the two poems that follow it, "Gigolo" and "Mystic." The "dead egg" of a man who speaks in it has a sense of everything rich and lovely in life but denied to him. The poem's empathy for his condition seems a reflex of the subtler agony, of fused passionate responsiveness and inability to sustain joy, that rides so much of Plath's writing.

Then, in "Gigolo," a different male persona is introduced—an irresistible lover for whom women are an aphrodisiac food, inseparable from the diet prescribed in folk-medicine for men desiring the greatest sexual potency:

Bright fish hooks, the smiles of women
Gulp at my bulk
And I, in my snazzy blacks,

Mill a litter of breasts like jellyfish.
To nourish
The cellos of moans I eat eggs—
Eggs and fish, the essentials,

The aphrodisiac squid.
My mouth sags,
The mouth of Christ
When my engine reaches the end of it.

The word "gigolo," of course, implies the exploitation of love; and the sense of exuberant sexuality (however gross) fades into a curious, unresolved imagery of obscene crucifixion through satiety. At the end of the poem, the glitter of male eroticism—for the poem does suggest that an exposé of masculine treachery and self-centeredness is at issue—is seen as in truth mere greed and narcissism:

. . . New oysters
Shriek in the sea and I
Glitter like Fontainebleau

Gratified,
All the fall of water an eye
Over whose pool I tenderly
Lean and see me.

A hovering connection is set up in these poems among something like a glut of self-love, the sense of inability to be in real touch with the world outside oneself, specifically masculine ego and power, and the vision of oneself reflected in an essentially murderous external cosmos. "Mystic," a sad poem that tries to reach out to everything happy and promising after disastrous experience (Dickinson again!), really cannot break out of the perspective established in its first stanza:

The air is a mill of hooks—
Questions without answer,
Glittering and drunk as flies
Whose kiss stings unbearably
In the fetid wombs of black air under pines in summer.

The fear of response to life's blandishments is as tangible in this stanza as the sensual alertness to them. "Mystic" is followed by "Kindness," which specifically rejects yielding to them—to "smiles" (as suspect a word as "glitter" in Plath's lexicon) and to sweetness of any kind. The poem repudiates love-overtures and the results of love: "You hand me two children, two roses." Yet, like "Child," it expresses distressed bewilderment about what is to become of children: "What is so real as the cry of a child?" Here and in the closing poem, "Edge," we come painfully near naked autobiographical confession. The marital resentments that had led Hughes and Plath to separate a few months earlier (October 1962, a note of Hughes's informs us), and the obviously anguished question of the children's future (normal enough in any case, but more so because of the suicidal pressure so relentlessly active in these poems), come right to the surface in the poems just mentioned.

Then, in "Words," the poet tries to place her art in objectified perspective. At this strategic point, nothing could have been more appropriate structurally, and yet it is an especially stunning achievement that Plath could have distanced things so at this juncture. The "words" she has written over the years helped her keep her turbulence in some order. Now she sees them in a complexly unwinding series of images for the power she exerted upon herself, the reverberations of her exercise of that power, the way that the reverberations become horses for her imagination and feelings, and the fact that her words have finally been unable to overcome her death-centeredness:

> Words dry and riderless,
> The indefatigable hoof-taps.
> While
> From the bottom of the pool, fixed stars
> Govern a life.

"Words" distances itself from the more privately confessional aspects of the other poems, while at the same time holding in desolate view those poems' interactions. It clinches our sense that we have been reading a sequence written *in extremis*. However tentative their formation, the poems transcend personal expression while carrying their suicidal set, with its highly controlled counternotes of tenderness and life-energy, into a state of impersonal, aesthetically realized "perfection." And the drastic pitch at which they exist, move, and interact, together with the astringent poetic discipline that allows no room for sentimental posturing or generalized proclamations, does away entirely with the problem of factitious identification with the victims of history and oppression. We may argue the question of Plath's right to assume such an identification as a point of logic; but in the poems the fusion of private and public agony, and of guilt and the sense of victimization, is a given of image and tone formation within the living structure. One might as well quarrel with Keats about whether or not Ruth's heart was truly sad, or with Milton about the existence of God.

Two other poems of relative objectification follow the remarkable "Words." These are "Contusion" and "Balloons." The former is a poem of pure death-obsession centered on an impersonal image. Its first line, "Color floods to the spot, dull purple," may be exactly reciprocal with the "red shred" image that forces itself on our attention at the end of "Balloons." The word "contusion" itself is not terribly frightening, but here its very formality and undramatic timbre are suggestive of barely detectable evil—something cancerous, a "doom mark" insidiously destroying its host.

"Contusion" moves with brilliantly morbid inevitability toward its closing images of death. After it, as if to drown it out, "Balloons" emerges in almost absolute contrast, its surface all sunny gaiety, full of exuberant humor. We are now in the world of the speaker's children. They have splendid balloons,

> Yellow cathead, blue fish—
> Such queer moons we live with
>
> Instead of dead furniture!
> Straw mats, white walls
> And these traveling
> Globes of thin air, red, green,
> Delighting
>
> The heart like wishes or free
> Peacocks blessing
> Old ground with a feather
> Beaten in starry metals.

It is a charming outpouring of happy sympathy, in beautiful distinction to what we find anywhere else in the "final group." And yet it does not altogether break the pattern of death-consciousness and strangeness. The poem has many oddly discordant notes. The balloons are called "oval soul-animals" in the first stanza. They are at least distant relatives to the Munich mannequins and the disquieting muses—"queer moons" too. The image of "dead furniture" suggests the heaviness of the daily world unless leavened by happy imagination. The lovely image of the peacock feather "beaten in starry metals" may be a subtle echo (admittedly the suggestion is farfetched, however) of the closing lines of "Words":

> From the bottom of the pool, fixed stars
> Govern a life.

But that funereal, fatalistic image is not the one in "Balloons," in any case. All the discordant notes we have mentioned suggest only very lightly that, as in "Child," the speaker is not simply a happy mother enjoying her children's delight in their playthings. Even the comic little vignette with which "Balloons" ends—of a child biting his pink balloon and then sitting back with "A red / Shred in his little fist"—carries only as much ironic implication as we are willing to allow it. It is the context of the whole group of final poems that leads one to give special weight to such an effect.

Consider again, for instance, the first line of "The Munich Mannequins," with its self-torn confession: "Perfection is terrible, it cannot have children." And consider the ambiguous lines later in that same poem—

> The absolute sacrifice.
> It means: no more idols but me . . .

The literal or symbolic image of dead children recurs throughout these poems, as elsewhere in Plath. "Totem" ends with a picture of flies that

> . . . buzz like blue children
> In nets of the infinite,
>
> Roped in at the end by the one
> Death with its many sticks.

But it is the final poem, "Edge," that stamps the image of the children in death on the sequence with absolutely cold, clear precision. We have been tentatively assuming that the order in the *Collected Poems* is the order of the final draft of a definite poem-sequence. Our purpose could hardly be a dogmatic one on this point, as we have already shown. It is simply to think about the text as given, remembering that the order of a sequence virtually always has its improvisatory side by the very nature of things. If, however, the order now proposed by Hughes, on the basis of dated manuscripts, is correct, and if the composition of these poems did indeed cease with "Edge," then the closing poem "solves" all the problems of the sequence at once. The speaker pictures the achievement of the perfection broached in terror in "The Munich Mannequins." She re-creates herself, dead (a marble image), and undoes the birth of her children, who now are dead along with her and have been "folded . . . back into her body"—

> . . . as petals
> Of a rose close when the garden

Stiffens and odors bleed
From the sweet, deep throats of the night flower.

"Words" provided a perfect ending for the sequence, in the *Ariel* version, from the standpoint of aesthetic distancing that nevertheless distils the essence of all that brute experience has piled into a tragic life. Here the distancing is more intimate and unbearably touching in its knowing self-deception. In the light of the vision in "Edge," at once morbid and pitifully tender, the happy scene in "Balloons" becomes a precarious, doomed moment, and the "red shred" left in the child's fist at the end seems a fearful prophecy.

Anne Sexton's "The Divorce Papers" (assuming, for the moment, the integrity of *its* text as given) lacks the ultimate astringency that would make everything in it, however literally rooted in the poet's life, count in the full poetic sense as do the Plath poems we have been discussing; yet at important intervals—as in the poems "The Red Dance"—the sequence does come close to a comparable achievement. It is made up of seventeen poems that, on the surface, depend heavily on the story of the poet's divorce and short-lived love-affair in 1973 and gravitation toward suicide the next year. One can trace the narrative's points of reference in the key dates alluded to in various poems. "The Wedding-Ring Dance," for instance, refers to her decision to obtain a divorce on 14 April 1973; "The Break Away," to her love-affair in the summer of 1973 (presumably the burden of the preceding poem "When the Glass of My Body Broke"); "The Love Plant," to that summer affair and to her realization in March 1974 that she hasn't been able to quell her feeling for her lover; and "Killing the Love" and "The Red Dance" to her success at last in doing so—and to the equation of that success with the triumph of the suicidal compulsion that now dominates the imagery. Beneath the surface of this autobiographical progression lies the psychological pressure of a confused, complex association of love with cruelty, rejection, and despondency. Thus, the first poem, "Where It Was At Back Then," proceeds to an apparently happy ending ironically undercut by the cruel images (implying a desire to castrate) that have gone before:

Husband,
last night I dreamt
they cut off your hands and feet.
Husband,
you whispered to me,
Now we are both incomplete.

Husband,
I held all four

in my arms like sons and daughters.
Husband,
I bent slowly down
and washed them in magical waters.

Husband,
I placed each one
where it belonged on you.
"A miracle,"
you said and we laughed
the laugh of the well-to-do.

Just how "well-to-do" they are the next seven poems reveal catastrophically. "The Wedlock" presents a picture of perfect incompatability, heavily seasoned with phrases and rhythms borrowed directly from Sylvia Plath—

Mr. Firecracker,
Mr. Panzer-man.
You with your pogo stick,
you with your bag full of jokes.

This poem is for the most part much more on the order of a jeering marital quarrel or wife's complaint than the first, which pointed directly to the deeper pressures and malaise which the sequence as a whole projects and contends with. Its concluding stanza, however, shifts tone and orientation completely and gets to the heart of the matter, a psychotic state that is felt as one's fate and that paralyzes all potentiality for sustained and nourishing love. In this stanza the Confessional dimension remains, but instead of the preceding nagging and screaming ("Suppertime I float toward you / from the stewpot / holding poems you shrug off / and you kiss me like a mosquito," etc.) we find a dream landscape foreshadowing the whole curve of the sequence:

When I'm crazy a daughter buys
a single yellow rose to come home by.
Home is our spy pond pool in the backyard,
the willow with its spooky yellow fingers
and the great orange bed where we lie
like two frozen paintings in a field of poppies.

A balance on the order of that in "The Wedlock" is sustained throughout the opening group of poems having to do with this bitterly unsatisfying relationship that is, in any case, blasted by the wife's psychosis. In this group the poems "Despair," "Bayonet," and "The Wedding Ring

Dance" stand out—the first because it explicitly and with more forceful concentration than most of "The Divorce Papers" presents the feeling of love as synonymous with despair and the love-relationship as murderous—"a railroad track toward hell"; the second because, as the title implies, that relationship is felt as a war in which the woman desires to use her powers as a "bayonet" to thrust into the man "as you have entered me" and transform him into "a sculpture," a white "object unthinking as a stone, / but with all the vibrations / of a crucifix"; and the third because the decision for divorce it announces ("the undoing dance") is in such a savage mood and goes far beyond rejection of the husband in its revulsion against the demands of love:

> I dance in circles holding
> the moth of the marriage,
> thin, sticky, fluttering
> its skirts, its webs.
> The moth oozing a tear,
> or is it a drop of urine?
> The moth, grinning like a pear,
> or is it teeth
> clamping the iron maiden shut?
>
> The moth,
> who is my mother,
> who is my father,
> who was my lover . . .

The downward, destructive, overexpanded movement of descent and horror in the opening group of eight poems is countered by a single poem that stands alone in the sequence. This is the next poem, "When the Glass of My Body Broke," an outcry of ecstatic joy at being awakened from sexual and emotional paralysis by a lover who "has picked me up and licked me alive" for the first time since "I was born a glass baby." The poem has its weaknesses; its vocabulary of love-fulfillment seems borrowed from D. H. Lawrence and pornographic literature ("hands that excite oblivion, / like a wind, / a strange wind / from somewhere tropic / making a storm between my blind legs, / letting me lift the mask of the child from my face"). But perhaps its very factitiousness at certain points prepares us for the reversal that follows immediately afterwards. And it does in large part breathe the sense of delight and release that even the illusion of love can arouse.

Then, though, the downward movement resumes. Because of the intervening poem of joy it seems all the fiercer, and the reference hereafter is less to the marital relation than to the new lover and to other lovers ("The Stand-Ins") who come and go, and to the sense of being stifled and unwillingly possessed by genuine love-experience and the memory of it ("The Love Plant"). Imagery of killing and stifling increases mightily in these poems, taking many morbid turns in the group of three poems following "When the Glass of My Body Broke." The title of "The Break Away" will suggest the violent torque of the first of these. Next, "The Stand-Ins" pulls the sequence up short with its hard repudiation of any need to regard male protestations and attachments sympathetically. Swastika and Cross are symbols of the killer-lover; so is Jesus. This is a poem of very mixed quality. Part of it falls into the most maudlin association, redeemed only by its helplessly and genuinely mad self-hypnotic aura, between "me" and the Auschwitz victims—one of the chief vulgarities, as we have suggested, of modern poetry, even in the case of Plath despite her supremely sustained artistry. But this dream-poem comes into its own at the end, in its curtly sardonic dismissal of the male hold and in its reversal of that well-known bit of masculine humor: "A woman is only a woman, but a good cigar is a smoke." Sexton writes:

> I woke.
> I did not know the hour,
> an hour of night like thick scum
> but I considered the dreams,
> the two: Swastika, Crucifix,
> and said: Oh well,
> it doesn't belong to me,
> if a cigar can be a cigar
> then a dream can be a dream.
> Right?
> Right?
> And went back to sleep
> and another start.

After the prolonged explanations and elaborations of "The Break Away," with its incidental revelation that one of the lovers' whimsies had been to call themselves "Two Camp Directors" (with a double suggestion of the pleasures of summer camp but also of the guilt of concentration camp officials), the relatively healthy-minded brusque realism of this passage is especially welcome. But the third poem of the group of reactions against the prospect of joy, "The Love Plant," is an even more agonized restatement of the need for repudiation. Then at last, in the final group of four poems, the entire perspective is fully revealed. Love must be destroyed because the underlying drive has always been toward self-liberation through suicide. "Killing the Love," the first poem in this group, begins the reorienting, but it is especially revealed in "The Red Dance," in which the speaker, for the first time, is only a distanced narrator who objectifies herself as a modern "Sappho" dancing along the edge of the Charles River, past and around all the tokens of the world's treacherous and repulsive nature and, in prospect, dreaming the vision of her immersion and obliteration in the river. The closing poem, "End, Middle, Beginning," takes a psychoanalytical turn at the start: "There was an unwanted child." This is felt as the great final revelation underlying the distrust and fear of love, especially since the sexual act that led to conception, forcing the child into life, was violently repudiated as an act of love from the child's point of view:

> There was an unwanted child.
> Aborted by three modern methods
> she hung on to the womb,
> hooked onto it
> building her house into it
> and it was to no avail,
> to black her out.

Thus the final poem begins, and then it moves relentlessly through a recapitulation of the unwanted child's progress through life's stages under the pressure of an unremitting sense of being stifled and silenced and emotionally inert. The progress is a reprise, in a different key, of all that has gone before in the sequence. Like "The Red Dance," the only other poem in the third person, it distances the hysteria-racked major movement of the sequence and converts it into a parable or allegory that belies and grimly parodies the story of the Sleeping Beauty. Having "grown fully, as they say," she is given a ring that she wore "like a root" in marriage, saying to herself, as "she lay like a statue in her bed," that "to be not loved is the human condition." A lucky encounter with love changes all this temporarily "by terrible chance," but then, "slowly, / love seeped away"—

> and she knew her fate,
> at last.
> Turn where you belong,
> into a deaf mute
> that metal house,
> let him drill you into no one.

The final poem (it would be the "final" one in its relation to the internal dynamics of the sequence even if placed elsewhere; that is, it would be the magnetic center around which the other poems gather and by which they spark off one another) achieves its harsh negative transcendence by pushing complaint into the realm of pure repudiation. It exploits the anticipations of psychoanalysis—to dive into unconscious memory again and again until the origins of one's pathology are clarified and confronted and liberation is achieved—but discovers only perpetual death-in-life as the ultimate "human condition." The more vivacious but equally doom-laden "The Red Dance" provides a reciprocal parable, of the aesthetic dimension of this condition: the dance toward the perfection of death, which is also a perfection of beauty. The full morbidity in this one poem matches that of Plath's "Words" and of certain other of her poems, notably "Contusion" and the slightly earlier "Death & Co." "The Red Dance" concludes:

> And the waters of the Charles were beautiful,
> sticking out in many colored tongues
> and this strange Sappho knew she would enter the lights
> and be lit by them and sink into them.
> And how the end would come—
> it had been foretold to her—
> she would aspirate swallowing a fish,
> going down with God's first creature
> dancing all the way.

The streams of tonality mingling in this sequence make a striking effort at equilibrium: a large and intense sensuality; a pervasive countersense of being imprisoned forever in a casing of repression; a struggle toward freedom and, simultaneously, a fear and hatred of it; enormous self-pity and self-castigation; ferocious resistance to male force, however much desired and needed, and rivalry with it both in male terms ("Bayonet") and in female terms ("The Red Dance"); and death-obsession that is inseparable from the creative, aesthetic instinct. "The Divorce Papers" is one extreme instance of the paradoxical gravamen of the confessional sequence: objectification and transcendence through the most intimate and vulnerable and particularized subjectivity.

Chapter Fifteen

Continuities, Post-Confessional and Eclectic

1. Ramon Guthrie's *Maximum Security Ward*

We have not exhausted the list of confessional sequences; we have tried, rather, to note the major instances, together with a few others that represent certain problems and challenges in this mode of sequence-formation. It may be that the confessional realized its best promise in *Life Studies,* and that most exemplars after this one repeat its successes at best and, at worst, force an unwelcome, poetically irrelevant emphasis in two ways. One way, insidiously pervasive, is to confuse autobiography and literal self-portraiture (including a succession of diary entries) with poetic art, which is forever a matter of language, sound, and evocative values, first and foremost. The other way is to assert a necessary universal symbolism in the details of a given protagonist's life and private ordeals, whether self-evidently present or not. Nevertheless, the working of psychological pressure, to which we have given so much attention, has been powerfully linked with the confessional dimension of sequences from the start.

It seems now that the burgeoning of the confessional mode was, historically, a necessary exploration of this essential dimension. It allowed poets the freedom to discover its limits. Also, it pressed our modern intimation that art is our decisive means of objectifying the life of inner awareness, in which form becomes the driving integrative force. The confusion in much confessional poetry, between anecdote and structure, and between private happenstance and true symbolic embodiment, is only a failure of artistic energy and realization.

At all events, the burgeoning of the confessional helped us see how indispensable an ingredient the specifically and idiosyncratically personal bearing of this affective element is. On the basis of Lowell's work and that of others, poets who continue writing sequences in what we may roughly describe as the "mainstream"—that is, the sort of affective and formal mixture we find in modern classics like *Song of Myself*, Yeats's sequences, *The Waste Land*, and the *Cantos*—have been encouraged to use the confessional ingredient more openly and centrally than their forerunners did. Foremost among these mainstream practitioners, we should list the somewhat neglected Ramon Guthrie, whose *Maximum Security Ward*, a sequence composed of forty-nine poems, focuses much of its attention on the poet's private sufferings, physical and emotional.[1] It is to that extent confessional, far more so than are the classic sequences just noted. And yet the balance is not finally in that direction; the confessional is absorbed into a dynamics that uses it without being dominated by it. The opening lines of the volume are an excellent introduction to its main tonal characteristics, including the absorption of its confessional element into a more encompassing lyrical structure:

> So name her Vivian. I, scarecrow Merlin—
> our Broceliande this frantic bramble of
> glass and plastic tubes and stainless steel—
> could count off such illusions as I have
> on a quarter of my thumbs.
>
> (. . . *even a postcard of Viollet-le-Duc's*
> *pensive chimera signed with her initial* . . .)

These lines open the first poem of the book, "Elegy for Mélusine from the Intensive Care Ward." We are engaged immediately by this quirky "Merlin" caught in the frantic confusion of the I. C. U. In the midst of personal disaster, he expresses himself with a touch of buffoonery and with keen imagination. Thus, figures of Arthurian romance are intertwined with the apparatus of modern medicine, with Notre Dame's famous gargoyle, with censorable cables, and with pseudo-anthropological-linguistic information in a splendid jumble of times, conditions, and places. Guthrie's diction is as flexible as his imagery: colloquial, wittily learned, tough-and-tender, self-ironic, and yearning. Throughout *Maximum Security Ward* we are never far from one or another tonality of an extremely volatile speaking voice through which this or that state of feeling is revealed: musing, questioning, preaching, praying, complaining,

1. Ramon Guthrie, *Maximum Security Ward 1964–1970* (New York: Farrar, Straus & Giroux, 1970; London: Sidgwick and Jackson, 1971).

cursing, exclaiming. At the same time, we are never far from song either, and the purely lyrical note seems as natural to the verse texture as all the other tones.

This lyrically evocative element is pervasive and the best possible corrective for any tendency to over-emphasize a single dramatic personality at the heart of the work: a hysterically distressed, invalided object of our pity. Even in the few lines we have quoted from "Elegy for Mélusine," one can see how much Guthrie relies on suggestive words, subtle rhythms, and sound echoes to provide a constantly lyrical undercurrent. The proper names conjure up a whole poetic world of romance and song, as does the phrase "pensive chimera"; and the speaker's longing for his cruel mistress may be personal but is also Petrarchan and medieval in its resonances. Further, this longing is used as a tonal refrain throughout "Elegy for Mélusine"; the words change, but the longing persists and the relevant stanzas are set off by spatial positioning and by type face. Even the fact that the poem is composed in stanzas, however varied in length, and in a familiar line that has a norm of four or five stresses, insinuates the decisive presence of the lyrical tradition in the poem's body and movement. This presence constitutes an impersonal pressure in the work's dynamics, transcending the confessional energy from the start.

The work's full vision is well worth this craftsmanship, and it too goes far beyond the literally personal references and private pressures that are part of the underlying emotional complex. There is something like a protagonist in the poems, but "he" is protean. He has many of Guthrie's attributes yet would hardly be Guthrie even if an illusory poetic figure could ever really replicate its maker. He refers to himself most frequently not as Guthrie or the poet but as Marsyas, the satyr who dared challenge Apollo to a test of musical skill. He is obviously *in extremis,* and in the course of the sequence a number of reasons for this parlous state are offered. First of all, he is desperately ill and probably dying. And then, he has a monumental sense of having missed out and failed. He has been treated callously by everyone, and by his mistress (Mélusine, so called after the French *fée* of medieval romance) in particular. He believes that, both as man and as artist, he has been totally defeated. Conventional religious faith is impossible for him; and his faith in man's goodness is at lowest ebb, for in our century man has proved so hideous that it is acutely embarrassing to call oneself human. Mortal suffering and extreme depression converge, sometimes comically ("bring on your Dead March with Muffled Drums / and Reversed Rifles and high-stepping young / Drum Majorettes with the minniest of Miniskirts"), sometimes with uncompromising starkness ("it is stamped with hot wire / on my breastbone").

Guthrie's general ideas and attitudes are not, of course, what gives *Maximum Security Ward* its power. What matters is the masterly evocation of a despairing, bewildered, wry condition of awareness, centered on a struggle against the sense of isolation from self, humanity, nature, and God. Imagination, intellect, and humor are the weapons of this poetic struggle with overwhelming pain, terror, and bitter knowledge. The epic effort is toward reorientation. The first third of the book—the 1000 lines and twenty poems of the first part—presents the full pressure of the dark vision it contends with. In the second part, twice as long, the first half mitigates this massed bleakness in important ways; and the rest of the sequence seems to forge a new perspective, still tragic but indefatigably striving to draw strength from reconceived religious, political, and aesthetic symbols. In context, this quest—really an imaginative flight through many realms of memory, history, myth, fantasy, and hallucination—becomes a terrific affirmation of humanity's spiritual endurance.

At the start, then, *Maximum Security Ward* faces into an intractable wilderness of hostile existence, internalized by the work's sheer nervous alertness. This complex confrontation is reflected in the way that, with the important exception of "Today Is Friday," the opening poems rely so heavily on grotesque farce and macabre fantasy. To return to the initial poem, the bulk of "Elegy for Mélusine" features a talking corpse who explains:

> grant me this: I *tried* to love life—
> tried my damnedest but just couldn't make it.
> Matter of acquired tastes you somehow can't acquire . . .

If this key poem is touching and elegiac in spite of its sardonic and farcical elements, the next poem goes into a wildly Joycean monologue by a demented inquisitor. The inquisitor is identified in the title, "Red-Headed Intern, Taking Notes"—a literal reference to the hospital setting in which the sequence is centered. But the language is so manic that it has clearly been transformed through hallucination into a hilarious parody of hospital routine and an expression of hysterical panic at the same time. "Speakless, can you flex your omohyoid," the "intern" asks, "and whinny ninety-nine?" And:

> No history of zombi-ism in the immediate family?
> And tularemia? No recent intercourse
> with a rabbit?
>   (Lash him firmly to the stretcher
>   and store him in the ghast house for the night.)

The parenthetical image at the end, with its helpless horror and its pun on "guest" and "ghastly," uncovers the grimness beneath the linguistic high-jinks. And then, suddenly, the sequence hurls us—in the next poem, "Today Is Friday"—to the depths of its particular hell of physical and mental suffering. This remarkable poem might well be the report of the "him" lashed to the stretcher and stored in the "ghast house" in that devil-driven parenthesis just quoted. Yet neither poem is exploitatively confessional despite the use of "I" in "Today Is Friday," which begins:

> Always it was going on
> In the white hollow roar
> you could hear it at a hundred paces if you listened closely
> and a hemisphere away if you didn't listen at all
> if you were paying no attention to it
> fixing your mind hard on something else
>   I will not hear it
>   I will not hear it
>   I

We noted earlier that "Today Is Friday" is an exception in the group of opening poems. It is unrelievedly stark, even more so than Donne's "A Nocturnall upon St. Lucie's Day," in which faith holds its own, however minimally, despite the total annihilation of self the poem professes and projects and relentlessly emphasizes. Guthrie's poem is relieved neither by any ultimate ground of religious certainty nor by the zany streak running through Part One as a whole. There is an implicit reference in the title to the Crucifixion, an allusion finely reciprocal with the parenthetical imagery—of a hallucinating patient's sense that he is being *lashed* to a stretcher and will be entombed in a ghostly place—closing the previous poem. "Today Is Friday" is completely given to an affect of extreme agony as an independent force. Its first stanza evokes this sense very purely, in terms of sound summed up in a central image at once concrete and abstract: "the white hollow roar." The evocation through sound-imagery is picked up in the second stanza, which converts it into a language of fused suffering, both physical and mental, such as one would experience under prolonged torture:

> Screaming it inwardly so hard it seemed
> your seminal vesicles must rupture with the strain
> you could hear it close at hand
> feel it crimping your nerve ends
> your brain pan buckling in its grip . . .

The poem presents the process of dying by way of this affective movement, as an assault on all the senses and on the most intimate parts of one's being. It is the most intense and concentratedly powerful poem in the book, and in an earlier draft[2] was placed at the beginning of the sequence. Perhaps Guthrie changed its position because he felt its drastic immediacy would render the rest of the book, with its wide emotional range, at once anticlimactic. Nevertheless, he did place it strategically early on, at a point where it centers attention *almost* at once on an impersonal imagery of sheer distress. Thus, the bizarrely humorous poems surrounding this one are put in the necessary perspective of the ultimate human victim, dying within the mechanical grip and "contracting" of an indifferent, active cosmic process while in the tormented possession of full consciousness:

Tangible
It is a great protracted
totally transparent cube
with sides and angles
perceptibly contracting against
eyeballs and nose and mouth and skin

It is always happening
It is always going on
When it gets tired of going on
maybe it will stop

"Today Is Friday" remains an implacable center of reference in the sequence as a whole, deepening every note of sadness and bitterness appearing elsewhere and putting any counter-notes into harsh, wry perspective. The two poems following it return to aspects of the literal hospital ward. "Via Crucis," for instance, describes its frightening realities, "straight out of Jacques Callot by Hogarth," from the victim-patient's viewpoint. The rude staff nurse is a "cloacal breathed, glad-handing ghoul" who refuses to fetch some water when requested; one hears "guggling / sounds of death," and sees the "new-bloodied bandages" of fellow-patients in the windowless room. In "Cadenza" we have juxtaposed a semi-humorous dialogue (but desperate in this context) between Abraham and a wary sacrificial ram, in the italicized opening stanza, and a small hospital vignette of a kindly night-nurse going off duty and offering to "see if I could do / sumpen about those pillows before I go." The first part is worth quoting as an instance of Guthrie's subtly mordant wit:

2. In the Dartmouth College Library Archives.

"My name is Marsyas,"
says ram
says ram to Abraham
caught in the thicket by the horns,
"My name is Marsyas."
Name of father? "None.
Look, I know an altar when I see one."
Name of mother? Come on, name of mother?
"All right. Hagar.
And don't make believe you never heard of her."

What would ordinarily be whimsical foolery with materials of Biblical and mythical lore becomes a tragic notation concerning a doomed life. The ram that Abraham sacrificed instead of Isaac turns out to have been his other son, Ishmael, whom he begot upon the slave Hagar. And so the neglected and outcast son of the Bible is identified with the Marsyas-figure, genius and ill-fated striver against a god, in whom the poet sees himself mirrored. A Sophoclean tragedy in miniature, but Guthrie leavens it by various virtuoso tricks: the nursery-rhyme beginning (sounding a bit like "Simple Simon and the Pieman"), the parody of official identification forms, and the street-wisdom, as it were, of the ram. The rhymes, exact, partial, and internal, are handled with a surface lightness that belies the way they emphasize an existential pathos and inherited fate.

The sixth poem, "Scene: A Bedside in the Witches' Kitchen," is transitional, beginning in the mood of hallucinatory zaniness we have already observed but then moving into a Poundian incantation (though maintaining Guthrie's own recognizable idiom) to female deities drawn from many mythologies. The incantation, sculptural in its envisioning of the goddesses and sensually alive to them ("cat-flanked Ishtar with the up-turned palms, / Rosmertha of the Gauls, with grief-gouged eyes / and rough-hewn cleft"), grows increasingly more personal and informal in its plea to them to "let me be never born." This poignancy blends quickly into another: a memory of Mélusine "in brambled upland meadows" and of "our walking there, arms pressed to ribs together." This ending opens up the floodgates of memory generally in the two ensuing poems, "Montparnasse . . ." and " 'Side by Side,' " which take us back to scenes a half-century earlier when the poet felt himself fully alive.

These are the first two poems that are openly and explicitly autobiographical in the way they present themselves. Thus, "Montparnasse" begins in a mood of elegiac recollection:

Montparnasse
that I shall never see again, the Montparnasse

> of Joyce and Pound, Stein, Stella Bowen,
> little Zadkine, Giacometti . . . all gone in any case,
> and would I might have died, been buried there.

The poems in this autobiographical mode scattered through the book evoke a range of experience, similar to Guthrie's own, that includes childhood poverty, active participation in both world wars, artistic life in Paris during the 1920s, and in general an active life beyond the academic one he spent at Dartmouth College as professor of French and comparative literature for some thirty-five years. "Montparnasse" contains a vivid vignette of Giacometti in the "dank, littered shed / that was his atelier." " 'Side by Side' " contains a slightly sardonic account of the other times, in wars and out, when the patient-protagonist avoided death. Further along in the sequence, "Fiercer Than Evening Wolves . . ." centers on his mother's wretched death in a charity ward. "There Are Those" contrasts the speaker's theological doubts with the certainties of two nostalgically remembered Irish friends. "Boul' Miche, May 1968" presents a vivid scene from the student riots in Paris and fixes on the instinctive bravery and self-sacrifice of the young. "Polar Bear" is based on a hospital episode, perhaps years earlier. Calmer than "Today Is Friday," it projects a sense of abandonment and freezing isolation through its central image, at once whimsical and touchingly immediate:

> That time coming out from under
> sodium pentathol my first words were,
> "I dreamt I was a polar bear
> that couldn't write poetry."
> Literally but to unhearing ears.
>
> Adrift upon that slab of floe
> under a slate sky
> his conic white
> snout swaying in unison
> with words that never came.

With its sad mingling of waking and dream worlds, of the ordinary reality of "a small deft nurse" and the anguished "white bewilderment" of the baffled bear, this poem goes far beyond Lowell's "Skunk Hour" in rendering a precise psychological state within a volatile atmosphere. Both poems find their tone through humorous anecdotes raised to high emotional pitch and charged with mythical resonance: a polar bear "groping for the tempo of a world / empty of both sense and sound"; skunks that "march on their soles up Main Street: / white stripes, moonstruck eyes' red fire / under the chalk-dry and spar spire / of the Trinitarian Church." However, Guthrie has the magical ability to make use of the autobiographical and confessional without sitting down on his own poem and blocking our view of it. And in general, the poems of autobiographical cast we have been noting serve to evoke a loved, regretted, and unresolved world of memory that is never again going to be experienced. Their affect is comparable to that of the massed memories in Pound's Canto 74, epitomized in the lines:

> we will see those old roads again, question,
>   possibly
> but nothing appears much less likely . . .

Let us say that the work's largest affective energies are those of absolute, intractable horror; of sources of emotional sustenance and also of self-doubting and depression in the past; and of creative will asserting itself, without heroic pretensions, despite the basic despair. It is in this context that the personal elements of *Maximum Security Ward* are deployed with aesthetic impersonality. Guthrie's sequence is perhaps the only triumph of just this sort that has yet been produced, given the fact that the earlier, "classic" sequences make far less use of unconcealed personal experiences and relationships.

We cannot linger over each poem in this deceptively articulated work, although indeed each poem has its place in the whole organic process ("deceptive" because the tone is so open and direct as to suggest at first reading a series of spontaneous, mostly unconnected lyrical or anecdotal forays). If we shift now to the eighteen poems that form the first half of Part Two, we find certain modifications of the stark alienation dominating Part One and expressed uncompromisingly in its penultimate poem, "Loin de Moi," with its pointed reaction to the Vietnam War:

> So now, at what by my watch, if they would give it back to me,
> must be about seven-thirty,
> I will not ask anything of any so-and-soing body
> in the world . . .
>   Certainly
> never to be human . . . The HUMAN RACE!
> No, not even for the laughs. The race of
> napalm Santa Clauses!
> Sheep herded by glib lies that greed concocts,
> he-harpies safely out of sight and sound
> cheerily showering some thousand tons of bombs
> on the innocent helpless to strike back,
> pointless despoilers and defilers of what
> might elsewise be a fairly pleasant world.
>
>   AND YET . . .

The "AND YET" opens the way to a major qualification, which comes in "Don't and Never Did," the opening poem of Part Two:

> All right, could I be Zeus, I still would choose
> rather to be human, to stand beside
> certain humans, even from afar. Or if,
> impervious as was never any god,
> I could be an ultimate grain of sand,
> my choice would stay the same.

At this point in the book, the assertion seems only sentimental despite some acerbic lines that follow. But by the eighteenth poem of the second part, "Not Dawn Yet," the intervening pieces, totaling some 1000 lines, have provided ample substance to give organic body to what is presented as Robert Desnos's dying meditation:

> . . . Having lived through
> to liberty was good. Being a poet,
> being a man was good.
> Did that mean life was good? Had been? Is?
> Allons, mon p'tit! Non, mais tu veux rire!
> But being human, yes, at times . . . at times
> when men are human.

"At times when men are human." The second part of *Maximum Security Ward* gives itself, precisely, to imagining what it is to be human. "Don't and Never Did" strips the contemplation of one's own death of much of the bitterness and mock-heroics of the preceding poems. The mood flattens to matter-of-fact pathos and acceptance:

> Don't shave and rouge and powder me . . .
> Let me look dead and tired and old,
> and no one look on me.

In this light, other poems begin to create a minimal, shoulder-shrugging affirmation of the value of *desiring* to be human even though we are what we are. "Icarus to Eve" concedes, amusedly, that there is something to the idea of evolutionary progress:

> (Without us, legs would still be fins.
> "Johnny! Don't you go too near that land.
> You want to get all dry?")

And "This Stealth" gives the concession a further turn:

> Human I never would have chosen to be,
> yet grant the poor bastard this: his lust, unlost
> for all frustration, to push his way beyond
> whatever he is.

In this group of poems we hear perhaps a remote echo of MacDiarmid's quarrels with himself about the human state, so passionate and crabbed and colloquial, just as Guthrie's addresses to his "gamut of goddesses" may recall MacDiarmid's sudden thrill of awe, at times, at the imagined sight of his "silken leddie" out of faerie and other intimations of dread supernatural presence. In any case, the "poor bastard" of this passage, striving to get beyond natural limits, embodies the best human possibilities we can hope for and is the type of the hero-saint in this work. The title of the next poem, "The Christoi," is the term chosen for such hero-saints, and in subsequent portrait-poems the sense of the real existence of these beings becomes concretely centered in a variety of vignettes—sometimes of heroic bearing, sometimes only partly so. In "The Christoi," the figures of Bach, Cézanne, Proust, Beethoven, Villon, and Rimbaud represent heroic achievement in art; but also numbered among the christoi or anointed ones are an Aurignacian cave artist, certain revolutionists, and, with a bow to Wallace Stevens, the "common man"—

> What did he mean (this insurance executive) by
> "The common man is the common hero?"

"Desnos" begins the series of "christoi" poems proper, which meditate on various models of that desirable, vulnerable humanity to which we should aspire. Here and in other poems Robert Desnos is seen as the prototypical common man and poet whose "life and words caught flame" in the hideous conflagration of World War II, and who died "all poet and hero and all saint." And so the vision of a former "slope-shouldered juggler-with-sounds" became powerful enough to provide the words engraved on the Mémorial de la Deportation, the "staunch anguished prow that cleaves the Seine." Guthrie translates that inscription as "hearts that hate war and beat for Liberty / to the very rhythm of the seasons and the tides / of day and night." Other christoi include Einstein, Burbank, and a failed painter called "Yorick." The danger of falling into moralistic discourse is a severe one in this part of the sequence, and one cannot say that Guthrie has skirted it entirely despite the strong infusion of humor (as in "And It Came to Pass," the Luther Burbank poem) and the intense reminders of the work's complexly desolate emotional center in "Good Friday," "The Surf," "The Dutch Head Nurse,"

"And the Veil of the Temple Was Rent," and "By the Watch." Nevertheless, the main sway of the work's purer tonalities prevails, particularly through the effect of these last-named poems. "Good Friday" and "The Surf" recall the context of the whole sequence in their suggestion of a triple descent into hell: Christ's, Lazarus's, and the poet's. Two were resurrected, but not the spirit at the center of *Maximum Security Ward:*

> My name is . . . My name is . . . hmm . . .
> You! You that were resurrected! Do you remember
> what I said my name was? My name is . . .
> Adam Icarus Marsyas Ishmael Merlin
>
> Mélusine, from this dank, jumbled death-bin,
> I cry out to you knowing well
> no answer ever will come. (Look goddamn it,
> you can write. You know the alphabet, you have hands,
> a sheet of paper, pens, pencils, a typewriter.
> You have written books. Or don't write. Sign your name.
> Or make an X—Mélusine, her sign.)

Among the other affective motifs we have been observing, thwarted love, in the abject, impatient, begging mode of these lines, surfaces repeatedly throughout the sequence. However, a pleasantly sensuous sexuality is kept alive alongside it as well. For instance, in "The Dutch Head Nurse," a vignette of The American Hospital in Paris, "a happy young Lapland nurse" is wistfully recalled:

> She was no reindeer rather a bright-eyed
> flicker-tailed ibex or chamois
> with nimble thighs that only fear of seeming senile
> kept me from stroking.

As one might expect with a satyr-hero like Marsyas, erotic feeling in many nuances pervades quite a few of the poems. It is treated humorously, irreverently, wonderingly, or nonchalantly and heartily, except in the disastrously forlorn yearning for Mélusine—the confessional ingredient that is important and basic and yet kept in balance by the rival tonalities. The stream of tragic feeling in the work as a whole is similarly counterbalanced without losing its centrality. Sexual loss, romantic hopelessness, swell this stream but merge with more universal distress. Thus, in "And the Veil of the Temple Was Rent," the traditional dying words of the christoi's namesake intertwine with the poet's despairing ones: "*Eli, Eli, lama sabachthani*" with "Nothing I have ever done . . .

was worth the doing." Then "By the Watch" recovers two key memories, both involving moments of instinctive violence—the first the killing of a harmless snake, the second the gunning down of German fighter pilots in World War I: "Death was a way that they and I had chosen." And the next two poems, "Death with Pants On" and "For Approximately the Same Reason Why a Man Can't Marry His Widow's Sister . . . ," take us to other war-scenes, in the Spanish Civil War and World War II. But the senseless slaying of the innocent snake evokes a special degree of remorse, understandable only if we realize its association with the murder of all those innocents "helpless to strike back" that arouses such fury in "Loin de Moi."

The closing third of *Maximum Security Ward* clarifies perspectives in the light of the magnificent and humble human beings intermittently celebrated as christoi. One of Guthrie's most successful poems, "Arnaut Daniel (circa 1190)," projects (typically) a moment when the troubadour is convinced that nothing he has done will endure. Monks are too dull to cherish and preserve his highly secular verse, and women too proud and fickle to admit he fathered their children. He shares a sexually oriented delight in things of the earth with Marsyas, who appears at the poem's start to recount his story in its mutually exclusive versions of triumph over and defeat by Apollo, and with the Romanesque sculptors whose work Daniel describes at the end. Death, to Arnaut, is not a matter of whether his poetry will immortalize him but of the loss of all of life's sweetness:

> . . . A couple of years from now
> it will be spring again
> and I shan't be around to see it.
> Willows budding in the meadow, gentians,
> jonquils in the woods, girls twittering
> in patches of sunlight by the river,
> giggling and squealing and hiking up their petticoats
> thigh-high to wade ankle-deep trickles in the field,
> and I not be there. . . .
> What I'll miss is girls stooping by the brook,
> picking cowslips, raising their arms
> to put them in their hair and show the sweet
> profiles shaping out with spring.

Further along in this section, we have various notations of pity and political militancy: identification with the innocently vulnerable; fear of nuclear holocaust, the ultimate betrayal whose sole survivor will be "Capt. Mephistopheles, U.S. Chaplain Corps"; and love for men and women who struggle as best they can against oppression and war. But a half-dozen poems after "Arnaut Daniel" the sequence returns to the realm of deeply inward experience in " 'Visse, Scrisse, Amó' " ("I lived, wrote, loved"—Stendhal's epitaph) and the four succeeding poems that conclude the sequence. In " 'Visse, Scrisse, Amó' " we have another descent into the world of the dead, as in "The Surf." But this time there is a peaceful, intimate communion with them, instead of another last-ditch cry to Mélusine from a "dank, jumbled death-bin." For the dead in this poem are now the christoi, particularly those artists who have made their visions of the world available to us:

> We have lived strangers' lives
> in depths and breadths of worlds they lived
> and died to make.

Thus, the book's early despair at the disparity between life as it is and life as it might be is powerfully counteracted here. In one of the earlier poems of Part Two, "Caduceus," the question had been raised (as an extension of Rimbaud's brilliant self-objectification "JE EST un autre"): "why not / JE SONT / beaucoup d'autres?" There the question seems facetious, part of a "naughty" parody of Stevens and other poets. Here it may be recollected by the reader as an indication of psychological possibility and a major means to mental stability. Having discovered, in " 'Visse, Scrisse, Amó,' " that the self is compounded of many selves, Marsyas confidently repeats his full roster of names in the next poem, " 'And the Evening and the Morning . . .' "[3] He is Marsyas, Merlin, Ishmael, Icarus, and Yorick; and if he cannot place himself in the ranks of the Titans (Rembrandt, the Romanesque sculptors, Einstein), he is nevertheless alongside them as one of the "lovers of liberty, beauty, justice." The rather uneven poem that follows, "Judgment Day," has a happy passage in which the kind of grace this positioning involves is shown:

> My name is Marsyas. I played a flute.
> Forget that silly challenge. I played it best alone,
> sitting on a rock or sprawled on banks of wolf's-foot,
> checkerberries.
> A chipmunk now and then would sit up and listen,
> a rabbit froze, ears flat along its back,
> after a while went on with nibbling.
> A bluejay cocked its head and gave a squawk.
> Once a box-turtle opened up and stretched
> its wattled neck in my direction.

3. Or *almost* the full roster. In "The Surf" he calls himself "Adam" as well.

> Nothing of an Orpheus about me. Not charmed,
> only at length reassured that this beast with
> its different kind of noise
> was as harmless as a nickering horse.

Thanks to the forty-six preceding poems, this simple, lovely scene is exceptionally evocative. Self-disgust and deep alienation are in abeyance here, where Marsyas shares the innocence of the harmless snake killed by one of his partial selves back in "By the Watch." The words "My name is Marsyas" link the passage to the Arnaut Daniel poem, which opened with the same words; the landscape, too, is reminiscent of the other poem, although here we are, obviously, in New England, not Old France. This is one of the few passages in the book with an American locale (except, of course, for the hospital). Perhaps Guthrie is lightly suggesting that he is content to be an American artist, quietly and modestly, of the present moment, without matching himself against the great European dead. At any rate, what comes through very purely is the feeling of simple joy: joy in the moment of rapport with animal innocence and in making one's own music as the birds make theirs, sharing the sounds one makes but with no need for public applause—the kind of self-contained isolation that can only be a good.

From this mood to that of the next-to-last poem, "The Making of the Bear," is a great leap. Both the passage just quoted and "The Making of the Bear" are acts of piety, but the latter is the work's deepest plunge into the primal power and innocence of an artistic vision dedicated to a nameless God, and to all that is mysterious and wonderful in creation. In this most remarkable of the portrait poems, the mind dreams into the sensibility of the paleolithic maker of a masterpiece hidden in the bowels of the earth; and the masterpiece takes on its own life, at once natural and supernatural:

> . . . When he began to breathe,
> I stopped and snuffed the wick, safe in his
> protection, slept. . . .
>
> Heft, strength, the saddle and the soles,
> the rambling appetite, fur, the rolling amble,
> the curious, investigating "Whoof!"
> the clatter of unretracting claws, the bear-play . . .
> the good
> bear-smell of being bears
> are what I had tried to make the flint say
> on the cavern wall.

Arnaut Daniel celebrated a "not too unlovely world." Marsyas found one among the New England checkerberries. Twenty thousand years ago another artist celebrated the mysteries of creation as well, and to him, as to the other two, public recognition was without importance:

> There
> in that total lack of light
> is where my bear is.
> No one will ever see him
> but he still
> is there.

The closing poem, "High Abyss," brings us millennia ahead to the present. We are at a performance of Beethoven's *String Quarter in C# Minor.* This sophisticated work is as much an expression of the religious and the artistic spirit as the cave painting of the bear:

> Delirious order
> of the march of suns and comets.
> In this expanse of tranquil ecstasy that made
> "cold tears of anguish and terror
> seep painfully through [Berlioz's] eyelids,"
> if I say "prayer," I have blown the word
> for ever use again.
> If I say "grief," what of the tremulous exultation,
> the clamorous glee, triumphant resignation?

As it soars into this stratosphere of emotion-laden transcendence, *Maximum Security Ward* becomes a carrier of the same spirit. At the same time, the poem seeks to fend off any impression of a claim to equality with Beethoven. In a passage borrowing its main image from Mark Twain's comic distinction between the right word and the word that is almost right (it is like the difference "between lightning and the lightning bug," he said), the thrilled humility engendered by the music is conveyed:

> I have come back having grasped perhaps as much
> as a lightning bug, clinging through a storm
> to a leaf's underside,
> might understand by fellow-feeling
> of the lightning stroke that in a single blast
> has ripped the elm trunk all its length.

Yet even Beethoven's ineffable achievement depends on men of less than titanic stature and on physical, not ineffable, materials to communicate it:

> Four sweating men are drawing horsehair
> across squills of lamb gut and silver wire—
> and give the resin credit too; it makes the squawk.

In its penultimate stanza, the poem comes as close as possible to naming the artist's nameless God, who has manifested himself as the interaction of a dead composer and living performers and audience by way of self-transcendent art. The listener has been rapt to a "locus beyond space-time continuum" and a "dawn beyond all limit of horizon"—the subjective reality we call religious awe. Its basis is man at his most humanly intense and creative, and yet with all his limitations heavy upon him:

> Four men bow woodenly, file off the stage
> taking their instruments, leaving the scores
> on racks behind them.

"The scores on racks behind them"—the final image of the sequence—are a reality and a metaphor recalling whatever it is the christoi can tangibly leave behind them, even unto the present work, Guthrie's sequence. Each such life-score, then, is the precipitated essence of a world of experience, passion, and "lust" to get beyond oneself and, "at times," be "human." The sequence ends with a moment of poise between a state of aesthetic transport at its purest and all the dark besetting realities that art copes with and even mobilizes into its own self-transcendence.

2. Ted Hughes's *Crow* and Galway Kinnell's *The Book of Nightmares*

Guthrie's *Maximum Security Ward* reflects the absorption into our later poetry of confessional assumptions (the need for a poetry of personal disaster). It confirms the validity of that absorption by bringing it into the orbit of the classic modern sequence without being swamped with exhibitionism in the process. At the same time, it marks the emergence in the 1970s of sequences infused with the horror of war. Not that the modern poetry of war did not exist before such a late date. It would be enough to mention Whitman and Hardy, let alone Owen, Sassoon, Edward Thomas, and many another poet of the two World Wars, to say nothing of Yeats's civil-war sequences and others we have touched on. But a new resurgence of such poetry developed in the wake of developments after World War II, coinciding with the reawakening of revolutionary attitudes relatively dormant after the Depression.

Many factors are involved: the continuing problem of the nuclear bomb; the Korean War; the revelations concerning the Holocaust and the related sense that the living are and will more and more become survivors; and most of all, apparently, for Americans at least, the Vietnam War. That war, to pick up from Guthrie's "Loin de Moi" again, has served as a focal point for all the revulsion against the human race and for much disillusionment and fascination with the essential savagery of existence. Recall Guthrie's lines about "the race of / napalm Santa Clauses" and the

> he-harpies safely out of sight and sound
> cheerily showering some thousand tons of bombs
> on the innocent helpless to strike back,
> pointless despoilers and defilers of what
> might elsewise be a fairly pleasant world.
>
> AND YET . . .

We have already observed the role of "AND YET" in opening the way toward coping with depressive horror in the development of *Maximum Security Ward.* While we can hardly measure the quality of a work by its "positiveness" of attitude, Guthrie's largeness of perspective, his lightly worn erudition, and his special range of sensibility and empathy are another matter. They enable him to keep his own worst insights into the foulness of things from destroying his openness to men and women and realities that are not at all foul—to maintain a humane balance in the midst of the frenzy his poems are sometimes driven by. In this respect it is interesting to compare his sequence with two others published a year after it appeared in 1970: Ted Hughes's *Crow* and Galway Kinnell's *The Book of Nightmares.*[4] These are sequences of a certain power and ambition, but they are limited in their human range and do not always ring true in their tonal emphasis.

Crow is particularly imbalanced, its bold energy called into service to celebrate the blackness that is one extreme of the modern psyche. Perhaps that is overstating the character of the poetry, whose attempt to project the terror of existence has a strongly sensationalist aspect. In one sense it is an almost unconscious reflex of a British generation too young

4. Ted Hughes, *Crow* (New York: Harper & Row; London: Faber & Faber, 1971; Galway Kinnell, *The Book of Nightmares* (Boston: Houghton Mifflin, 1971).

to have taken part in combat yet highly aware of others' war experience and of the whole tragic pattern of modern European history. Hughes, like his wife, Sylvia Plath, arrogates to himself the disillusionment and burnt-out sense of history that we associate with Nazi Germany's victims and with East European experience. But whereas she assimilates all this into the sense of a free-floating demonic, suicidal, unbearably sensitive intelligence, Hughes goes out for a brutal verbal kill: an experiment in symbolic sadism and (perhaps) self-laceration. Sexually charged, technologically inspired murderousness is the major pressure on sensibility in this sequence of 66 poems. In "Crow's Account of the Battle," for instance, we have nothing like the human immediacy of a Wilfred Owen (instinctively responded to by Guthrie), but rather a sort of metaphysics concretized in images of war's atrocity and yet left hideously impersonal:

> The cartridges were banging off, as planned,
> The fingers were keeping things going
> According to excitement and orders,
> The unhurt eyes were full of deadliness.
> The bullets pursued their courses
> Through clods of stone, earth and skin,
> Through intestines, pocket-books, brains, hair, teeth
> According to Universal laws
> And mouths cried "Mamma"
> From sudden traps of calculus,
> Theorems wrenched men in two,
> Shock-severed eyes watched blood
> Squandering as from a drain-pipe
> Into the blanks between the stars. . . .
> Reality was giving its lesson,
> Its mishmash of scripture and physics,
> With here, brains in hands, for example,
> And there, legs in a treetop.

Compare this surrealist improvisation, put together as for a jigsaw-puzzle—and for shock-value but without precise feeling—with the second and third stanzas of Owen's "Dulce et Decorum Est":

> Gas! GAS! Quick, boys!—An ecstasy of fumbling,
> Fitting the clumsy helmets just in time;
> But someone still was yelling out and stumbling
> And flound'ring like a man in fire or lime . . .
> Dim, through the misty panes and thick green light,
> As under a green sea, I saw him drowning.

> In all my dreams, before my helpless sight,
> He plunges at me, guttering, choking, drowning.[5]

How far from this compassionately sensuous evocation, or from Whitman's evocation of the sea-battle and its aftermath in sections 35 and 36 of *Song of Myself*, is Hughes's "reality-lesson" here. The denizens of his mechanized world are fragments of bodies grotesquely scattered in a mad vision of a society gone completely out of touch with the individual human meaning of a life. This is a sort of totalitarian poetry, in the sense that it attempts to convey total horror and to imply that the cost of recognizing it is total madness: terror, guilt, remorse, and uncontrollable disgust. The empathy of a Whitman or an Owen is almost sentimental self-indulgence in such a context, as if one needs to harden oneself against over-response to suffering in a universe whose only law is cruelty. In this universe existence itself is a crime. Birth, nursing, play are but means of draining life from others; love-making is haunted by war's bestialities and anaesthetizes instead of revitalizing. "Criminal Ballad" tells us:

> And when he clasped his first love belly to belly
> The yellow woman started to bellow
> On the floor, and the husband stared
> Through an anaesthetized mask
> And felt the cardboard of his body

We may find a remote echo here of Owen's "Greater Love" ("Red lips are not so red / As the stained stones kissed by the English dead"), which offers as a shocking paradox a displacement of the intensity of love-desire by that of the agony of the wounded and dismembered on the battlefield. But in Hughes's poems the shock itself is past; the transference of intensity has not only taken place long ago but been so prolonged that feeling has been neutralized in the process. In the world of "Criminal Ballad" ordinary family life is hopelessly impossible, and the "silly songs and the barking" of those innocents, dogs and children, are drowned by machine-gun fire and the cries of the tortured and the insane ("a screaming and laughing in the cell"). The scene is infernal: "the woman of complete pain rolling in flame / Was calling to him all the time / From the empty goldfish pond." The "him" referred to is normal man—guilty, as Yeats would have put it, of "the crime of birth." But here the source of guilt is neither original sin nor existential shame but

5. *The Collected Poems of Wilfred Owen* (London: Chatto & Windus, 1963), p. 55.

the heritage of war as the condition of human existence, with the added element of our modern weaponry and mass-warfare. The dehumanization of whole societies is implied, in an atmosphere of pure hysteria:

> And when he began to shout to defend his hearing
> And shake his vision to splinters
> His hands covered with blood suddenly
> And now he ran from the children and ran through the house
> Holding his bloody hands clear of everything
> And ran along the road and into the wood
> And under the leaves he sat weeping
>
> And under the leaves he sat weeping
>
> Till he began to laugh

Crow as a whole oscillates between projections of the darkness at the heart of creation (the first nine poems, in fact, are a dark parody of Genesis) and suggestions of the essential human spirit, faint, distorted, or broken as it may be. The image of Crow, the monstrous bird dominating virtually every poem, serves as a magnetic center for epic humor and self-assertion, for vulnerability in the unequal contest in nature (rather than in the mythical realm) between bird and man, and for the amorality natural in beasts but not, supposedly, in man. One's general impression of Crow is of a grotesque figure in an animated cartoon, but one that can alter from poem to poem in almost any aspect of meaning or behavior, and can be either a grim (or comically awkward *and* grim) predator or something like the burden of metaphysical self-consciousness by which man questions or denies his own validity. Thus, "Crow Tyrannosaurus" ends neatly with an allegory of man the "trap-sprung" slaughterer losing all essential human feeling (hard-won at best):

> Crow thought "Alas
> Alas ought I
> To stop eating
> And try to become the light?"
>
> But his eye saw a grub. And his head, trapsprung, stabbed.
> And he listened
> And he heard
> Weeping
>
> Grubs grubs He stabbed he stabbed
> Weeping
> Weeping
>
> Weeping he walked and stabbed

Thus came the eye's
roundness
the ear's
deafness.

One sees here the counter-effort to imitate Blake's sense of universal pathos; and perhaps there is an imitation of Blake's surface toughness as well in the obvious irony. The poem shares a dangerous ambiguity with certain other key-poems in the sequence that we shall speak of shortly. As a whole, *Crow* lacks the qualitative movement of a true lyric sequence. It tends, rather, to be an anthology of fragments from an imaginary narrative tradition concerning what the subtitle calls "the Life and Songs of the Crow." The suggestion is of material still so much in the oral tradition that its events and personalities have not yet been fixed into any one consistent mold. Combined with this sophisticated mimicry of a primitive mythology in process of formation, there is a structuring by way of massing sets of attitudes related to characteristic concerns of modern sensibility. The first of these is a conviction of the arbitrariness of philosophical and religious or theological systems. The second has to do with the disastrous laws of physical and instinctual nature—a post-Spenglerian, post-Freudian version of Original Sin, encouraged by the darkest aspects of history since World War I and the blows dealt our faith in human goodness. At the center of the horror, somehow, lie the contradictions of sexuality and love, as suggested in the contradictory imagery of "Fragment of an Ancient Tablet": "Above—her brow, the notable casket of gems. / Below—the belly with its blood-knot." To the disasters of our evolution and our psychic makeup must be added that created by language, our greatest invention, which has led to the deceptions and antihuman impersonality of abstract thought. The emotional contexts in which these attitudes and philosophical bearings are presented are primarily irony, guilt, violent hilarity, and, very occasionally, a slightly suspect tender pity. The pity is suspect because close to something like relish in the overwhelming human predicament of so much suffering and defeat (to say nothing of death) within which our insect-tiny possibilities contrive to survive: "littleblood . . . sucking death's mouldy tits."

The poems of *Crow* fall into seven main groups, with a certain amount of overlapping. The first six poems present a "Crow-gony," or cosmogony centered on the main symbol, that echoes Genesis and primitive creation-incantations. The next two focus on the sexual predicament; and then poems 9–16 turn to the terrors of history—prolonged, it is implied, by the modern wars, the Holocaust, and the coming nuclear Armageddon. Poems 17–26 are survival-poems, as if the sequence had emerged from the rubble of the destructive history embodied in all that has so far been presented. This group has a strong psychological and theological orientation, and introduces as man's post-original sin his self-betrayal, by language, into impersonal abstractness: the subordination to powerful words of the instinct to make a meaning of an individual life. This is the subjective dimension of total war and the triumph of the State conceived as a mindless instrument of mass-murder. A large block of poems, the next sixteen, now explores the sense of guilt associated with this betrayal and with our subjection to the tyranny of the word—conceptions not unconnected with Freud's in *Civilization and Its Discontents*—in terms both humanly personal (especially in poems 27–33) and archetypal (especially in poems 34–42). Poems 43–50 provide lyrical, metaphysical, and mythopoeic afterbeats; poems 51–68 face into the ultimate bleakness with a hard laughter (shades of Heraclitus, Nietzsche, Yeats, and Brecht); and the final sixteen poems provide a sort of reprise of the points of contact with the stark disaster of existence, the faint notes of possibility and sweetness that persist, and the complexity and mixed vulnerability and savagery of the human condition.

Clearly, the sequence suffers from its overloading, occasioned by a desire to hammer in the allegorical implications of its Laurentian primitivism. The terms of such primitivism have long been familiar, and the effort here is to make it new by intensifying the implicit savagery and fatalism, partly by a redundancy that echoes history itself. The result is a kind of rhetoric of clamoring insistence, in which sheer bulk of materials is deployed to press the issue. The idea is to hit out with main strength from the very beginning and then keep going, with tonal variations along the way to break the monotony of a single blackly ironic mood and to develop the nuances and counter-notes. But oddly enough, the overridingly drastic pounding of the initial poems blocks off the possibility of anything like a gradual opening into subjective realization of the horror. This is odd because, in a *shorter* sequence, Hughes's method would work better. Rapidly developing juxtapositions and contrasts within, as it were, *hearing distance* of one another, would almost immediately relate the initial impact to opposite or mediating or quite independent affects. A quick succession of tonal shifts, and their interaction, create in the classic sequences a sense of gradual realization because of the modulations, quiet echoings, and delayed connections that make themselves felt almost subliminally at first.

Although *Crow* does not achieve this sort of structure, it does approach it in its own way. Underneath the curiously rhetorical and mythical-historical ordering that culminates in a mass of poems presenting modern guilt and chaos and finally attempting an emotively ordered reprise, we do have something like lyrical movement at points along the way. The first poem, "Two Legends," implies the character of this almost buried inner movement. The first "legend" revises primitive creation myths to suggest the birth of a death-fraught being that is entirely "black" within and without: the oversaturation and almost polemical insistence here set the basic key of feeling for the whole book:

> Black was the without eye
> Black the within tongue
> Black was the heart
> Black the liver, black the lungs
> Unable to suck in light
> Black the blood in its loud tunnel
> Black the bowels packed in furnace
> Black too the muscles
> Striving to pull out into the light
> Black the nerves, black the brain
> With its tombed visions
> Black also the soul, the huge stammer
> Of the cry that, swelling, could not
> Pronounce its sun.

The second "legend" then, without yielding a bit of the dominating "black" vision, nevertheless introduces images of action, effort, color (but oxymoronically, as in Milton's "darkness visible"), and even (again, in purely negative terms) transcendence:

> Black is the wet otter's head, lifted.
> Black is the rock, plunging in foam.
> Black is the gall lying on the bed of the blood.
>
> Black is the earth-globe, one inch under,
> An egg of blackness
> Where sun and moon alternate their weathers
>
> To hatch a crow, a black rainbow
> Bent in emptiness
>   over emptiness
>
> But flying

So the urge toward transcendence is established through an energy of language here, a shadow-language of anti-triumphant triumph, an "empty" promise that yet remains a promise. Just before the end of the sequence, the "two legends" of the opening poem are balanced off by "Two Eskimo Songs." These both repeat the tragically conceived cosmogony of the opening group of poems, though in altered terms. They are cruel songs: one about the origin of sexual love and pain and of song itself in the knowledge of death; the other about the desire of "water" to "live," and its discouragement because of the misery of all that has already achieved life and because of the alien and forbidding nature of the rest of reality. After "Two Eskimo Songs" the book ends with "Littleblood," to which we have already referred:

> O littleblood, hiding from the mountains in the mountains.
> Wounded by stars and leaking shadow
> Eating the medical earth.
>
> O littleblood, little boneless little skinless
> Ploughing with a linnet's carcase
> Reaping the wind and threshing the stones.
>
> O littleblood, drumming in a cow's skull
> Dancing with a gnat's feet
> With an elephant's nose with a crocodile's tail.
>
> Grown so wise grown so terrible
> Sucking death's mouldy tits.
>
> Sit on my finger, sing in my ear, O littleblood.

We have hardly advanced here beyond the "crow, a black rainbow / Bent in emptiness / over emptiness // But flying" at the close of "Two Legends." Out of the deep centuries has come the tiniest mite of surviving creative energy, wounded and contaminated by all it has passed through, bearing the memory of all forms of life in the long evolutionary process: sole hope, and how limited, for the future! This is elegiac poetry presented as celebration, and in the wake of the previous poem in which we see "water," after it has given up its effort to achieve life—and given up its useless weeping at last as well—lying "at the bottom of all things // Utterly worn out   utterly clear." This poem, second of the "Two Eskimo Songs," is called "How Water Began To Play," with the apparent suggestion that the pure play of the mind, whether in art or otherwise, begins with abnegation of the effort to overcome the unmanageable horror of fatality. But we need not read such literal interpretation into the poem, despite the title's obvious invitation to do something of the sort; we may simply go along with the feeling of an effort abandoned, and a residue of exhaustion and clarification through letting-go left behind, with whatever the further connotations of "play" may be in this context. ("Littleblood," of course, with its serious whimsy and dwindled vision of song after cosmic loss and death, seems reciprocal with "How Water Began To Play.")

At certain points, as in the two poems just discussed, the indispensable personal note enters the sequence and quickens it with an emotional idiom beyond the rasping violence and ironies that seek to conquer by main force. Among further examples, we would especially cite "Owl's Song," "Crow's Undersong," "Dawn's Rose," "Glimpse," and, most particularly, "That Moment." In the first of these Hughes gives a strong indication that, in substituting a singing owl for his usual "Crow" (but without actually naming the "He" of the poem except in the title), he is "really" talking about himself:

> He sang
> How the swan blanched forever
> How the wolf threw away its telltale heart . . .
>
> He sang
> How everything had nothing more to lose
>
> Then sat still with fear . . .

"Dawn's Rose," though it offers crow-images ("a crow talking to stony skylines," "Desolate is the crow's puckered cry"), does so in lower-case and without any sense of an immense cartoon-bird of horrendously symbolic import. In this poem those images are only elements of a heavily painful imagery of dawn that brings reminders of our weakness and woe rather than joy. "Crow's Undersong" speaks to a private memory and vision, of the appearance of the female principle as the timid embodiment of a dream of love and generation and hope that has brought humanity as far as it could come despite the impossibility of ultimate victory. "She cannot come all the way," says the poem at the start; moreover, to complicate the sad relationship between human hopes and human defeat,

> If there had been no hope she would not have come
>
> And there would have been no crying in the city
>
> (There would have been no city)

More piercing than these poems in their impact, however, are "Glimpse" and "That Moment." The very brevity of the former poem makes it crowd its bitter vision into view without benefit of explanation:

> "O leaves," Crow sang, trembling, "O leaves—"
>
> The touch of a leaf's edge at his throat
> Guillotined further comment.

Nevertheless
Speechless he continued to stare at the leaves
Through the god's head instantly substituted.

How quickly here we are given, first, the desire to be in touch with all that's beautiful, alive, and growing; then, the stark reminder of the dangerousness of reality; and finally the confusion of the ideal and the terror-ridden that is inseparable from the contemplation of the real world, a contemplation magnetically irresistible. In this poem we see almost the whole dilemma that *Crow* confronts.

Not quite the whole of it, though. As we noted earlier on in comparing this sequence with *Maximum Security Ward,* the humane empathy and emphasis that Guthrie refracts throughout his poems, for the most part avoiding the trap of sentimentality, is largely absent from *Crow.* The closest Hughes comes to it is in the interesting poem "That Moment," with its surrealist picture of the final human moment after the last, all-destructive war, when life will begin again only on the grossest, most desperate terms:

When the pistol muzzle oozing blue vapor
Was lifted away
Like a cigarette lifted from an ashtray

And the only face left in the world
Lay broken
Between hands that relaxed, being too late

And the trees closed forever
And the streets closed forever

And the body lay on the gravel
Of the abandoned world
Among abandoned utilities
Exposed to infinity forever

Crow had to start searching for something to eat.

The poem is one of the few in the sequence, together with "Crow's Undersong" and some others, that risk directly human images which project tragic predicament in a world of mere survival. Its reverberations carry far in the sequence, where it appears quite early (the tenth poem). But its connections with other poems are indicative without being decisive. The distribution of the more intensively lyrical notations in *Crow* is, in general, almost haphazard rather than forming a pure stream on its own throughout the work and disciplining the rest of the structuring thereby.

Galway Kinnell's *The Book of Nightmares* is a much more closely knit work than *Crow*. Composed of ten poems, each in seven sections, it is dominated by a meditating sensibility that appears completely consistent and localized in time and place. Personal reminiscence looms large in this sequence as a major ordering device. Although a variety of locales are recalled, the speaker is presented most frequently in three of them only: in the mountains, at home (in bed with his wife or comforting his infant daughter Maud), or in the swaybacked bed of a disastrously haunted room in the Xvarna Hotel, the "Hotel of Lost Light." The nightmares of the title are strongest in this last place, in the bed where a drunk has died and death seems hideously close. Unlike *Crow*, which offers many moments of power, however ugly, Kinnell's book consistently projects a sensibility far more vulnerable than Hughes's. There is nothing in it to compare with the affirmation, in "Examination at the Womb-Door," that Crow is for the moment "stronger than death" despite the fact that "*Death* . . . owns the whole rainy, stony earth."

Kinnell is closer to the Hughes of "Criminal Ballad" or "Crow's Account of the Battle," poetry of terror and loss, but his feeling is more immediate. His sixth poem, "The Dead Shall Be Raised Incorruptible," is the key to the intense malaise with which the sequence copes. As compared, for instance, with the keen but distanced grief and pathos of Hughes's "That Moment," this poem is marked by more direct preoccupation with the literal scene and outrage of war:

> A piece of flesh gives off
> smoke in the field—
>
> carrion,
> caput mortuum,
> orts,
> pelf,
> fenks,
> sordes,
> gurry dumped from hospital trashcans:
>
> *Lieutenant!*
> *This corpse will not stop burning!*

The war is specifically the Vietnam War. The mad speaker of the second part tells of obsessively shooting down the "little black pajamas jumping / and falling," whether or not they are "friendlies." The third section satirizes a television commercial implying that the human body and its odors and imperfections are disgusting. In context, the implication is that such disgust enables us to ignore the insult and destruction we visit upon bodies in war; and the last line of this brief section, "*We shall not all sleep, but we shall all be changed,*" connects this "sanitary" turning away from the body's realities with Christian denial of the actuality of physical death. Then the fourth section shifts the key of satire to that of François Villon, whose "Testament" is imitated in very modern terms that indict modern Christianity for the behavior of its professed communicants: a "trespass on earth" that extends from genocide to the possible extermination of the whole human race. (As one of Villon's best translators, Kinnell catches the wry and biting mockery of that great master if not his concentrated lyrical brilliance.) The fifth section narrows down to imagery of a rotting body in a ditch, but the final two sections press toward a balance—not cheerful but made harshly insistent—between a morbidity imposed by the pain and mutilation of the flesh and the persistence of spirit. Part 6 presents a gruesome scene, almost as if in a fantasy of Hell, of a soldier on fire, with his neck broken, running and "thinking" with a mad elation:

> *I ran*
> *my neck broken I ran*
> *holding my head up with both hands I ran*
> *thinking the flames*
> *the flames may burn the oboe*
> *but listen buddy boy they can't touch the notes!*

This is mortal flesh *in extremis* yet beyond itself, a blended realistic and visionary scene. The seventh section preserves some of the feeling but dissipates itself by introducing a variety of images and literary echoes to suggest the broader range of reference of the sequence as a whole. (Kinnell's symmetrical pattern of section divisions forces him, here as elsewhere, to contrive a purely external balance with corresponding sections in other poems.)

This sixth poem is of primary importance in controlling the book's essentially confessional tendency. It provides an emotional center rooted in distress and anger and shared pain and yet having no direct relationship to the poet's private life. So too the third poem, "The Shoes of Wandering," and the fifth, "In the Hotel of Lost Light," focus on the human predicament from vantage points outside the usual circumstances of his experience. In the former poem, he has entered the world of the neglected and the displaced, wearing used shoes from the Salvation Army ("shoes strangers have died from," "steppingstones / of someone else's wandering") and sleeping in the Xvarna Hotel:

> I draw the one,
> lightning-tracked blind

in the narrow room under the freeway, I put off
the shoes, set them
side by side
by the beside, curl
up on bedclothes gone stiff
from love-acid, night-sweat, gnash-dust
of tooth, and lapse back
into darkness.

The fifth poem returns to this scene, beginning:

In the left-
hand sag the drunk smelling of autopsies
died in, my body slumped out
into the shape of his, I watch, as he
must have watched, a fly
tangled in mouth-glue, whining his wings . . .

and ending:

The foregoing scribed down
in March, of the year Seventy,
on my sixteen-thousandth night of war and madness,
in the Hotel of Lost Light, under the freeway
which roams out into the dark
of the moon, in the absolute spell
of departure, and by the light
from the joined hemispheres of the spider's eyes.

Poems III, V, and VI focus on war and political vileness and on lives of extreme poverty and degradation (lives like flies trapped in spider-webs). These are placed in contrast with the closeups of the births of the poet's children, his tender feeling for them, and his marital and other love-relationships in "Under the Maud Moon" (Poem I), "Dear Stranger Extant in Memory by the Blue Juniata" (Poem IV), "Little Sleep's-Head Sprouting Hair in the Moonlight" (Poem VII), "The Call across the Valley of Not-Knowing" (Poem VIII), and "Lastness" (Poem X). The descriptions of childbirth are at once entranced and healthy minded. They stand in absolute contrast to the squeamishness and self-deceptive, uncaring pseudo-spirituality that prevent so many of us from looking directly at suffering in and out of war. Were the three outward-reaching poems omitted from the sequence, however, the work would strike us as a sometimes touching and mystically alert act of self-absorption, with an eccentric supernaturalist twist. The intrusion of "sacred" phrasing that transfers to the protagonist's every act, the sequence's every notation, an aura of Holy Writ, begins at the start of the first poem, "Under the Maud Moon":

> On the path,
> by this wet site
> of old fires—
> black ashes, black stones, where tramps
> must have squatted down,
> gnawing on stream water,
> unhouseling themselves on cursed bread,
> failing to get warm at twigfire—
>
> I stop,
> gather wet wood,
> cut dry shavings, and for her,
> whose face
> I held in my hands
> a few hours, whom I gave back
> only to keep holding the space where she was,
>
> I light
> a small fire in the rain.

This speaker is his own Holy Land, worshiping the sacred soil of wherever he happens to be. Despite the shorter lines, small echoes of Whitman play about the rhythm and phrasing, combined oddly with an Olson-like syntax in the second stanza—another sort of echo that is emphasized in the third section of this poem with its parallel "who"-constructions:

> And she who is born,
> she who sings and cries,
> she who begins the passage . . .

The "she" is the baby Maud, whose birth is celebrated, literally described, and sanctified in the ensuing three sections—e.g.:

> and she skids out on her face into light,
> this peck
> of stunned flesh
> clotted with celestial cheesiness, glowing
> with the astral violet
> of the underlife. . . .

"Celestial cheesiness" is a welcome bit of humor—one hopes with a pinch of intentional self-irony in it as well. For the exalted sentimentality continues to the very end of this opening poem, served by an occasionally glittering lyricism that places father and daughter on a cosmically holy stage within the vast, ordinarily indifferent cold reaches of time and space:

> For when the Maud moon
> glimmered in those first nights,
> and the Archer lay
> sucking the icy biestings of the cosmos,
> in his crib of stars . . .

The poem takes a self-elegiac turn at the very end: a prayer that, after the deaths of her parents, Maud will hear the "spectral" voice of her father calling to her from this book and from the whole past ("everything that dies"),

> And then
> you shall open
> this book, even if it is the book of nightmares.

"Under the Maud Moon" is echoed in the closing poem, "Lastness," which celebrates the birth of Maud's younger brother Fergus similarly ("when he was born . . . / . . . he opened / his eyes; his head out there all alone / in the room" amid "the ninth-month's / blood splashing beneath him"). Its self-elegiac dimension is echoed, too, in the seventh poem, "Little Sleep's-Head Sprouting Hair in the Moonlight," in a passage that begins like a popular song:

> If one day it happens
> you find yourself with someone you love
> in a café at one end
> of the Pont Mirabeau, at the zinc bar

From the standpoint of its confessional aspect, the sequence culminates in the eighth poem, "The Call across the Valley of Not-Knowing." Despite the symmetry of the opening and closing birth-poems, their sentimentalities and mystical pretensions make for a more contrived than convincing affirmation: wistful and with a certain charm and ambition to command the universe by casting a little spell over it. Poem VIII is more adult; it confronts the intractable in the form of two difficult realms of experience. One is an unresolved love-relationship, begun and encouraged despite the knowledge from the start that life's "necessities" would not permit the protagonist to sustain it—in other words, a relationship fraught with passion, self-indulgence, and early betrayal. The other is the illusion, illustrated here by the vicious sheriff whose hand held the protagonist's as if lovingly while he fingerprinted him, of an affectionate reaching-out by all that's inhuman and impersonal.

But Kinnell does not permit the drastically negative to prevail in any one poem. Intermixed with these contexts, which parallel those out of which Guthrie and Hughes write, are scenes of earthly *and* spiritual marital reciprocity that make a special claim. In them a certain ease is announced: the authority of a richly prolific life attuned to all that is good and desirable despite the negative and intractable aspects of existence. The rhetoric here ("and the genitals sent out wave after wave of holy desire / until even the dead brain cells / surged and fell in god-like, androgynous fantasies") demands great tolerance, as do the sections elsewhere in the sequence that seek to evoke a sense of divine forces working in the visceral, unconscious life of animals (and babies, and loving folk).

The sequences of Guthrie, Hughes, and Kinnell share a need to sum up the driving terror underlying reality—the terror revealed most crushingly in the man-made world of war. They also share the counterneed to resist the irresistible and to contrive ways of reversing the terror. The situation of *Maximum Security Ward* makes this complex perspective seem more natural than do the more forced horrific and benign visions of *Crow* and *The Book of Nightmares*, and the subjective range and depth—the copiousness—of its associative memory, imagination, and open humanity of feeling allow it the freest affective play while keeping the varied streams of its movement in order.

3. Psychological Pressure and the Continuing Main Stream (H. D., Dylan Thomas, Laura Riding, Muriel Rukeyser, Adrienne Rich, and Others)

Poetry that mingles the confessional, the political, and the meditative has a long, protean history—one can trace it back to *The Prelude* and earlier. Even if one started with *The Prelude*, the case would be very strong for emphasizing the importance of psychological pressure in such a work's structure: its dependence on key moments of experience, each with its own complex affect (the boat-stealing scene and its depressive aftermath, the ecstatic ice-skating scene, the strange suddenness of the long-expected yet suddenly accomplished achievement of a goal in the "Simplon Pass" section, the encounter with the blind beggar on a crowded London street, etc.). If Wordsworth were writing in our century, there seems little doubt that *The Prelude* would be compressed into one or more sequences whose separate units were centered on these moments and juxtaposed without their narrative and discursive links. That is, the finished work would be closer to the separate units out of which it was forged in the first place. And it can be argued that many a work since then—take as one instance, arbitrarily, Melville's "After the Pleasure Party"—embodies a poetic struggle to bring the pressure of explosive psychological forces (in this instance, sexual ones) into the foreground.

The whole direction of modern poetry, with its cultivation of intense subjectivity and radiant centers of affect, has been toward becoming a poetry of psychological pressure and its attempted encompassment. We have examined a fair number of masterpieces and works of lesser but genuine interest in observing the uneven progress and many dimensions of this evolution. It is useful, as well, to see how certain writers pick up, and take for granted, methods of the sequence developed by the masters—Olson, for instance, in the wake of Pound and Williams, or Sexton and Snodgrass in Lowell's. And it is useful to recall that the growth of the sequence has been fed by so many works of quality that a huge volume—an encyclopaedia—would be needed to take proper note of them all.

Thus, one has only to reach back almost blindly to touch sequences and near-sequences written earlier in the century: efforts over which we might well have lingered. Such examples would include H. D.'s first volume, *Sea Garden* (1916), and works of the 1920s and 1930's like Laura Riding's "Forgotten Girlhood," "To a Loveless Lover," "Echoes," and "Rhythms of Love," and Dylan Thomas's "Altarwise by Owl-Light."[6] All these poets mined veins of some importance, yet were too idiosyncratic or elusive, in the main, for others to follow readily.

Of the works just mentioned, H. D.'s *Sea Garden* presents the most purely lyrical and passionately immediate modulation toward a sequence. Its preoccupation with a difficult beauty wrung out of terror, loss, and harsh adversity summons up a long tradition of Romantic desolation with a heroic twist, and is connected with comparable emotional stances in Pound and Williams. Such pieces as "Sea Rose," "Evening," "Loss," and "Pear Tree," some of them important exemplars of Imagist poetics, are in the vanguard of their moment. At the same time, they are touchingly backward-looking: toward the nineties and the pre-Raphaelites, and toward Dickinson in certain moods associated with the Brownings and even with Meredith. "Cities," the concluding poem of *Sea Garden*, brings the volume into the arena of social criticism; here

6. See H. D. (Hilda Doolittle), *Collected Poems* (New York: New Directions, 1953); *The Poems of Laura Riding* (New York: Persea Books, 1980); and Dylan Thomas, *The Poems*, ed. Daniel Jones (London: Dent, 1971).

H. D. transfers her bruising effort to sustain her vision of beauty amid every violating pressure into a strained sense of ghostly survival. The poem may be considered a precursor of *The Waste Land*, but no more so than truly powerful earlier poems by Blake, Wordsworth, and others. Like the second poem of the volume—the beautiful, ambiguous "The Helmsman"—"Cities" points the way to H. D.'s hermetic, mystical later sequences and suggests an impenetrable, solipsistic private realm of ultimate emotional resort.[7]

Dylan Thomas's sequence of ten riddling, cosmically symbolic sonnets, which according to Vernon Watkins he intended to expand into a much larger work,[8] is one of the sports of our poetry. Like so much of his other writing, it fuses Rabelaisian exuberance, obsession with the physiology of sex and procreation, Joycean word-play, and mesmerization by the death-bound journey of individual lives. There is no surface confessionalism in these sonnets—except in the farfetched sense of their projection both of bawdy vitality and of personal pity and grief for general doom. There is nothing like Lowell's easy anecdotal slouch at the start of a poem, imitable by all and sundry (the sound and general pitch of ordinary speech, quite apart from what is done with them, poetically, in Lowell's best work), in a sequence that begins:

> Altarwise by owl-light in the half-way house
> The gentleman lay graveward with his furies . . .

There is, though, everything here to delight, intrigue, and puzzle us into becoming attuned to the metaphoric fling of associative and emotional movement that follows. When, a few lines further on in this first sonnet that puts Genesis and the genitals into manic alignment (as indeed the Biblical tale of Eden does more fearsomely), we come to the lines "Old cock from nowheres and the heaven's egg, / With bones unbuttoned to the half-way winds," we do not think: "Dylan's being an exhibitionist again." He *is*, of course, but meanwhile the metaphors have been carrying us to cosmic and mythical realms, both Christian and pagan (to say nothing of Freudian)—so that the sexual act becomes a vast allegorical panorama on the order of certain late medieval and early Renaissance paintings. So does its opposite, death; and so the second sonnet begins: "Death is all metaphors, shape in one history"—and then goes on to illustrate with a plethora of metaphors for the life-in-death and death-in-life paradox that is our condition: "hemlock-headed in the wood of weathers," as the closing line has it.

7. See H. D., *Trilogy* and *Hermetic Definition* (Cheshire: Carcanet Press, 1973 and Oxford: Carcanet Press, 1972 respectively).

8. Dylan Thomas, *Letters to Vernon Watkins*, ed. Vernon Watkins (New York: New Directions, 1957), pp. 13–14.

Thomas's sonnets pile much of the burden of this paradoxical insight onto an archetypal, many-faceted universal Crucifixion imagery. Sheer fatality and its instrument, our great enemy time, are what our sexual energy heroically contends with in life's epic struggle to persist. As the sonnets advance, their marvelous metaphors shooting tragic rockets of defiant or wondering or boastful or misery-laden vision against the pressing conviction of life's pointlessness, the realizations become more painful, the assertions more insistent. Sonnet VIII concentrates the sense of life as crisis, and of desperate will acting on behalf of human meaning (the Crucifixion motif—that is, the central sensibility, as in *Song of Myself,* re-enacting the mission of Christ), into an epitome of the whole sequence:

> This was the crucifixion on the mountain,
> Time's nerve in vinegar, the gallow grave
> As tarred with blood as the bright thorns I wept;
> The world's my wound, God's Mary in her grief,
> Bent like three trees and bird-papped through her shift,
> With pins for teardrops is the long wound's woman.
> This was the sky, Jack Christ, each minstrel angle
> Drove in the heaven-driven of the nails
> Till the three-coloured rainbow from my nipples
> From pole to pole leapt round the snail-waked world.
> I by the tree of thieves, all glory's sawbones,
> Unsex the skeleton this mountain minute,
> And by this blowclock witness of the sun
> Suffer the heaven's children through my heartbeat.

Despite the racketing forward energy of the lines, so many of them unbroken and enjambed as well, and of the highly active metaphors, the affect here is of failed effort, suffering, and pity in the face of all our sexual and spirtual energies. In Sonnet IX, which ends with the line "And rivers of the dead around my neck," the heaviness of the negative is almost overwhelming; and the closing sonnet is at first a whimsical apology for a hopeless effort by "the tale's sailor from a Christian voyage" and then a prayer that mortal human dreams of a recovered Eden prevail after all beyond the serpent-venom's murderous power. We can see in these final lines how much both Lowell and Stevens learned from Thomas—the one poet from the driving rhythm that heaps up their metaphors so passionately, the other from their emotionally hovering, brooding paradoxicality (lessons of modifiable technique, however, rather than of idiosyncratic style):

> Green as beginning, let the garden diving
> Soar, with its two bark towers, to that Day

When the worm builds with the gold straws of venom
My nest of mercies in the rude, red tree.

It seems unlikely that Thomas could really have expanded this sequence very much without falling into hopeless redundancy, not only because of repeating the same basic verse-form but also because the work's intensity is already threatened by the need for more and more metaphoric ingenuity. His ten densely packed sonnets seem more than enough to contain a single curve of affective movement, in which complex contradictory elements are at work, without dissipating its major force. In Thomas's case, his very ingenuity and copiousness would soon have militated against the affective accumulation he achieves here. Needless to say, neither his ingenuity nor his fertile imagination, with its leaps of organically related imagery, can be readily imitated.

Riding may have had a considerable subterranean influence, but of course such an influence would hardly be very noticeable. Like Dickinson, however, she may have affected later female poets by the implied confessionalism of so much of her poetry. (We say "later poets" because, soon after her *Collected Poems* appeared in 1938, when she was only thirty-seven, Riding renounced all future composing of poems, on theoretical grounds—discussed in the introduction to the new edition of her poems we have referred to earlier, and in statements reprinted in its appendix—that need not, despite their intrinsic interest, concern us here.) Her short sequences are allusive and secretive, and often deliberately cool, in tone. The earliest given in her carefully designed book has a Sitwell-like bounce, with its nursery-rhyme meters and tone. Its title, "Forgotten Girlhood," suggests the world of children's play, but also their fluid existence—a realm of change, forgetting, and half-awareness of the world's hardships and lures: the threat and promise of sex and of grownup domesticity. Some parts of the sequence are of the stuff that helped make the Beatles famous much later on; take for instance the poem "All the Way Back," which begins:

Bill Bubble in a bowler hat
Walking by picked Lida up.
Lida said "I feel like dead."
Bubble said
"Not dead but wed."
No more trouble, no more trouble,
Safe in the arms of Husband Bubble.

A rocking-chair, a velvet hat,
Greengrocer, dinner, a five-room flat,
Come in, come in,

Same old pot and wooden spoon,
But it's only soup staring up at the moon.

This sort of writing is unusual in Riding, however. The other sequences we have mentioned, like most of her work, are more sober in tone. The first of them, "To a Loveless Lover," is probably too consistent in perspective over its brief stretch of four short sections to be more than a modulation toward a sequence. It begins obliquely, with a riddling tone very different from Thomas's explosive metaphors that are riddles, too, but in context are seen finally as surprising leaps of expression. In Riding, instead, we have the sense of portentously indirect speech—a distanced suggestion of a difficult, intimate feeling:

How we happened to be both human,
Of the material of the machine . . .

The one original substance is one.
Two is two's destruction.
But love is the single word wherein
The double murder of the machine
Is denied
In one suicide . . .

Then, suddenly, the private disturbance beneath this mystifying surface is partly exposed, and with it the bitterness lurking in these mysteriously half-abstract images. The next lines explain:

Long very long ago,
A time unthinkable,
We loved each other.

Greet an old doubt
With contemporary conviction—
Lest going you give me lovelessness
And the accursed courage for a close.

Though lacking the hectic feverishness of the invective of disillusioned love in poets like Plath and Sexton, "To a Loveless Lover" clearly anticipates them. The rest of the semi-sequence veers back and forth between distanced coolness that grows increasingly sardonic at the same time and unresolved passion and hurt that are fairly directly expressed. In the fourth and final section these opposites are merged and the feeling swings around against the supposed speaker herself. Presumably the wretched lover has regretted his behavior and asked forgiveness; however that may be (and, very properly, we are not told), the reoriented tonality here does give the piece something of the structure of a sequence:

> The cycle of revenge comes round,
> Your expiation ties in me.
>
> Mercy, mercy for me
> Who would only suffer,
> Who would never sin.
> The righteous are transfixed
> While sinners are swept round to judgement.
> Mercy, mercy for me where I stand
> A bigot of forgiveness.

The 26-part "Echoes" (many short fragments) and the seven-part "Rhythms of Love" are closer to genuine sequences. Both give far subtler, prismatic insights into feelings of love-relationship and pride; and a sadness pervades them, kept free of pathos by their restraint and rigor of language. The delicate modulations of the former poem are worth study though we but mention them here—partly to show that the poetry of feminine sensibility, politicized by a more recent feminist poet like Adrienne Rich, has antecedents in its delphic and deliberately self-contradictory aspects. "Echoes" is exquisitely alert from the start:

> Since learning all in such a tremble last night—
> Not with my eyes adroit in the dark,
> But with my fingers hard with fright,
> Astretch to touch a phantom, closing on myself—
> I have been smiling.

Then, a surprise—the next fragment exudes bemused complicity with males:

> Mothering innocents to monsters is
> Not of fertility but fascination
> In women.

The astringency of these lines will suggest the extraordinary, Dickinsonian toughmindedness—or acid purity—of Riding's work generally. One more instance:

> Forgive me, giver, if I destroy the gift!
> It is so nearly what would please me,
> I cannot but perfect it.

These short, riddling sequences ("Echoes" is less than five pages long despite its 26 parts) remind us of the phenomenon we may call the mini-sequence, which often relies strongly on a riddling element to suggest powerful emotions at work. An interesting example is Donald Davie's "After an Accident" (1964),[9] a work that has to do with a car-crash in which Davie and his wife narrowly escaped being killed. In this five-part work, the first poem presents the shock of the experience ("Steep shadows and the purple / Hole in my darling's head") and the vision of Death as a mirror image of oneself; the second compares the thoughts of husband and wife after the accident; and the next three play on states of curiously detached realization. The melancholy timbre of the phrasing bathes the whole in an ambience of twilit acceptance and somber gratitude—a narrative converted into its subjective dimensions. A comparable process occurs in J. V. Cunningham's *To What Strangers, What Welcome,*[10] made up of a series of fifteen short poems that are essentially lyrical epigrams built around an implied little tale of love recaptured, then lost, by the Pacific. At the heart of the sequence is a bleakness often signaled by a poem's opening lines: "I drive Westward. Tumble and loco weed / Persist" (Poem 1) and "In a few days now when two memories meet / In that place of disease, waste, and desire" (Poem 3) are typical examples. In this sequence a humiliated and lonely sensibility is presented nakedly—witness the ending of Poem 4:

> We neither give nor receive:
> The unfinishable drink
> Left on the table, the sleep
> Alcoholic and final
> In the mute exile of time.

There are touching counter-notes of envisioned joy and purity ("Innocent to innocent, / One asked, What is perfect love?") and notes of astringent clarity to forestall lugubriousness ("Good is what we can do with evil"; "A premise of identity / Where the lost hurries to be lost"). Yet of course these notes, like the sense of gratitude and of widened understanding in "After an Accident," only enhance the poignancy.

Davie's and Cunningham's mini-sequences, like Riding's, are indications of the absorption of the sequence-form into the general poetic con-

9. Donald Davie, *Collected Poems 1950–1970* (New York: Oxford University Press, 1972), pp. 160–63.

10. *To What Strangers, What Welcome: A Sequence of Short Poems* (Denver: Swallow, 1964); reprinted in *The Collected Poems and Epigrams of J. V. Cunningham* (Chicago: Swallow Press; London: Faber, 1971), pp. 91–106.

ception of lyrical structure. Poets who might once have told a little tale, or a big one, surrounding it with purely lyrical passages or lyrically suffused meditation—or ending the poem with such effects—now reassemble the emotional centers and aspects of an important experience or observation of sustained character to get directly to an affective structure. Not always, of course. Seamus Heaney's "Glanmore Sonnets" (1979),[11] for instance, fall between structured narrative and dynamic structure. This group of ten sonnets seeks to connect communion with a place and its childhood associations and the relationship of a husband and wife—their sense of themselves and the natural world around them and of his poetry as a re-creation and repossession of all the meanings involved.

We may end our discussion (for end it we must, at last) with two poets who have centered sequences on inner psychic states but are serious political poets as well: Muriel Rukeyser and Adrienne Rich. As a promising young poet of the 1930s, Rukeyser readily absorbed styles and perspectives from Eliot, Lawrence, and Auden particularly and began writing sequences early in her career. Her "The Book of the Dead" (1938) is a twenty-poem sequence dealing with silicosis victims in West Virginia and the investigation of mine conditions by a subcommittee of the House of Representatives.[12] Much of the sequence is a setting of documents and hearings to free verse; some of it is relatively abstract and rhetorical, like MacDiarmid's more didactic poetry and leftwing political oratory of the period (as well as its reflex in the poems of Auden, MacNeice, and Day Lewis in England); and there is even a touch of Masters in some of the local portraits. After a transitional sequence called "Lives"—begun in *A Turning Wind* (1939)—in which she tried to merge inner states of her own with those of selected humanly and artistically courageous figures, she shifted in her *Elegies* (1938–45) to a more subjective mode represented in most of her later work.[13] The *Elegies* and her brief "Waterlily Fire" sequence in her volume of the same title (1962) are sufficiently representative of her achievement. The beginning of her "First Elegy. Rotten Lake," with its Waste Land imagery of loss and hope, mingles Eliot-notes with diction reminiscent of both Crane and Auden. These echoes, however, do not prevent effects of genuine sensuousness in the second stanza and of confessional openness in the third, despite the Freudian and Marxian burden those stanzas carry. The stanzas are a typical Rukeyser mixture:

11. Seamus Heaney, *Field Work* (London: Faber and Faber, 1979), pp. 33–42.

12. Muriel Rukeyser, *U.S. 1* (New York: Covici-Friede, 1938), pp. 9–72.

13. Muriel Rukeyser, *Elegies* (Stuttgart-Bad Cannstatt: Walter Cantz for New Directions, 1949). The 1938–45 dates of composition are supplied by Muriel Rukeyser, *Waterlily Fire: Poems 1935–1962* (New York: Macmillan, 1962), our text of reference for quotation here.

As I went down to Rotten Lake I remembered
the wrecked season, haunted by plans of salvage,
snow, the closed door, footsteps and resurrections,
    machinery of sorrow.

The warm grass gave to the feet and the stilltide water
was floor of evening and magnetic light and
reflection of wish, the black-haired beast with my eyes
    walking beside me.

The green and yellow lights, the street of water standing
point to the image of that house whose destruction
I weep, when I weep you.   My door (no), poems, rest,
    (don't say it!) untamable need.

Under—or above—the derivativeness and tendentiousness, a pure lyricism can be heard, seldom allowed to have its own way; and also a pure personalism: a sense of brave and difficult struggle epitomized in the first stanza of "Ninth Elegy. The Antagonists":

> Pieces of animals, pieces of all my friends
> prepare assassinations while I sleep.
> They shape my being, a gallery of lives
> fighting within me, and all unreconciled.
> Before them move my waking dreams, and ways
> of the spirit, and simple action.   Among these
> I can be well and holy.   Torn by them I am wild,
> smile, and revenge myself upon my friends
> and find myself among my enemies.
> But all these forms of incompleteness pass
> out of their broken power to a place
> where dream and dream meet and resolve in grace.

The "pure lyricism" we have mentioned is hardly sustained; it appears in the first two lines, fades away, and surfaces again for an instant in line 8, not too convincingly. And of course the stanza makes haste to assert a "positive" view as soon as it realizes it is giving away the truth about inner anxieties and self-betrayals. The *Elegies* as a whole present this same complex of momentary bits of melody and glimpses of suffering on the one hand, and smothering expanses of facile affirmation on the other. In the specific contexts of the ten poems that make up the sequence, this complex manifests itself in relation to various political, aesthetic, or psychological problems—reflected in such titles as "Third Elegy. The Fear of Form," "Fourth Elegy. The Refugees," and "Eighth Elegy. Children's Elegy," as well as in more generally suggestive titles. These are poems of the years when the Spanish Republic fell, the forces of war gathered and exploded, and the full terror of the Nazi period was burned into the modern mind. The *Elegies* reflect a mystical democratic rhetoric of the period ranging from liberal humanitarianism to Popular Front and Communist visionary vistas of social transformation. The rhetoric had its genuine basis in horror at fascist savagery, compassion for refugees and bombed-out civilian populations, residual pacifism born of the experience of World War I, and hopes for a better postwar world. Muriel Rukeyser, like Archibald MacLeish and many other poets, found it difficult to sort out the heartfelt language of social idealism and political oratory from the diction of an astringent emotional precision. Even poets like Wallace Stevens and Marianne Moore lose their way in the sincere rhetoric of his "Soldier, there is a war" passage, at the end of "Notes Toward a Supreme Fiction," and of her "In Distrust of Merits." (If only sincerity were all!)

Rukeyser's "Waterlily Fire" represents her later sequence-style at its best. It is relatively compressed, consisting, like *The Waste Land* or one of the *Quartets*, of five sections: independent yet linked and interacting poems. Perhaps—we are not talking of ultimate quality—it is most like "Little Gidding." It is centered on two specific events: one of them the 1958 fire in New York's Museum of Modern Art that destroyed a large *Waterlilies* canvas by Monet; the other a 1961 demonstration by Rukeyser and others against a nuclear air-raid drill, when they refused to take shelter at the warning sound of sirens. These events, explained by the poet in her notes to the sequence, can be compared with Eliot's scene after the all-clear signal following an air-raid; and even, perhaps, with his stunning impression, at the start of "Little Gidding," of snow on hedgerows and bright sunshine everywhere on a cold winter day as "midwinter spring"—an illusion of fused seasons suggesting a passionately desired state of existing beyond time. Something like the latter impression is apparently intended at the opening of "Waterlily Fire":

> Girl grown woman   fire   mother of fire
> I go the stone street turning to fire.   Voices
> Go screaming.   Fire   to the green glass wall.
> And there where my youth flies blazing into fire
> The   dance   of sane and insane images, noon
> Of seasons and days.   Noontime of my one hour.
>
> Saw down the bright noon street the crooked faces
> Among the tall daylight in the city of change.
> The scene has walls   stone   glass   all my gone life.

Rukeyser's notes explain that "Before the Museum . . . was built, I worked for a while in the house that then occupied that place. On the day of the fire, I arrived to see it as a place in the air." Here the problem is the opposite one to that presented in *Elegies:* the imposition of a set of private associations without finding a language adequate to evoke it (rather than the smothering of the genuinely evocative in rhetoric). The *aperçu* stated in the note, of the once-familiar house, now gone, still hovering as a projection of memory in the burning space, is more evocative than the passage of poetry it accounts for. The fragmented utterances and portentous long pauses (indicated by the spacing) seek to force a mystical time-sense like that of "Little Gidding" without sufficient presentative basis. One even needs the note to realize (despite the repetition of the word "fire" four times) that we are to respond to a situation in which a building is on fire—let alone a museum full of glorious paintings that also happens to have replaced another building important in the poet's memory and sense of her own identity. There is a sense of disorder, a distraught and disoriented state of mind, and irreparable loss; and yet nothing in this passage of hysterical excitement suggests much beyond incoherence.

There *is* more, though. In the sequence as a whole we have something like a rhetoric of secret emotions—hints and outcries. The poetry becomes utterance by a singing sybil, the poet who has come through stages of a difficult life into the present crisis in which all is destruction and she needs to be reawakened through love. The fire in the Museum of Modern Art and the social acceptance of the possibility of nuclear war are felt as cognate with the poet's predicament—powerful motifs and insights, in the abstract, that the poem touches on without being able to give enough substance to. Not one really sharp clear image of the burning or burned paintings illuminates the poem; and the brief passage on the anti-nuclear protest is perhaps complacent, although it reflects a loneliness of brave spirits too:

> This moment in a city, in its dream of war.
> We chose to be,
> Becoming the only ones under the trees
> when the harsh sound
> Of the machine sirens spoke. There were these two men,
> And the bearded one, the boys, the Negro mother feeding
> Her baby. And threats, the ambulances with open doors.
> Now silence. Everyone else within the walls. We sang.
> We are the living island,
> We the flesh of this island, being lived,
> Whoever knows us is part of us today.

Poetry such as Rukeyser's is part of the dense matrix of writing of the age. It has its moral fervor, its personal vulnerability and commitment, its urge for transcendence—even when, as so often, it does not provide those moments of keenest realization that, taken together, are precisely the organic body of transcendence. Because of the work of the masters, such poets have fallen into the sequence form as the expected one for the most serious poetry. This, surely, is an aspect of the curious situation wherein a genre has developed and provided models even while so little note has been taken of its existence—*except in practice.* But it is difficult to write a successful sequence by any rigorous poetic standards; every demand associated with the most accomplished lyric poetry on a smaller scale must be made on its separate parts, and there remains the problem of the direction and reciprocity of these parts within the larger structure.

The work of Adrienne Rich invites considerations similar to those we have observed in Rukeyser. Here again we have a poet of intensity and lyric feeling, often derivative, often affecting the authority of the didactic orator not only when her writing is oracular on matters political and ideological but also when it deals with the intimately personal. Her "The Phenomenology of Anger" (1972)[14] is both more belligerently assertive and more dynamically alive in detail than Rukeyser's work is generally; but the sybilline tone is unmistakable and probably learned from the older, more generous-spirited poet. The closing section (part 10) might easily have been by Rukeyser:

> 10. how we are burning up our lives
> testimony:
> the subway
> hurtling to Brooklyn
> her head on her knees
> asleep or drugged
>
> la vía del tren subterráneo
> es peligrosa
>
> many sleep
> the whole way
>
> others sit
> staring holes of fire into the air
>
> others plan rebellion:
> night after night
> awake in prison, my mind

14. Adrienne Rich, *Diving into the Wreck: Poems 1971–1972* (New York: Norton, 1973), pp. 27–31.

licked at the mattress like a flame
till the cellblock went up roaring

Thoreau setting fire to the woods

Every act of becoming conscious
(it says here in this book)
is an unnatural act

These oracular utterances center on a scene in a subway car, one of many such scenes in our poetry and fiction. The best-known, probably, is the section called "The Tunnel" in Hart Crane's *The Bridge*, a much more fully developed closeup, with overheard conversations and the wild, hideously imagined sight of Poe's "head swinging from the swollen strap" and his body "smoking along the bitten rails"—altogether a more powerful hell-scene. Despite the active character of lines 3–12, and the sad suggestiveness of lines 5–8, the passage in Rich's poem is largely oratorical. It begins with a fairly obvious thought about the waste of our lives and energies, quotes a warning sign that of course also suggests the oppression of so many workingclass lives and the possibility of taking an "underground," revolutionary way of struggle, and then offers a number of images of revolutionary thought and struggle. It ends, finally, with an ambiguously ironic note, perhaps in part a sneer against Freudian theory concerning the self-alienating evolution of human consciousness. As the rest of the sequence shows, the bearing is revolutionary; the revolution envisioned is that of the female against imprisonment by the perversions of male power. The sequence plays with fantasies of fighting violence with violence, murder with murder:

4. White light splits the room.
Table. Window. Lampshade. You.

My hands, sticky in a new way.
Menstrual blood
seeming to leak from your side.

Will the judges try to tell me
Which was the blood of whom?

5. Madness. Suicide. Murder.
Is there no way out but these?
The enemy, always just out of sight
snowshoeing the next forest, shrouded
in a snowy blur, abominable snowman
—at once the most destructive
and the most elusive being

gunning down the babies at My Lai
vanishing in the face of confrontation.

The prince of air and darkness
computing body counts, masturbating
in the factory
of facts.

Clearly, the surrealist dream-picture of Section 4 is the deepest indication of the ultimate, innermost premises of the sequence. The "you" addressed at the end of the second line becomes, in succeeding sections, the male companion (husband, lover, distorted sexual complement) whom the sequence is in painful process of rooting out of its emotional center. The relationship is seen in a new, divisive "white light" in which everything is clarified malignly. The need to kill "you"—whether literally and personally or as the personification of the old relationship—is projected in the images of "my hands, sticky in a new way" and of "menstrual blood / seeming to leak from your side." In dream or reverie, "I" have murdered "you" and our sexual roles have thereby been reversed and, presumably, canceled. The equation of menstruation with victimization is not merely implied; the two closing lines suggest that the male is guilty of wounding the female into menstrual flow (an exact reciprocal of some of the more violent sexual grotesqueries in *Crow*). No doubt the imagery can be rationalized into a feminist argument that men have exploited women because of their specifically female physiology and functions, while at the same time making women feel guilty and inferior because of them. Few would contest the position. But the psychological pressure revealed in this section is of a primal hatred and resentment, mitigated by notes of moral confusion and of a lost world of sharing (as if one had changed the original atmosphere of quietly living and working together by turning an unbearably bright spotlight on in a room; or had suddenly taken as a personal affront the reciprocal sexual differences in a love-relationship). The powerful feeling here—"if I stabbed you, the blood flowing from your side would be the precise equivalent of what happens to me all the time"—is beyond rational argument, something on the order of war-propaganda that plays on bestial archetypes. And, in fact, the next section moves into just such archetypes for the male enemy. War atrocities in Vietnam, for instance, are alluded to not (as in Rukeyser) as our national crimes and shared guilt, but as male diabolism and male sexual gratification.

The movement of the ten-part sequence pivots on these two sections, which face into the violent rejection of past relationships and conceptions of the self that constitutes the major effort of the work. The effort begins obliquely in Poem 1, with a bit of wryly humane meditation at the start that soon gives way to a feminist complaint:

> 1. The freedom of the wholly mad
> to smear & play with her madness
> write with her fingers dipped in it
> the length of a room
>
> which is not, of course, the freedom
> you have, walking on Broadway
> to stop & turn back or go on
> 10 blocks; 20 blocks
>
> but feels enviable maybe
> to the compromised
>
> curled in the placenta of the real
> which was to feed & which is strangling her.

The language here is somewhat ambiguous, but one soon notices that the "wholly mad" figure at the start and the unborn, placenta-strangled figure at the end are both female. The suggestion, therefore, is that the untrammeled "you" of the second stanza, moving freely in the outside world at whatever risk, is male; but the typically "compromised" woman, protectively nourished to the point of stifling, is not allowed to be born into total freedom and so may envy the madwoman's freedom of expression. The adventurous tone of the first stanza dwindles into an instrument of sly argument, although the closing image of aborted maturing has a genuine resonance of frustration. Poetically, the succession of affects is, first, an outburst of uninhibitedly manic infantilism; second, a slight monotone of static; and third a nagging, resentment-laden insistence. When the next poem then shifts to the first person and loads itself with images of inability to rekindle a fire or cause a rose to bloom anew or release the burning intensities of tenderness and desire, the generalized frustration evoked at the end of Poem 1 re-emerges with its energy now immediate and personal and charged with sexual yearning.

The monotone of doctrinaire static returns in Poem 3, which is filled with images of maledom's strange, hard domain that possess the unrequited womanly soul of the previous poem. It is an interim of "self-hatred" before the outbursts of Poems 4 and 5. After these climactic outbursts, which turn the hatred outward, the murderous feeling is qualified in Section 6 into a dream of purifying the enemy rather than killing him. The qualification may be humorously self-corrective, but the context makes this unlikely. It is moralistic reorientation rather than self-irony that leads to the assertion:

. . . When I dream of meeting
the enemy, this is my dream:

white acetylene
ripples from my body
effortlessly released
perfectly trained
on the true enemy

raking his body down to the thread
of existence
burning away his lie
leaving him in a new
world; a changed
man

Once this righteously reorienting science-fiction imagery has exculpated the speaker of true murderousness, Poem 7 returns heartily to its song of hatred toward a particular man and toward male insensitivity, male false depth of humanity, the male body, and male destructiveness generally. The phrase "I hate" is used as a refrain introducing three lines, and the private complaint underlying the rhetoric throughout the sequence is brought specifically, in italics, into the open:

Last night, in this room, weeping
I asked you: *what are you feeling?*
do you feel anything?

Shades of "A Game of Chess" in *The Waste Land!* Needless to say that here the object of the assault (who, a few lines earlier, has been told: "I hate you," "I hate the mask you wear," and "I hate your words") is not quoted as offering any reply. Obviously, there is nothing for him to say that could be believed. The next poem (Section 8) gains some witty distancing by counterposing images of the present state of our lives as a "temporary" phase, of which a "dogeared earth" and "wormeaten moon" are appropriate symbols, and a dreamt-of ideal state, in which women and men are "gaily / in collusion with green leaves, stalks, / building mineral cities, transparent domes"—

a conspiracy to coexist
with the Crab Nebula, the exploding
universe, the Mind—

Section 8 serves a self-exculpatory function like that of Section 6—a drawing back from extreme statement that is also a preparation for another drastic shift. And that shift comes in Section 9, which turns its back on all relationships of love and fellow-feeling with men. This penultimate poem of the sequence is its second climax:

> 9. "The only real love I have ever felt
> was for children and other women.
> Everything else was lust, pity,
> self-hatred, pity, lust."
> This is a woman's confession.
> Now, look again at the face
> of Botticelli's Venus, Kali,
> the Judith of Chartres
> with her so-called smile.

For all the straightforwardness of the quoted declaration in the first four lines, the passage is essentially enigmatic. *It* does not declare for lesbianism, although that would seem an important aspect of the sequence's implied polemic. "This is a woman's confession" does not literally mean "all women's confession"; and yet the passage is trivial if that is not the meaning. The "confession," with its expression of self-contempt for feelings in the past, resonates with anguish and with qualified tenderness, but goes beyond these tones only if felt to present a proof of universal womanly reality—that aspect of any woman's repressed need ("curled in the placenta . . . which is strangling her") to open herself to the love of other women. The closing sentence, too, is enigmatic, like the faces of Venus, Kali, and Judith, whose expressions are not described in any case. Nor does the poem suggest specific associations of these figures, whether of power, beauty, or ecstasy, although only Venus' name will carry any such associations for most readers. It is left to the implied polemic of the whole sequence to point the central passion of Poem 9 underlying its "confession" and its allusion to powerful female figures out of three traditions. Then, in Poem 10, the implied sexual-political argument is picked up and amplified, combining with notes drawn from all the previous sections of the sequence. Thus, the weary woman in the subway car is somehow the victim of a male-dominated world, a dangerous world whether one accepts it as it is or struggles against it. Women's lives are being burned away pointlessly (see the fire-imagery in Poem 1); and women's dreams are full of imagined fire—which will break loose on its own sooner or later, like Thoreau's campfire that accidentally set trees burning. Losing control is the danger that must be risked, just as the achievement of full self-awareness by women is necessary despite the fact that it may lead them into unpredictable ways considered "unnatural" (like lesbianism, no doubt).

We have lingered a bit over "The Phenomenology of Anger" because it so clearly represents a continuing problem of the sequence. It is obvious to any poet of ability that the form provides an opportunity to set affective units side by side without providing a smooth surface continuity of discourse, drama, or narration. The form encourages the poet to work through these centers of lyrical affect in an exploratory way, rather than trying to contrive them so as to force the sequence in an arbitrary direction. Ideally, they create their own subjective direction and reveal the sequence's complex of psychological pressures driving it toward its final structural dynamics. As we have suggested, the task of sustaining such a structure with the necessary rigor is a difficult one, imperfectly understood by many poets, and even the greatest can lapse into discourse when something like a hard, gemlike flame is needed.

A work like "The Phenomenology of Anger" or "Waterlily Fire" has its inner affective life, in the one instance having to do with the awakening to hatred and a painful vision of freedom and love; and in the other hysteria in the face of the destructive principle and the sheer effort to cope and to maintain a courageous identity. But in both instances the didactic and rhetorical predominate and diminish the affective purity and authority of the work. The very way in which Poem 10 of Rich's sequence *illustrates* every *point* made or suggested in the preceding poems is an index of the fact that to an important degree affective structure has been subordinated to a code of symbolic indications, with the emotion residing in the referent rather than riding the language. As we have seen, this is not the whole truth about this sequence. It has some strong affective moments in its own right. But to a noticeable degree there is a tendentious dimension: the use of inadequate, unmemorable phrasing and rhythms in the hope that the *unwritten* aspect of the work, its dependence on set emotional and political expectations, will carry the day anyway. Morally or politically coded sequences, ambivalent because their strongest moments undercut the code, exist in large numbers in current poetry—in one sense as the underside of the triumph of the modern poetic genre most suitable for epic and heroic undertakings.

Index

Books and sequences are indexed by title, not author. Individual poems are listed by sequence; otherwise (with the exception of anonymous works), by author.

Copyrights and Acknowledgments

*Selections from works by the following poets and authors were made possible by the kind permission of themselves, their respective publishers, agents, or literary executors:

Conrad Aiken: by permission of Brandt & Brandt Literary Agents

W. H. Auden: by permission of Random House, Inc., for selections from *The Collected Poetry of W. H. Auden,* copyright 1945 by W. H. Auden; from *The English Auden,* © 1977 by Edward Mendelson, William Meredith, and Monroe K. Spears, Executors of the Estate of W. H. Auden; from *W. H. Auden: Collected Poems,* © 1976 by Edward Mendelson, William Meredith, and Monroe K. Spears, Executors of the Estate of W. H. Auden. By permission of Faber and Faber Ltd. for selections from *The English Auden: Poems, Essays and Dramatic Writings 1927–1939* and *Collected Poems*

John Berryman: by permission of Farrar, Straus and Giroux, Inc., and Faber and Faber Ltd., for selections from *The Dream Songs,* copyrights © 1959, 1962, 1964, 1965, 1966, 1967, 1968, 1969 by John Berryman

Paul Blackburn: by permission of Joan M. Blackburn for excerpts from his *Pure Vidal: Translations*

Basil Bunting: by permission of Oxford University Press for excerpts from *Collected Poems,* © Basil Bunting 1978

Austin Clarke: by permission of Wake Forest University Press and The Dolmen Press from *Selected Poems,* edited by Thomas Kinsella, copyright 1976 by Nora Clarke

Hart Crane: by permission of Liveright Publishing Corporation and Laurence Pollinger Limited, for selections from *The Complete Poems and Selected Letters and Prose of Hart Crane,* edited by Brom Weber, copyright 1933, © 1958, 1966 by Liveright Publishing Corporation; by permission of David Mann for quotations from Hart Crane's uncollected works

Emily Dickinson: by permission of the publishers and the Trustees of Amherst College for selections from *The Poems of Emily Dickinson,* edited by Thomas H. Johnson, Cambridge, Mass.: The Belknap Press of Harvard University Press, copyright 1951, © 1955, 1979, by the President and Fellows of Harvard College; and by permission of Little, Brown & Co. from *The Complete Poems of Emily Dickinson*, edited by Thomas H. Johnson, copyright 1914, 1929, 1935, 1942 by Martha Dickinson Bianchi; copyright © 1957, 1963 by Mary L. Hampson

T. S. Eliot: by permission of Harcourt Brace Jovanovich and Faber and Faber Ltd. for selections from *Collected Poems 1909–1962;* copyright © 1943, 1963, 1964 by T. S. Eliot; copyright 1971 by Esme Valerie Eliot

Allen Ginsberg: by permission of City Light Books for selections from *Kaddish & Other Poems*, copyright © 1961 by Allen Ginsberg

Ramon Guthrie: by permission of Dartmouth College, Hanover, New Hampshire

Thomas Hardy: by permission of the Macmillan Publishing Co., Inc. for selections from *Complete Poems* edited by James Gibson

Geoffrey Hill: by permission of Andre Deutsch Limited for selections from *Mercian Hymns*

A. E. Housman: by permission of Holt, Rinehart and Winston, Publishers, and the Society of Authors as the literary representative of the Estate, and Jonathan Cape Ltd. as publishers, for selections from *The Collected Poems of A. E. Housman*, copyright 1939, 1940, © 1965 by Holt, Rinehart and Winston; copyright © 1967, 1968 by Robert E. Symons

Ted Hughes: by permission of Harper & Row, Publishers, and Faber and Faber Ltd. for selections from *New Selected Poems* copyright 1982 by Ted Hughes and *Crow*, copyright 1971 by Ted Hughes

David Jones: by permission of Faber and Faber Ltd. for selections from *Anathémata*

Galway Kinnell: by permission of Houghton Mifflin Company for selections from *The Book of Nightmares* copyright © 1971 by Galway Kinnell

Thomas Kinsella: by permission of Wake Forest University Press and Peppercanister Press for selections from *Poems 1956–1973* and *Peppercanister Poems 1972–1978*, both copyright © 1979 by Thomas Kinsella

Robert Lowell: by permission of Farrar, Straus, and Giroux, and Faber and Faber Ltd., for excerpts from *Life Studies*, copyright © 1956, 1959 by Robert Lowell, and from *Day by Day*, copyright 1975, 1976, 1977 by Robert Lowell

Hugh MacDiarmid: by permission of the Macmillan Publishing Co., Inc. for excerpts from *Collected Poems,* © Christopher Murray Grieve 1948, 1962

Edgar Lee Masters: by permission of Ellen C. Masters

John Montague: by permission of Wake Forest University Press and the Dolmen Press for excerpts from *The Rough Field,* 3rd edition, copyright © 1979 by John Montague

Charles Olson: by permission of Corinth Books and the estate of Charles Olson for selections from *The Maximus Poems,* copyright © 1960 by Charles Olson, published by Jargon Books in association with Corinth Books and by permission of the estate of Charles Olson for "Maximus to Gloucester, Letter 27 [withheld]"

Sylvia Plath: by permission of Harper & Row Publishers and Olwyn Hughes for selections from *The Collected Poems—Sylvia Plath,* edited by Ted Hughes, copyright © 1960, 1965, 1971, 1981 by the Estate of Sylvia Plath

Ezra Pound: by permission of New Directions Publishing Corporation and Faber and Faber Ltd. for excerpts from *Personae,* copyright 1926 by Ezra Pound; from *The Collected Early Poems of Ezra Pound,* copyright © 1976 by the Trustees of the Ezra Pound Literary Property Trust—all rights reserved; from *The Cantos of Ezra Pound,* copyright 1934, 1937, 1948 by Ezra Pound; copyright © 1972 by The Estate of Ezra Pound

Laura Riding: by permission of Persea Books, Inc. for excerpts from *The Poems of Laura Riding,* copyright © 1938, 1980 by Laura (Riding) Jackson

Adrienne Rich: by permission of W. W. Norton Company, Inc. for excerpts from *Diving into the Wreck, Poems 1971–1972,* copyright © 1973 by W. W. Norton & Company, Inc.

Muriel Rukeyser: by permission of C. Mitchell Douglas from *Waterlily Fire Poems 1935–1962,* copyright 1935, 1938, 1939, 1944, 1948, 1949, 1950, 1951, 1952, 1954, 1955, 1956, 1957, 1958, 1960, 1962 by Muriel Rukeyser

Anne Sexton: by permission of Houghton Mifflin Company and the Sterling Lord Agency, Inc., for selections from *45 Mercy Street,* copyright © 1976 by Linda Gray Sexton and Loring Conant, Jr., Executors of the Estate of Anne Sexton

W. D. Snodgrass: by permission of Alfred A. Knopf, Inc. from *Heart's Needle,* copyright © 1959 by W. D. Snodgrass

Wallace Stevens: by permission of Alfred A. Knopf, Inc., and Faber and Faber Ltd. for selections from *The Poem at the End of the Mind:*